THE EQUATIONS OF
RADIATION HYDRODYNAMICS

GERALD C. POMRANING

DOVER PUBLICATIONS, INC.
Mineola, New York

Bibliographical Note

This Dover edition, first published in 2005, is an unabridged republication of the work originally published by Pergamon Press, Oxford and New York, in 1973.

Library of Congress Cataloging-in-Publication Data

Pomraning, G. C. (Gerald C.)
  The equations of radiation hydrodynamics / Gerald C. Pomraning.
    p. cm.
  Originally published: Oxford ; New York : Pergamon Press, 1973, in series: International series of monographs in natural philosophy ; v. 54.
  Includes bibliographical references and index.
  ISBN 0-486-44599-2 (pbk.)
  1. Heat—Transmission. 2. Hydrodynamics. I. Title.

QC320.P63 2005
536'.3—dc22

2005051881

Manufactured in the United States of America
Dover Publications, Inc., 31 East 2nd Street, Mineola, N.Y. 11501

Contents

Contents

List of Tables

List of Figures

List of Figures

Preface

This book is concerned with the propagation of thermal radiation through a fluid, and the effect of this radiation on the hydrodynamics describing the fluid motion. The term "thermal radiation" means electromagnetic radiation of atomic, as opposed to nuclear, origin. Such radiation is generally emitted by matter in a state of thermal excitation, thus accounting for the designation of the radiation as thermal. The energy density of this type of radiation in an enclosure whose walls are maintained at a constant and uniform temperature is given by the well-known Planck formula. More generally, however, the energy distribution of the radiation field is not described by the Planck function. Under certain rather unrestrictive conditions, the state of the radiation can be described by a kinetic or transport equation, referred to historically as the equation of radiative transfer. In large part, this book concentrates on various formulations of the equation of transfer describing the propagation of thermal radiation.

The importance of thermal radiation in physical problems increases as the temperature is raised, primarily because the radiation energy density associated with a Planck distribution varies as the fourth power of the temperature. At moderate temperatures (say, thousands of degrees Kelvin), the role of the radiation is primarily one of transporting energy by radiative processes. At higher temperatures (say, millions of degrees Kelvin), the energy and momentum densities of the radiation field may become comparable to or even dominate the corresponding fluid quantities. In this case, the radiation field significantly affects the dynamics of the fluid. Hydrodynamics with explicit account of the radiation energy and momentum contributions constitutes the charter of "radiation hydrodynamics". Such considerations find their practical application in the understanding of certain astrophysical and nuclear weapons effects phenomena.

A brief outline of the contents of this book is as follows: The first chapter introduces the fundamental concepts required to formulate a kinetic description of electromagnetic energy transport. Chapter II then derives in various forms and geometries the equation of transfer containing the basic physics needed for the vast majority of work in radiation hydrodynamics. Also discussed in this chapter is the validity of the equation of transfer as a description of electromagnetic energy transport, and the concepts of local thermodynamic equilibrium and induced or stimulated emission and scattering of radiation. Chapter III considers various approximate formulations of radiative transfer, concentrating on approximate treatments of the angular and frequency variables. Chapters IV through VI consider the incorporation of three additional physical effects, often neglected, into the equation of transfer. These are polarization, refraction and dispersion, and the relativistic effects of the fluid motion.

Chapter VII gives the reader a brief introduction into the microscopic physics associated with the equation of transfer, namely the physics of the absorption, emission, and scattering of radiation on the atomic level. Chapter VIII describes at some length the description of Compton and inverse Compton scattering (the scattering of radiation from free electrons in motion) as it enters the equation of transfer. The final item discussed, in Chapter IX, is the hydrodynamic description of the fluid, including effects due to the presence of the radiation field. The equations of hydrodynamics, in Eulerian, modified Eulerian, and Lagrangian forms, are given in both the relativistic case and in the nonrelativistic limit. The discussion is restricted, however, to an ideal fluid (no viscous or heat conduction effects). One appendix is included on the Lorentz transformation of the equation of transfer.

As is clear from this brief outline, radiative transfer is stressed in this book at the expense of hydrodynamics. In part, this reflects the interests of the author. However, it is also true that once one accepts ideal fluid dynamics as the description of the fluid transport, the hydrodynamics, even with the inclusion of relativistic and radiation effects, is relatively straightforward. The difficult aspect of hydrodynamics is the calculation of the transport coefficients of viscosity and thermal conduction, effects which are ignored (by definition of an ideal fluid) in ideal fluid hydrodynamics. In general, this neglect introduces a small error for problems in the radiation hydrodynamic regime.

The intent of this book is to give a detailed discussion, derive from first principles, and display in full form all of the equations pertinent to the specialized field of radiation hydrodynamics. Specifically not discussed are analytic and numerical techniques which have been developed for the solution of these equations. For references in this area, the reader is referred to the bibliography given at the end of the book. Because of its rather narrow scope, this book is probably not well suited for use as the sole textbook in a course on physical gas dynamics, for example. However, it is hoped that it will be useful as a supplement to more standard texts on radiative transfer and hydrodynamics. The detail given in this book should be useful to students as well as researchers actively engaged in radiative transfer and radiation hydrodynamic work.

It is traditional to end a preface such as this with an expression of gratitude to those persons who have been instrumental in seeing a book through to its final manuscript. It is a pleasure to continue this tradition here. In the first place, I would like to thank Dr. P. M. Campbell for his substantial help in the writing of Chapter VII and the assembling of the bibliography. In both instances Dr. Campbell must be considered as a co-author, although I alone am responsible for any errors or omissions. Secondly, a special thanks is due Miss Jo Van Elverdinghe who did all of the typing. With a combination of fortitude, patience, and skill, she produced a superb manuscript. Finally, I would like to offer a sincere thank you to Dr. ter Haar for allowing this book to be part of his distinguished series on natural philosophy.

La Jolla, California   G. C. P.

I

Definitions and Basic Concepts
in Radiative Transfer

---

1. Introduction

In this chapter we consider the basic concepts needed to describe the radiation field and its interaction with matter. The distribution function for photons is introduced and various angular moments of this distribution function which have physical significance are defined. We also discuss the interaction of the radiation field with matter through the processes of absorption, scattering, and emission of photons.

The primary intent of this first chapter is to lay the groundwork for the derivation of the equation of radiative transfer, to be considered in Chapter II.

2. The Radiation Field

We consider the energy in the radiation field to be carried by point, massless particles called photons. With each photon we associate a frequency ν such that the energy E of the photon is $h\nu$, where h is Planck's constant. It is known that a massless particle has momentum E/c, where c is the vacuum speed of light, and hence the momentum of a photon is $h\nu/c$. Between collisions with matter, a photon is supposed to travel in a straight line with speed c and with no change in frequency.

At any time t, six variables are required to specify the position of the photon in phase space, namely three position variables and three momentum variables. We denote the three position variables by the vector $\mathbf{r}$. In radiative transfer work it is conventional to use, rather than the three momentum variables, three equivalent variables. These are the frequency ν and the direction of travel of the photon $\mathbf{\Omega}$ (it requires two angular variables to specify $\mathbf{\Omega}$). In terms of these variables we define the distribution function f

$$f \equiv f(\mathbf{r},\, \nu,\, \mathbf{\Omega},\, t), \tag{1.1}$$

such that

$$dn = f\, d\mathbf{r}\, d\nu\, d\mathbf{\Omega}. \tag{1.2}$$

Here dn is the number of photons (at time t) at space point $\mathbf{r}$ in a differential volume element $d\mathbf{r}$, with frequency ν in a frequency interval $d\nu$, and traveling in a direction $\mathbf{\Omega}$ in a solid angle element $d\Omega$. As a concrete example, let us represent $\mathbf{r}$ by the Cartesian coordinates x, y, z and Ω by a polar angle θ (measured with respect to any fixed direction in space) and a corresponding azimuthal angle φ. Then

$$d\mathbf{r} = dx\,dy\,dz, \tag{1.3}$$

$$d\Omega = \sin\theta\,d\theta\,d\varphi = d\mu\,d\varphi, \tag{1.4}$$

where $\mu \equiv \cos\theta$. [A minus sign is omitted in Eq. (1.4) since by conventional definition μ runs from -1 to 1, but when θ runs from 0 to π, μ runs from 1 to -1.] Then Eq. (1.2) becomes, with full display of the arguments of the distribution function,

$$dn = f(x, y, z, \nu, \mu, \varphi, t)\,dx\,dy\,dz\,d\nu\,d\mu\,d\varphi. \tag{1.5}$$

Of course, one is not restricted to use x, y, z, μ, and φ as the phase space variables. Any other representation of the vectors $\mathbf{r}$ and Ω is acceptable and, in fact, in many cases desirable to take full advantage of the symmetry which a particular situation may involve.

It is conventional in radiative transfer to introduce $I \equiv I(\mathbf{r}, \nu, \Omega, t)$, the "specific intensity" of radiation, in place of the distribution function f. The definition of the specific intensity is

$$I(\mathbf{r}, \nu, \Omega, t) = ch\nu\,f(\mathbf{r}, \nu, \Omega, t). \tag{1.6}$$

The physical interpretation of I is contained in the relationship

$$dE = I(\mathbf{r}, \nu, \Omega, t)\cos\theta\,d\nu\,d\Omega\,d\sigma\,dt. \tag{1.7}$$

Here dE is the amount of radiant energy in $d\nu$ centered at ν, traveling in a direction Ω confined to a solid angle element $d\Omega$, which crosses, in a time element dt, an area $d\sigma$ oriented such that θ is the angle which the direction Ω makes with the normal to $\sigma(\cos\theta = \Omega\cdot\mathbf{n})$. Figure 1.1 displays this interpretation schematically. If the specific intensity is independent of Ω at a point, it is said to be "isotropic" at that point. If the intensity is independent of both $\mathbf{r}$ and Ω, the radiation field is said to be "homogeneous" and isotropic.

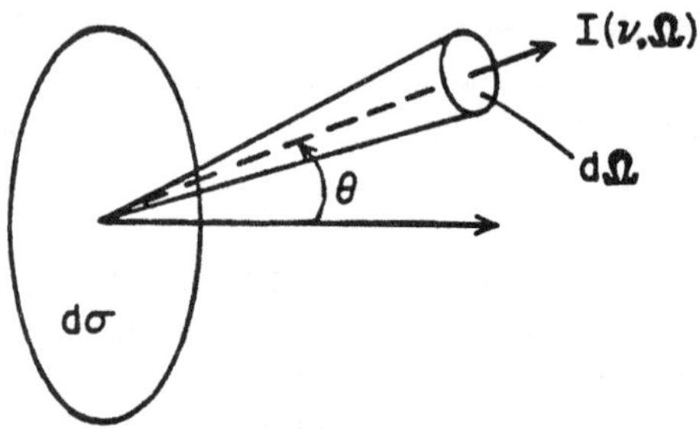

FIG. 1.1. Geometric interpretation of the specific intensity.

The most important example of a homogeneous, isotropic radiation field is that which coexists with matter in complete thermodynamic equilibrium at a temperature T. In this case the specific intensity of radiation is given by the Planck function $B \equiv B(\nu, T)$

$$I = B = \frac{2h\nu^3}{c^2}(e^{h\nu/kT} - 1)^{-1}, \tag{1.8}$$

where k is the usual Boltzmann constant.

Three quantities of physical interest are the energy density, the radiative flux, and the pressure tensor associated with the radiation field. These correspond to various angular moments of the specific intensity.

Considering first the energy density $u \equiv u(\mathbf{r}, t)$, we find, from the definitions of the distribution functions f and I,

$$u = \int_0^\infty dv \int_{4\pi} d\Omega\, h v f = \frac{1}{c} \int_0^\infty dv \int_{4\pi} d\Omega\, I. \tag{1.9}$$

If I is given by the Planck function, the corresponding energy density is

$$u = \frac{1}{c} \int_0^\infty dv \int_{4\pi} d\Omega\, B = \frac{4\pi}{c} \int_0^\infty dv\, \frac{2h v^3}{c^2} (e^{h v / kT} - 1)^{-1}. \tag{1.10}$$

This integral can be carried out by setting $x = h v / kT$ and expanding the denominator in an infinite series according to

$$u = \frac{8\pi k^4 T^4}{h^3 c^3} \int_0^\infty dx\, x^3 (e^x - 1)^{-1}$$

$$= \frac{8\pi k^4 T^4}{h^3 c^3} \int_0^\infty dx\, x^3 \sum_{n=1}^\infty e^{-nx} = \left(\frac{48\pi k^4}{h^3 c^3} \right) T^4 \sum_{n=1}^\infty \frac{1}{n^4}. \tag{1.11}$$

Since the above sum is equal to $\pi^4/90$, we find

$$u = aT^4, \tag{1.12}$$

where a, the radiation constant, is given by

$$a = \frac{8\pi^5 k^4}{15 h^3 c^3}. \tag{1.13}$$

This result is often written

$$u = \frac{4\sigma}{c} T^4, \tag{1.14}$$

where $\sigma = ac/4$ is called the Stefan–Boltzmann constant.

The second quantity of physical interest is a vector quantity called the radiative flux. It is defined as the rate of energy flow per unit area across a surface. Consider a surface element oriented perpendicular to the x axis. The flow of energy across this surface, per unit area, is the x component of the radiative flux $F_x \equiv F_x(\mathbf{r}, t)$ and is given by

$$F_x = \int_0^\infty dv \int_{4\pi} d\Omega\, c h v \Omega_x f = \int_0^\infty dv \int_{4\pi} d\Omega\, \Omega_x I, \tag{1.15}$$

where Ω_x is the projection of Ω along the x axis. In general the radiative flux is a vector quantity $\mathbf{F} \equiv \mathbf{F}(\mathbf{r}, t)$ and we write

$$\mathbf{F} = \int_0^\infty dv \int_{4\pi} d\Omega \, \mathbf{\Omega} I. \tag{1.16}$$

This is nothing more than a shorthand notation for a vector with cartesian components

$$\left.\begin{aligned}
F_x &= \int_0^\infty dv \int_{4\pi} d\Omega \, \Omega_x I, \\[2mm]
F_y &= \int_0^\infty dv \int_{4\pi} d\Omega \, \Omega_y I, \\[2mm]
F_z &= \int_0^\infty dv \int_{4\pi} d\Omega \, \Omega_z I.
\end{aligned}\right\} \tag{1.17}$$

For any isotropic specific intensity, the radiative flux is clearly zero. In particular, if $I = B$, the Planck function, we have $\mathbf{F} = 0$. This is, of course, the expected result since in complete thermodynamic equilibrium there is no net flow of radiative energy in any direction.

Finally, we consider the pressure tensor due to radiation. In the kinetic theory of gases the pressure is defined as the rate of momentum flow across a surface. The same definition applies to the radiation field. The component $p_{xy} \equiv p_{xy}(\mathbf{r}, t)$, for example, is defined as the rate of y momentum flow per unit area through a surface perpendicular to the x axis. Since a photon carries y momentum in the amount $h\nu\Omega_y/c$, we have

$$p_{xy} = \int_0^\infty dv \int_{4\pi} d\Omega (c\Omega_x)(h\nu\Omega_y/c)f = \frac{1}{c}\int_0^\infty dv \int_{4\pi} d\Omega \, \Omega_x\Omega_y I. \tag{1.18}$$

In general we write

$$\mathbf{p} = \mathbf{p}(\mathbf{r}, t) = \frac{1}{c}\int_0^\infty dv \int_{4\pi} d\Omega \, \mathbf{\Omega}\mathbf{\Omega} I. \tag{1.19}$$

The nine quantities defined by Eq. (1.19) are the components of the pressure tensor. Written out explicitly, we have

$$\left.\begin{aligned}
p_{xx} &= \frac{1}{c}\int_0^\infty dv \int_{4\pi} d\Omega\,\Omega_x\Omega_x I, & p_{xy} &= \frac{1}{c}\int_0^\infty dv \int_{4\pi} d\Omega\,\Omega_x\Omega_y I, & p_{xz} &= \frac{1}{c}\int_0^\infty dv \int_{4\pi} d\Omega\,\Omega_x\Omega_z I, \\[2mm]
p_{yx} &= \frac{1}{c}\int_0^\infty dv \int_{4\pi} d\Omega\,\Omega_y\Omega_x I, & p_{yy} &= \frac{1}{c}\int_0^\infty dv \int_{4\pi} d\Omega\,\Omega_y\Omega_y I, & p_{yz} &= \frac{1}{c}\int_0^\infty dv \int_{4\pi} d\Omega\,\Omega_y\Omega_z I, \\[2mm]
p_{zx} &= \frac{1}{c}\int_0^\infty dv \int_{4\pi} d\Omega\,\Omega_z\Omega_x I, & p_{zy} &= \frac{1}{c}\int_0^\infty dv \int_{4\pi} d\Omega\,\Omega_z\Omega_y I, & p_{zz} &= \frac{1}{c}\int_0^\infty dv \int_{4\pi} d\Omega\,\Omega_z\Omega_z I.
\end{aligned}\right\} \tag{1.20}$$

It can be seen that the radiation pressure tensor is symmetric, i.e. $p_{jk} = p_{kj}$. Also, since

$\Omega_x^2 + \Omega_y^2 + \Omega_z^2 = 1$, we have, for an arbitrary radiation field, the trace relationship

$$p_{xx} + p_{yy} + p_{zz} = \frac{1}{c} \int_0^\infty dv \int_{4\pi} d\Omega I = u. \tag{1.21}$$

This leads to the definition of the average pressure as one-third of the energy density, i.e.,

$$\bar{p} = \tfrac{1}{3}(p_{xx} + p_{yy} + p_{zz}) = \tfrac{1}{3}u. \tag{1.22}$$

For an isotropic specific intensity, one finds

$$p_{xx} = p_{yy} = p_{zz} = \bar{p} = \tfrac{1}{3}u, \tag{1.23}$$

with all off diagonal terms zero, i.e.,

$$p_{xy} = p_{xz} = p_{yz} = 0. \tag{1.24}$$

In particular, for a Planck distribution one has

$$p_{xx} = p_{yy} = p_{zz} = \bar{p} = \tfrac{1}{3}aT^4. \tag{1.25}$$

At the other extreme from isotropy, if at a point in space and time photons are only traveling parallel to the x axis (the specific intensity is a Dirac delta function in the x direction), then at this point

$$p_{xx} = u, \tag{1.26}$$

with the other eight components of the pressure tensor identically zero.

It is interesting to note that in forming u, $\mathbf{F}$, and p we have constructed tensors of increasing rank by multiplying the specific intensity by 1, Ω, and $\Omega\Omega$, respectively, and integrating over all frequency and angle. One could continue this procedure to form tensors of higher rank, but the resulting quantities have little physical significance.

3. Interaction of the Radiation Field with Matter

Having introduced the basic quantities associated with a description of the radiation field, we now turn to a brief discussion of the three basic interactions between photons and matter, namely absorption, scattering, and emission.

We first consider absorption. As a photon travels through matter there is a certain probability that it will interact with the material and disappear, i.e., it will be absorbed. To describe this process quantitatively, we introduce the "macroscopic absorption coefficient" or, more simply, the absorption coefficient $\sigma_a \equiv \sigma_a(\mathbf{r}, v, t)$. This coefficient is defined such that the probability of a photon being absorbed in traveling a distance ds is given by

$$\text{probability of absorption} = \sigma_a(v)\, ds. \tag{1.27}$$

The absorption coefficient in general depends upon the frequency v of the photon as well as space and time since the material properties are in general functions of $\mathbf{r}$ and t. It is noteworthy that we have not indicated an angular dependence, i.e., a dependence upon Ω, for the absorption coefficient. This is a consequence of an assumption that, for a given path length ds, the probability of absorption is independent of the direction of travel of the photon. This implies, as we shall discuss in more detail in the next chapter, that the matter has no preferential direction as one finds, for example, in a crystal.

It is sometimes convenient to consider σ_a to be the product of two terms according to, suppressing the $\mathbf{r}$ and t dependences,

$$\sigma_a(\nu) = \varrho \varkappa(\nu), \tag{1.28}$$

where ϱ is the mass density of the material and $\varkappa(\nu)$ is called the "mass absorption coefficient". Alternately, one often writes

$$\sigma_a(\nu) = N\mu_a(\nu), \tag{1.29}$$

where N is the atomic density and $\mu_a(\nu)$ is called the "microscopic absorption coefficient".

Similar to absorption, a photon can undergo scattering interactions with matter. The "scattering coefficient" $\sigma_s \equiv \sigma_s(\mathbf{r}, \nu, t)$ is defined in analogy to the absorption coefficient such that

$$\text{probability of scattering} = \sigma_s(\nu)\, ds. \tag{1.30}$$

As with absorption, it is assumed that this probability is independent of the direction $\mathbf{\Omega}$. In a scattering event a photon does not disappear as in absorption, but continues to exist with another direction of travel and frequency, in general. That is, the scattering interaction serves to change the photon's characteristics ν' and $\mathbf{\Omega}'$ to a new set of characteristics ν and $\mathbf{\Omega}$. To quantitatively describe the scattering event, one requires a probabilistic statement concerning this change. This leads to the definition of the "differential scattering coefficient" $\sigma_s(\mathbf{r}, \nu' \rightarrow \nu, \mathbf{\Omega}' \cdot \mathbf{\Omega}, t)$ such that the probability of a photon being scattered from ν' to ν contained in $d\nu$, and from $\mathbf{\Omega}'$ to $\mathbf{\Omega}$ contained in $d\mathbf{\Omega}$, in traveling a distance ds is given by

$$\text{probability} = \sigma_s(\nu' \rightarrow \nu, \mathbf{\Omega}' \cdot \mathbf{\Omega})\, d\nu\, d\mathbf{\Omega}\, ds. \tag{1.31}$$

It is important here to note that rather than writing $\mathbf{\Omega}' \rightarrow \mathbf{\Omega}$ as an argument of σ_s to denote the transfer from direction $\mathbf{\Omega}'$ to direction $\mathbf{\Omega}$, we have written $\mathbf{\Omega}' \cdot \mathbf{\Omega}$. This dot product is just the cosine of the scattering angle and by using this single argument, rather than $\mathbf{\Omega}'$ and $\mathbf{\Omega}$ separately, we have assumed that the scattering probability depends only upon the scattered angle. This assumption is consistent with those concerning the lack of angular dependence of the absorption and scattering coefficients. If one integrates the differential scattering coefficient over all exit frequencies and angles, one obtains the scattering coefficient defined by Eq. (1.30). That is,

$$\sigma_s(\nu') = \int_0^\infty d\nu \int_{4\pi} d\mathbf{\Omega}\, \sigma_s(\nu' \rightarrow \nu, \mathbf{\Omega}' \cdot \mathbf{\Omega}) = 2\pi \int_0^\infty d\nu \int_{-1}^1 d\mu_0 \sigma_s(\nu' \rightarrow \nu, \mu_0). \tag{1.32}$$

The last equality follows by choosing $\mathbf{\Omega}'$ as the z axis for the $\mathbf{\Omega}$ integration, and letting $\mu_0 = \mathbf{\Omega}' \cdot \mathbf{\Omega}$.

Often it is convenient to decompose the differential scattering coefficient into the product of two terms according to

$$\sigma_s(\nu' \rightarrow \nu, \mathbf{\Omega}' \cdot \mathbf{\Omega}) = \sigma_s(\nu')K(\nu' \rightarrow \nu, \mathbf{\Omega}' \cdot \mathbf{\Omega}), \tag{1.33}$$

where the kernel $K \equiv K(\mathbf{r}, \nu' \to \nu, \boldsymbol{\Omega}' \cdot \boldsymbol{\Omega}, t)$ has the interpretation of a probability distribution denoting the probability of scattering to ν and $\boldsymbol{\Omega}$, given that a scattering event has occurred at ν' and $\boldsymbol{\Omega}'$. K is properly termed a probability distribution in that it is normalized to unity

$$\int_0^\infty d\nu \int_{4\pi} d\Omega K(\nu' \to \nu, \boldsymbol{\Omega}' \cdot \boldsymbol{\Omega}) = 2\pi \int_0^\infty d\nu \int_{-1}^1 d\mu_0 K(\nu' \to \nu, \mu_0) = 1, \tag{1.34}$$

as follows from Eqs. (1.32) and (1.33). If there is no frequency change upon scattering, the kernel K contains a Dirac delta function in frequency

$$K(\nu' \to \nu, \boldsymbol{\Omega}' \cdot \boldsymbol{\Omega}) = K(\boldsymbol{\Omega}' \cdot \boldsymbol{\Omega})\delta(\nu' - \nu), \tag{1.35}$$

and the scattering is said to be "coherent". It can be seen from Eq. (1.34) that $K(\boldsymbol{\Omega}' \cdot \boldsymbol{\Omega})$ is normalized according to

$$\int_{4\pi} d\Omega K(\boldsymbol{\Omega}' \cdot \boldsymbol{\Omega}) = 2\pi \int_{-1}^1 d\mu_0 K(\mu_0) = 1. \tag{1.36}$$

If the kernel $K(\nu' \to \nu, \boldsymbol{\Omega}' \cdot \boldsymbol{\Omega})$ is independent of the scattering angle

$$K(\nu' \to \nu, \boldsymbol{\Omega}' \cdot \boldsymbol{\Omega}) = \frac{1}{4\pi} K(\nu' \to \nu), \tag{1.37}$$

the scattering is termed "isotropic". From Eq. (1.34) it follows that $K(\nu' \to \nu)$ is normalized as

$$\int_0^\infty d\nu K(\nu' \to \nu) = 1. \tag{1.38}$$

If the scattering is both coherent and isotropic, one has the simplest kernel, namely

$$K(\nu' \to \nu, \boldsymbol{\Omega}' \cdot \boldsymbol{\Omega}) = \frac{1}{4\pi} \delta(\nu' - \nu). \tag{1.39}$$

For many problems of interest, Eq. (1.39) is a reasonable approximation to the actual kernel.

The "total interaction coefficient", denoted by $\sigma_t = \sigma_t(\mathbf{r}, \nu, t)$ or simply $\sigma = \sigma(\mathbf{r}, \nu, t)$, is given by

$$\sigma(\nu) = \sigma_a(\nu) + \sigma_s(\nu). \tag{1.40}$$

The symbol $\tilde{\omega} = \tilde{\omega}(\mathbf{r}, \nu, t)$ generally denotes the probability of scattering given that a collision has occurred, i.e.,

$$\tilde{\omega} = \frac{\sigma_s}{\sigma_a + \sigma_s} = \sigma_s/\sigma. \tag{1.41}$$

The case of no absorption, $\tilde{\omega} = 1$, is referred to as the "conservative" or "perfect scattering" case.

Another quantity which is often used to characterize the interaction between the radiation field and matter is the "mean free path" of a photon. If a photon is traveling in a homogeneous medium of interaction coefficient $\sigma(\nu)$, how far, on the average, will this photon stream before suffering a collision? This distance is called the mean free path and denoted by $\lambda = \lambda(\nu)$.

The Equations of Radiation Hydrodynamics

To derive the relationship between λ and σ, we need anticipate a special case of the equation of transfer to be discussed in detail in the next chapter. From the definition of the total interaction coefficient σ, we have

$$dN(\text{collided}) = -N\sigma\, ds, \qquad (1.42)$$

where N represents the number of photons in a directed beam of radiation and ds is an element of path length along this beam. Hence for this simple situation the equation of transfer is simply

$$\frac{dN}{ds} = -N\sigma, \qquad (1.43)$$

which has as its solution

$$N = N_0 e^{-\sigma s}. \qquad (1.44)$$

That is, N_0 photons initially in a beam will be reduced exponentially to N photons in the beam after traveling a distance s. The number of photons which collide in a path length element ds is then given by

$$|dN| = N_0 e^{-\sigma s}\, \sigma\, ds. \qquad (1.45)$$

This number of photons, namely $|dN|$, have traveled a distance s before suffering a collision, and hence the average distance $\bar{s}$, or λ, to a collision is just s averaged over $|dN|$, i.e.,

$$\bar{s} = \lambda = \frac{\displaystyle\int_0^\infty ds(N_0\sigma e^{-\sigma s})s}{\displaystyle\int_0^\infty ds(N_0\sigma e^{-\sigma s})}. \qquad (1.46)$$

This gives

$$\lambda = \lambda(\nu) = 1/\sigma(\nu). \qquad (1.47)$$

That is, in a homogeneous medium the mean free path, which in general is frequency dependent, is just the inverse of the total interaction coefficient. If the material properties are functions of space and time, the mean free path in turn depends upon these variables. In this more general situation, Eq. (1.47) is then taken as a definition of the mean free path, i.e.,

$$\lambda = \lambda(\mathbf{r},\, \nu,\, t) = 1/\sigma(\mathbf{r},\, \nu,\, t). \qquad (1.48)$$

Similarly, one can define absorption and scattering mean free paths as

$$\lambda_a(\mathbf{r},\, \nu,\, t) = 1/\sigma_a(\mathbf{r},\, \nu,\, t), \qquad (1.49)$$

$$\lambda_s(\mathbf{r},\, \nu,\, t) = 1/\sigma_s(\mathbf{r},\, \nu,\, t). \qquad (1.50)$$

These mean free paths obey the inverse addition rule

$$\lambda^{-1} = \lambda_a^{-1} + \lambda_s^{-1}. \qquad (1.51)$$

The final item we shall consider in this chapter is the "emission" of photons. We have already discussed the interaction of photons with matter which in the case of absorption destroys the photon and in the case of scattering changes its direction and frequency. In neither case are photons created. How then are photons introduced into the system? One way this can be accomplished is by shining light into the material through its bounding surface. We discuss this in the next chapter in connection with the equation of transfer.

The other possibility is that photons are born in the matter, through the process of spontaneous emission. That is, all materials spontaneously emit photons characteristic of the state of the matter. We quantify this source by introducing the function $q \equiv q(\mathbf{r}, v, t)$ such that the number of photons emitted per unit time and volume at frequency v in dv and direction Ω in $d\Omega$ is given by

$$\text{photons emitted} = q(\mathbf{r}, v, t)\, dv\, d\Omega. \tag{1.52}$$

This source of photons, as shown by the arguments of q, is taken to be independent of Ω. As previously stated, this follows from the assumption that the matter has no preferential direction. The frequency dependence of $q(v)$ will be considered in more detail in the next chapter.

II

The Equation of Transfer

1. Introduction

The equation of transfer, also called the transport equation, is the mathematical statement of the conservation of photons. Our purpose in this chapter is to formulate this conservation law in various forms. Initially, we derive the so-called integro-differential form of the equation of transfer. We give two alternate derivations which, of course, lead to the same result. Following a statement of the boundary conditions and a display of the equation in various geometries, we cast the equation of transfer into integral form, with particular emphasis on the isotropic scattering case. We reduce the general integral form to that appropriate in the three one dimensional geometries (plane, cylinder, sphere). A variational characterization of the equation of transfer is also introduced. We end the chapter with discussions of (1) the inclusion of induced emission and scattering (since photons are bosons) in the equation of transfer; (2) the concept of local thermodynamic equilibrium which simplifies the source term in the equation of transfer; and (3) the validity of the equation of transfer as a description of the transport of electromagnetic energy.

2. The Equation of Transfer—An Eulerian Derivation

In analogy to the terminology used in hydrodynamics, we present in this section what might be termed an Eulerian derivation of the equation of transfer.

Let the vector $\mathbf{r}$ be described by the Cartesian coordinates x, y, z and the vector $\mathbf{\Omega}$ be described by a polar angle θ, measured with respect to a fixed axis in space (such as the z axis), and a corresponding azimuthal angle φ. If we introduce $\mu \equiv \cos\theta$, then

$$d\mathbf{r} = dx\, dy\, dz, \tag{2.1}$$

$$d\mathbf{\Omega} = \sin\theta\, d\theta\, d\varphi = d\mu\, d\varphi. \tag{2.2}$$

We consider a six dimensional "cube", fixed in space, of dimensions Δx, Δy, Δz, Δv, $\Delta\mu$, $\Delta\varphi$. The number of photons within this cube at time t is given by

$$\text{number of photons} = f(\mathbf{r}, v, \mathbf{\Omega}, t)\, \Delta x\, \Delta y\, \Delta z\, \Delta v\, \Delta\mu\, \Delta\varphi, \tag{2.3}$$

where $f(\mathbf{r}, v, \mathbf{\Omega}, t)$ is the distribution function introduced in the last chapter. The arguments of f are average arguments; i.e., v means an average value of v in Δv, and similarly for the arguments $\mathbf{r}$ and $\mathbf{\Omega}$. Eventually we will let Δv go to the differential interval dv, $\Delta\mathbf{r}$ go to $d\mathbf{r}$, and $\Delta\mathbf{\Omega}$ go to $d\mathbf{\Omega}$, so the exact definition of these averages is not important.

Now, the time rate of change of the number of photons in this six dimensional cube is given by

$$\text{change} = \frac{\partial}{\partial t}[f(\mathbf{r}, v, \mathbf{\Omega}, t)\,\Delta V] = \Delta V \frac{\partial}{\partial t} f(\mathbf{r}, v, \mathbf{\Omega}, t), \tag{2.4}$$

where we let ΔV denote the volume of the cube $\Delta V = \Delta x\,\Delta y\,\Delta z\,\Delta v\,\Delta\mu\,\Delta\varphi$. The volume element ΔV can be taken outside the time derivative in Eq. (2.4) since it is a volume element fixed in space, not dependent upon time or any other variables. This time rate of change is due to five separate processes; namely, the net rate of streaming of photons out of the cube through the bounding surface, and the rate of four processes occurring within the cube: (1) absorption, (2) scattering from v, $\mathbf{\Omega}$ to all other frequencies and directions (outscattering), (3) scattering to v, $\mathbf{\Omega}$ from all other frequencies and directions (inscattering), and (4) emission (the source of photons). We consider each of these five contributions separately.

We first consider the rate of streaming through the surfaces of the cube perpendicular to the x axis. We have, for the net rate of photons leaving the cube through these two surfaces,

$$(\text{streaming})_x = \dot{x} f(\mathbf{r}, v, \mathbf{\Omega}, t)\,\Big|_x^{x+\Delta x} \Delta y\,\Delta z\,\Delta v\,\Delta\mu\,\Delta\varphi, \tag{2.5}$$

where $\dot{x}$ denotes the x component of the photon velocity ($\dot{x} = c\Omega_x$) and $\Delta y\,\Delta z\,\Delta v\,\Delta\mu\,\Delta\varphi$ is the appropriate surface area. Anticipating that Δx will approach the differential dx, we can rewrite Eq. (2.5) as

$$(\text{streaming})_x = \Delta V \frac{\partial}{\partial x}[\dot{x} f(\mathbf{r}, v, \mathbf{\Omega}, t)]. \tag{2.6}$$

Similar results follow for the flow from the cube in the other five directions (y, z, v, μ, φ), and the total net rate of streaming from the cube is given by

$$\text{streaming} = \left[\frac{\partial}{\partial x}(\dot{x}f) + \frac{\partial}{\partial y}(\dot{y}f) + \frac{\partial}{\partial z}(\dot{z}f) + \frac{\partial}{\partial v}(\dot{v}f) + \frac{\partial}{\partial \mu}(\dot{\mu}f) + \frac{\partial}{\partial \varphi}(\dot{\varphi}f)\right]\Delta V, \tag{2.7}$$

where $f \equiv f(\mathbf{r}, v, \mathbf{\Omega}, t)$. A term such as $\dot{\varphi}$ in Eq. (2.7) deserves a bit of explanation. It is the time rate of change of the angle φ, describing the azimuthal direction of the photon, as the photon streams along its path.

The rate of absorption within the six dimensional cube is the product of the number of photons in the cube, $f\Delta V$, and the probability of absorption per photon and per unit time. This probability follows from the discussion in Chapter I and is just the product of the absorption coefficient $\sigma_a(\mathbf{r}, v, t)$ and the rate of path length generation, which is just the photon speed c. Hence the rate of absorption within the cube is given by

$$\text{absorption} = c\sigma_a(\mathbf{r}, v, t) f(\mathbf{r}, v, \mathbf{\Omega}, t)\,\Delta V. \tag{2.8}$$

Similar arguments give the contributions due to scattering in the cube. We find for the time rate of outscattering and inscattering within the vol ume element ΔV

$$\text{outscattering} = c\Delta V \int_0^\infty dv' \int_{4\pi} d\Omega' \sigma_s(\mathbf{r}, v \to v', \boldsymbol{\Omega} \cdot \boldsymbol{\Omega}', t) f(\mathbf{r}, v, \boldsymbol{\Omega}, t), \qquad (2.9)$$

$$\text{inscattering} = c\Delta V \int_0^\infty dv' \int_{4\pi} d\Omega' \sigma_s(\mathbf{r}, v' \to v, \boldsymbol{\Omega}' \cdot \boldsymbol{\Omega}, t) f(\mathbf{r}, v', \boldsymbol{\Omega}', t), \qquad (2.10)$$

where $\sigma_s(\mathbf{r}, v' \to v, \boldsymbol{\Omega}' \cdot \boldsymbol{\Omega}, t)$ is the differential scattering coefficient introduced in Chapter I. The integrals over v' and $\boldsymbol{\Omega}'$ in Eq. (2.9) account for the fact that we need sum the scattering from $v, \boldsymbol{\Omega}$ to all other frequencies and directions $v', \boldsymbol{\Omega}'$. Similarly, these integrations are required in Eq. (2.10) to account for scattering from all $v', \boldsymbol{\Omega}'$ to the frequency and direction of interest $v, \boldsymbol{\Omega}$. The integrations in Eq. (2.9) can be performed, if desired, since the distribution function in the integrand is independent of the integration variables. This yields

$$\text{outscattering} = c\Delta V \sigma_s(\mathbf{r}, v, t) f(\mathbf{r}, v, \boldsymbol{\Omega}, t), \qquad (2.11)$$

where $\sigma_s(\mathbf{r}, v, t)$ is the total scattering coefficient as defined by Eq. (1.32).

The final item in the balance equation we need consider is the emission of photons. In view of the definition of the function $q \equiv q(\mathbf{r}, v, t)$ given in the last chapter, the rate of production of photons within the cube under consideration is simply

$$\text{emission} = q(\mathbf{r}, v, t)\, \Delta V. \qquad (2.12)$$

As pointed out earlier, this emission rate is independent of $\boldsymbol{\Omega}$.

The equation of transfer follows by equating the sum of these five terms, Eqs. (2.7) through (2.10) and (2.12), with appropriate signs to designate a loss or gain, to the overall rate given by Eq. (2.4). Cancelling the $\Delta V = \Delta x\, \Delta y\, \Delta z\, \Delta v\, \Delta \mu\, \Delta \varphi$ common to all terms, and letting ΔV approach a differential element to be consistent with the passage from Eq. (2.5) to (2.6), we find

$$\frac{\partial f(v, \boldsymbol{\Omega})}{\partial t} + \frac{\partial(\dot{x}f)}{\partial x} + \frac{\partial(\dot{y}f)}{\partial y} + \frac{\partial(\dot{z}f)}{\partial z} + \frac{\partial(\dot{v}f)}{\partial v} + \frac{\partial(\dot{\mu}f)}{\partial \mu} + \frac{\partial(\dot{\varphi}f)}{\partial \varphi}$$

$$= q(v) - c\sigma_a(v) f(v, \boldsymbol{\Omega}) + c \int_0^\infty dv' \int_{4\pi} d\Omega' \sigma_s(v' \to v, \boldsymbol{\Omega}' \cdot \boldsymbol{\Omega}) f(v', \boldsymbol{\Omega}')$$

$$- c \int_0^\infty dv' \int_{4\pi} d\Omega' \sigma_s(v \to v', \boldsymbol{\Omega} \cdot \boldsymbol{\Omega}') f(v, \boldsymbol{\Omega}). \qquad (2.13)$$

For notational simplicity we have dropped all arguments $\mathbf{r}$ and t, as well as v and $\boldsymbol{\Omega}$ in the streaming terms, in Eq. (2.13). Now $\dot{\mu} = \dot{\varphi} = 0$, since these angles are measured with respect to fixed axes in space, and a photon is assumed to travel in a straight line between collisions. Further, $\dot{v} = 0$ since photons stream with no change in frequency. Finally, $\dot{x} = c\Omega_x$, $\dot{y} = c\Omega_y$, and $\dot{z} = c\Omega_z$, since photons stream with speed c in the direction $\boldsymbol{\Omega}$. Thus Eq. (2.13) simplifies to

$$\frac{\partial f(v, \boldsymbol{\Omega})}{\partial t} + c\boldsymbol{\Omega} \cdot \nabla f(v, \boldsymbol{\Omega}) = q(v) - c\sigma_a(v) f(v, \boldsymbol{\Omega})$$

$$+ c \int_0^\infty dv' \int_{4\pi} d\Omega' [\sigma_s(v' \to v, \boldsymbol{\Omega}' \cdot \boldsymbol{\Omega}) f(v', \boldsymbol{\Omega}') - \sigma_s(v \to v', \boldsymbol{\Omega} \cdot \boldsymbol{\Omega}') f(v, \boldsymbol{\Omega})]. \qquad (2.14)$$

If we introduce the specific intensity of radiation, $I = ch\nu f$, in place of f in Eq. (2.14), we find the equation of transfer in the conventional form

$$\frac{1}{c}\frac{\partial I(\nu, \Omega)}{\partial t} + \Omega \cdot \nabla I(\nu, \Omega) = S(\nu) - \sigma_a(\nu)\, I(\nu, \Omega)$$

$$+ \int_0^\infty d\nu' \int_{4\pi} d\Omega' \left[\frac{\nu}{\nu'}\sigma_s(\nu' \to \nu, \Omega' \cdot \Omega)\, I(\nu', \Omega') - \sigma_s(\nu \to \nu', \Omega \cdot \Omega')\, I(\nu, \Omega)\right]. \quad (2.15)$$

In this equation we have defined the function $S = S(\mathbf{r}, \nu, t)$ as

$$S(\mathbf{r}, \nu, t) = h\nu q(\mathbf{r}, \nu, t), \qquad (2.16)$$

which is obviously the rate of energy emission due to spontaneous processes. Sometimes one sees Eq. (2.15) with the integrals in the outscattering term carried out. This form of Eq. (2.15) is

$$\frac{1}{c}\frac{\partial I(\nu, \Omega)}{\partial t} + \Omega \cdot \nabla I(\nu, \Omega) = S(\nu) - \sigma(\nu)\, I(\nu, \Omega)$$

$$+ \int_0^\infty d\nu' \int_{4\pi} d\Omega' \frac{\nu}{\nu'}\sigma_s(\nu' \to \nu, \Omega' \cdot \Omega)\, I(\nu', \Omega'), \qquad (2.17)$$

where $\sigma(\mathbf{r}, \nu, t)$ is the total interaction coefficient (the sum of the absorption and scattering coefficients).

Equation (2.16) or (2.17) is clearly an integro-differential equation which, because of its partly differential character, requires both spatial and temporal boundary conditions. Before discussing these conditions, however, we consider an alternate derivation of this important result.

3. The Equation of Transfer—A Lagrangian Derivation

In this section we give what might be called a Lagrangian derivation of the equation of transfer, borrowing this term from the field of hydrodynamics.

As in the last section, we choose a coordinate system such that in a six dimensional volume element we have

$$\text{number of photons} = f(\mathbf{r}, \nu, \Omega, t)\, \Delta x\, \Delta y\, \Delta z\, \Delta \nu\, \Delta \mu\, \Delta \varphi. \qquad (2.18)$$

As this packet of photons travels in matter, its number would be conserved except for the processes of absorption, scattering, and emission. We have just computed in the last section the effects of these processes, and we can immediately write the balance equation

$$\frac{d}{dt}[f(\nu, \Omega)\, \Delta V] = q(\nu)\, \Delta V - c\sigma_a(\nu)f(\nu, \Omega)\, \Delta V$$

$$+ c\Delta V \int_0^\infty d\nu' \int_{4\pi} d\Omega'\, [\sigma_s(\nu' \to \nu, \Omega' \cdot \Omega)f(\nu', \Omega')$$

$$- \sigma_s(\nu \to \nu', \Omega \cdot \Omega')f(\nu, \Omega)], \qquad (2.19)$$

where we have again dropped all arguments $\mathbf{r}$ and t for simplicity and set $\Delta V = \Delta x\,\Delta y\,\Delta z\,\Delta v\,\Delta\mu\,\Delta\varphi$. Before proceeding with the development, some discussion of Eq. (2.19) is required. In the first place, the time derivative here is a total derivative, taken along the natural path of the streaming packet of photons. That is, $d(f\Delta V)/dt$ means the difference between the value of $f\Delta V$ at $s+ds$ and its value at s (where $ds = c\,dt$ is an element of length along the photon path) divided by the transit time dt. This difference in values comes, in general, from both an explicit time dependence of $f\Delta V$ and an implicit time dependence through the other variables involved. This total time derivative is the so-called Lagrangian time derivative of hydrodynamics. One can also, in terms of d/dt, define a total path length derivative as

$$\frac{d}{ds} = \frac{1}{c}\frac{d}{dt}.$$

$$(2.20)$$

Such a derivative is often used in transport theory discussions in lieu of d/dt.

Secondly, the volume element $\Delta V = \Delta x\,\Delta y\,\Delta z\,\Delta v\,\Delta\mu\,\Delta\varphi$ is not fixed in space as in the last section, but travels along with the packet of photons. Its size varies in time in just such a way that at any instant of time it encompasses the photons of interest. Hence $d\Delta V/dt \neq 0$, but the change in ΔV with time must be calculated. However, the angles $\theta = \cos^{-1}(\mu)$ and φ are measured with respect to fixed axes in space, just as in the last section.

With this discussion of the precise meaning of the terms in Eq. (2.19), we proceed with the development of the equation of transfer from this Lagrangian point of view. The elementary rule for differentiating a product gives, for the left hand side of Eq. (2.19),

$$\frac{d}{dt}(f\Delta x\,\Delta y\,\Delta z\,\Delta v\,\Delta\mu\,\Delta\varphi) = \Delta x\,\Delta y\,\Delta z\,\Delta v\,\Delta\mu\,\Delta\varphi\,\frac{df}{dt}$$

$$+f\,\Delta y\,\Delta z\,\Delta v\,\Delta\mu\,\Delta\varphi\,\frac{d\Delta x}{dt}$$

$$+\text{(five similar terms)},$$

$$(2.21)$$

or, introducing ΔV,

$$\frac{d}{dt}(f\Delta V) = \Delta V\,\frac{df}{dt}+f\Delta V\left[\frac{1}{\Delta x}\,\frac{d\Delta x}{dt}+\frac{1}{\Delta y}\,\frac{d\Delta y}{dt}\right.$$

$$\left.+\frac{1}{\Delta z}\,\frac{d\Delta z}{dt}+\frac{1}{\Delta v}\,\frac{d\Delta v}{dt}+\frac{1}{\Delta\mu}\,\frac{d\Delta\mu}{dt}+\frac{1}{\Delta\varphi}\,\frac{d\Delta\varphi}{dt}\right].$$

$$(2.22)$$

Considering $\Delta x \equiv x_2 - x_1$, we have

$$\frac{d\Delta x}{dt} = \frac{d}{dt}(x_2 - x_1) = \dot{x}\bigg|_{x=x_2} - \dot{x}\bigg|_{x=x_1} = \frac{\partial\dot{x}}{\partial x}\,\Delta x.$$

$$(2.23)$$

Similar results follow for $d\Delta y/dt$, $d\Delta z/dt$, etc. Equation (2.22) then becomes

$$\frac{d}{dt}(f\Delta V) = \Delta V\,\frac{df}{dt}+f\Delta V\left[\frac{\partial\dot{x}}{\partial x}+\frac{\partial\dot{y}}{\partial y}+\frac{\partial\dot{z}}{\partial z}+\frac{\partial\dot{v}}{\partial v}+\frac{\partial\dot{\mu}}{\partial\mu}+\frac{\partial\dot{\varphi}}{\partial\varphi}\right].$$

$$(2.24)$$

Recognizing that $f \equiv f(x, y, z, v, \mu, \varphi, t)$ we find that the use of the chain rule of differentiation to compute df/dt allows Eq. (2.24) to be written

$$\frac{d}{dt}(f\Delta V) = \Delta V \left[\frac{\partial f}{\partial t} + \dot{x}\,\frac{\partial f}{\partial x} + \dot{y}\,\frac{\partial f}{\partial y} + \dot{z}\,\frac{\partial f}{\partial z} + \dot{v}\,\frac{\partial f}{\partial v} + \dot{\mu}\,\frac{\partial f}{\partial \mu} + \dot{\varphi}\,\frac{\partial f}{\partial \varphi} \right]$$
$$+ f\Delta V \left[\frac{\partial \dot{x}}{\partial x} + \frac{\partial \dot{y}}{\partial y} + \frac{\partial \dot{z}}{\partial z} + \frac{\partial \dot{v}}{\partial v} + \frac{\partial \dot{\mu}}{\partial \mu} + \frac{\partial \dot{\varphi}}{\partial \varphi} \right] . \tag{2.25}$$

Combining the two bracketed terms in Eq. (2.25) we obtain our final result for the Lagrangian time derivative of $f\Delta V$, namely

$$\frac{d}{dt}(f\Delta V) = \Delta V \left[\frac{\partial f}{\partial t} + \frac{\partial(\dot{x}f)}{\partial x} + \frac{\partial(\dot{y}f)}{\partial y} + \frac{\partial(\dot{z}f)}{\partial z} + \frac{\partial(\dot{v}f)}{\partial v} + \frac{\partial(\dot{\mu}f)}{\partial \mu} + \frac{\partial(\dot{\varphi}f)}{\partial \varphi} \right]. \tag{2.26}$$

Use of this result in Eq. (2.19) gives, upon cancellation of the ΔV common to all terms,

$$\frac{\partial f(v, \mathbf{\Omega})}{\partial t} + \frac{\partial(\dot{x}f)}{\partial x} + \frac{\partial(\dot{y}f)}{\partial y} + \frac{\partial(\dot{z}f)}{\partial z} + \frac{\partial(\dot{v}f)}{\partial v} + \frac{\partial(\dot{\mu}f)}{\partial \mu} + \frac{\partial(\dot{\varphi}f)}{\partial \varphi}$$
$$= q(v) - c\sigma_a(v)f(v, \mathbf{\Omega}) + c\int_0^\infty dv' \int_{4\pi} d\Omega' [\sigma_s(v' \rightarrow v, \mathbf{\Omega}' \cdot \mathbf{\Omega})f(v', \mathbf{\Omega}')$$
$$- \sigma_s(v \rightarrow v', \mathbf{\Omega} \cdot \mathbf{\Omega}')f(v, \mathbf{\Omega})]. \tag{2.27}$$

This result is identical to Eq. (2.13) of the last section. Taking account of the fact that photons stream in straight lines with no change in frequency, and introducing the specific intensity I in place of the distribution function f, we are once again led to the integro-differential form of the transport equation

$$\frac{1}{c}\frac{\partial I(v, \mathbf{\Omega})}{\partial t} + \mathbf{\Omega} \cdot \nabla I(v, \mathbf{\Omega}) = S(v) - \sigma_a(v)\,I(v, \mathbf{\Omega})$$
$$+ \int_0^\infty dv' \int_{4\pi} d\Omega' \left[\frac{v}{v'}\sigma_s(v' \rightarrow v, \mathbf{\Omega}' \cdot \mathbf{\Omega})\,I(v', \mathbf{\Omega}') - \sigma_s(v \rightarrow v', \mathbf{\Omega} \cdot \mathbf{\Omega}')\,I(v, \mathbf{\Omega}) \right]. \tag{2.28}$$

Both the derivation just given and the Eulerian derivation of the previous section are equally rigorous derivations of the equation of transfer. One or the other may appeal to any given reader. We felt it useful to give two derivations because of the different viewpoints they involve, and the importance of the result. Almost all of neutral particle transport work is based on this equation or equations (either approximate or exact) derived from this result.

4. The Boundary Conditions on the Equation of Transfer

Since the transport equation is a first order differential equation in space and time, boundary conditions in both these variables are required.

We assume that the system of interest, which is arbitrary in composition and shape, is non-re-entrant and characterized by a volume V and surface area A. By non-re-entrant we mean that any photon which leaves the body will not re-enter it through another part of the surface. If the body is re-entrant, we enclose it in a non-re-entrant hypothetical surface (such as a spherical shell) and consider the system to be bounded by the imposed, rather than the real, surface. The new system is then non-re-entrant, but consists, in part, of vacuum ($\sigma = S = 0$).

On physical grounds we know that it is sufficient to specify the specific intensity at all points on the surface A in the incoming direction. For a non-re-entrant surface, this implies the boundary condition

$$I(\mathbf{r}_s, \, \nu, \, \mathbf{\Omega}, \, t) = \Gamma(\mathbf{r}_s, \, \nu, \, \mathbf{\Omega}, \, t), \quad \mathbf{n} \cdot \mathbf{\Omega} < 0, \tag{2.29}$$

where Γ is a specified function of all its arguments, $\mathbf{r}_s$ is a point on the surface, and $\mathbf{n}$ is an outward normal vector at this point.

An important special case of Eq. (2.29) is the so-called "vacuum" or "free surface" boundary condition

$$I(\mathbf{r}_s, \, \nu, \, \mathbf{\Omega}, \, t) = 0, \quad \mathbf{n} \cdot \mathbf{\Omega} < 0. \tag{2.30}$$

This merely states that no photons enter the system through its bounding surface.

In the time variable we assume that the range of interest is $0 < t < \infty$, and hence we need specify the initial condition at $t = 0$. We write this temporal boundary condition as

$$I(\mathbf{r}, \, \nu, \, \mathbf{\Omega}, \, 0) = \Lambda(\mathbf{r}, \, \nu, \, \mathbf{\Omega}), \tag{2.31}$$

where Λ is a specified function of its three arguments.

The equation of transfer, Eq. (2.28), together with the boundary conditions, Eqs. (2.29) and (2.31), completely specify the radiative transfer problem, at least within the limits of the physics assumed thus far. In radiative transfer work the main task is to simplify this very complex equation to the point of being able to obtain a solution without losing the essential physics in the process.

5. The Equation of Transfer in Various Coordinate Systems

In this section we consider the form the general equation of transfer takes in specific coordinate systems. In all cases the essential point is to interpret $\mathbf{\Omega} \cdot \nabla I$ in Eq. (2.17) as a directional derivative in the $\mathbf{\Omega}$ direction. That is, we write

$$\mathbf{\Omega} \cdot \nabla I = \frac{\partial I}{\partial s}, \tag{2.32}$$

where s is a length along $\mathbf{\Omega}$. We use a partial derivative here to distinguish between this directional derivative and the total path derivative d/ds introduced in Eq. (2.20). In particular, the s derivative in Eq. (2.32) is taken with time and frequency held constant.

We first consider the simplest case, that of a plane system which is infinite in the x and y directions. The specific intensity then depends only upon the single spatial coordinate z and the single angular coordinate θ, the angle between $\mathbf{\Omega}$ and the z axis. (We have assumed that any photon distribution incident upon the slab faces is independent of x and y and can

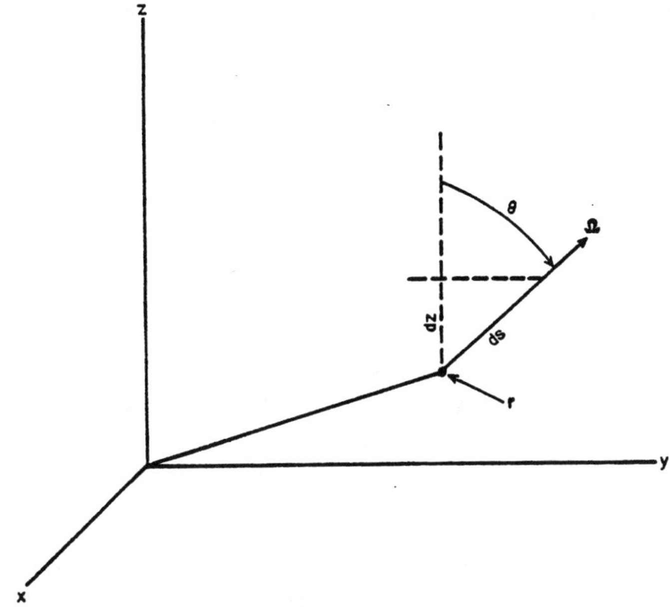

FIG. 2.1. Plane geometry.

be described by this single angular variable.) This situation is shown in Fig. 2.1. Introducing $\mu \equiv \cos\theta$ we have, since $I \equiv I(z, \mu)$,

$$\frac{\partial I}{\partial s} = \frac{\partial I}{\partial z}\left(\frac{dz}{ds}\right) + \frac{\partial I}{\partial \mu}\left(\frac{d\mu}{ds}\right). \tag{2.33}$$

From the figure we clearly see that

$$\frac{dz}{ds} = \cos\theta \equiv \mu; \quad \frac{d\mu}{ds} = 0. \tag{2.34}$$

Hence in this geometry Eq. (2.17) becomes

$$\frac{1}{c}\frac{\partial I(\nu, \mu)}{\partial t} + \mu\frac{\partial I(\nu, \mu)}{\partial z} + \sigma(\nu)\,I(\nu, \mu) = S(\nu)$$

$$+ \int_0^\infty d\nu' \int_{4\pi} d\Omega' \,\frac{\nu}{\nu'}\,\sigma_s(\nu' \to \nu, \Omega'\cdot\Omega)\,I(\nu', \mu'), \tag{2.35}$$

where $I \equiv I(z, \nu, \mu, t)$.

One can also simplify the scattering term in plane geometry. We expand the angular dependence of the differential scattering coefficient in the Legendre polynomials $P_n(\Omega'\cdot\Omega)$ according to

$$\sigma_s(\nu' \to \nu, \Omega'\cdot\Omega) = \sum_{n=0}^\infty \frac{2n+1}{4\pi}\,\sigma_{sn}(\nu' \to \nu)\,P_n(\Omega'\cdot\Omega). \tag{2.36}$$

The Equations of Radiation Hydrodynamics

The $\sigma_{sn}(\nu' \to \nu)$ are the expansion coefficients which follow from the orthogonality property of the Legendre polynomials as

$$\sigma_{sn}(\nu' \to \nu) = 2\pi \int_{-1}^{1} d\mu_0 \sigma_s(\nu' \to \nu, \mu_0)\, P_n(\mu_0). \tag{2.37}$$

To proceed further we make use of the so-called addition formula for the Legendre polynomials

$$P_n(\mathbf{\Omega'} \cdot \mathbf{\Omega}) = P_n(\mu)P_n(\mu') + 2\sum_{m=1}^{n} \frac{(n-m)!}{(n+m)!}\, P_n^m(\mu)\, P_n^m(\mu')\, \cos m(\varphi - \varphi'). \tag{2.38}$$

Here $\theta \equiv \cos^{-1}(\mu)$ and φ are polar and azimuthal angles respectively defining the direction $\mathbf{\Omega}$, and similarly μ' and φ' define $\mathbf{\Omega'}$. The $P_n^m(\mu)$ in Eq. (2.38) are the associated Legendre polynomials. For a general discussion of Legendre polynomials and the addition formula, see, for example, Erdelyi *et al.* (1953a). Equations (2.36) and (2.38) allow us to write

$$\int_{4\pi} d\Omega' \sigma_s(\nu' \to \nu, \mathbf{\Omega'} \cdot \mathbf{\Omega})\, I(\nu', \mu') = \int_{0}^{2\pi} d\varphi' \int_{-1}^{1} d\mu' \sigma_s(\nu' \to \nu, \mathbf{\Omega'} \cdot \mathbf{\Omega})\, I(\nu', \mu')$$

$$= \sum_{n=0}^{\infty} \frac{2n+1}{4\pi}\, \sigma_{sn}(\nu' \to \nu) \int_{-1}^{1} d\mu' I(\nu', \mu') \int_{0}^{2\pi} d\varphi' \left[P_n(\mu)\, P_n(\mu') \right.$$

$$\left. + 2\sum_{m=1}^{n} \frac{(n-m)!}{(n+m)!}\, P_n^m(\mu)\, P_n^m(\mu')\, \cos m(\varphi - \varphi') \right]. \tag{2.39}$$

Now, because of the periodicity of the cosine function

$$\int_{0}^{2\pi} d\varphi'\, \cos m(\varphi - \varphi') = 0, \qquad m = 1, 2, \ldots, \tag{2.40}$$

and hence Eq. (2.39) simplifies to

$$\int_{4\pi} d\Omega' \sigma_s(\nu' \to \nu, \mathbf{\Omega'} \cdot \mathbf{\Omega})\, I(\nu', \mu') = \sum_{n=0}^{\infty} \frac{2n+1}{2}\, \sigma_{sn}(\nu' \to \nu)\, P_n(\mu) \int_{-1}^{1} d\mu' P_n(\mu')\, I(\nu', \mu').$$

$$\tag{2.41}$$

Use of Eq. (2.41) in Eq. (2.35) yields the final result for the equation of transfer in plane geometry,

$$\frac{1}{c}\frac{\partial I(\nu, \mu)}{\partial t} + \mu \frac{\partial I(\nu, \mu)}{\partial z} + \sigma(\nu)\, I(\nu, \mu) = S(\nu)$$

$$+ \sum_{n=0}^{\infty} \frac{2n+1}{2}\, P_n(\mu) \int_{0}^{\infty} d\nu' \frac{\nu}{\nu'}\, \sigma_{sn}(\nu' \to \nu) \int_{-1}^{1} d\mu' P_n(\mu')\, I(\nu', \mu'), \tag{2.42}$$

where $I \equiv I(z, \nu, \mu, t)$. In practice, the infinite sum in Eq. (2.42) can generally be truncated after only a few terms. This is true for two reasons. In the first place, the differential scattering coefficients of interest are generally well represented by a few terms in the Legendre expansion, and, secondly, the solution of the equation of transfer is relatively insensitive to the higher terms in the expansion.

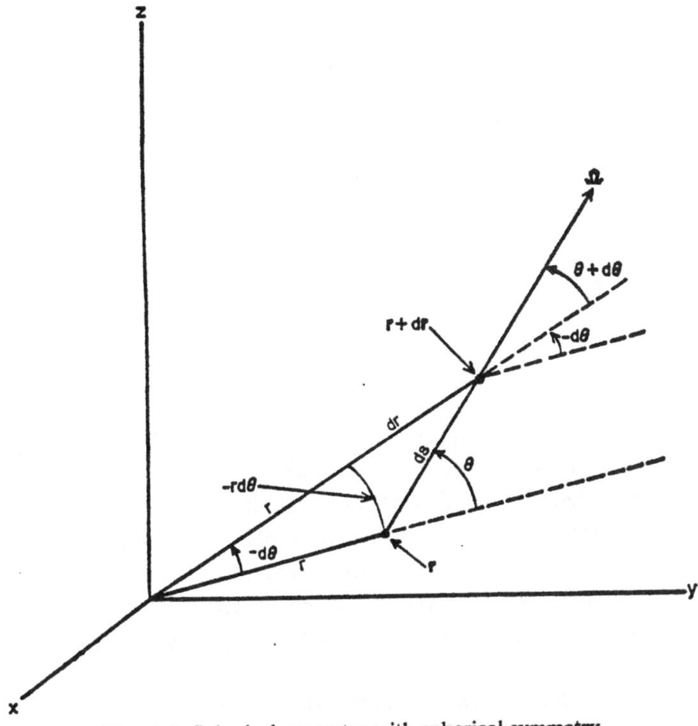

FIG. 2.2. Spherical geometry with spherical symmetry.

We now consider the slightly more difficult case of spherical geometry with complete spherical symmetry. In this situation the specific intensity again depends upon only one spatial variable, the distance from the origin r, and one angular variable, the angle θ between r and Ω. We define $\mu \equiv \cos \theta$ and consider $I \equiv I(r, \mu)$. Then

$$\frac{\partial I}{\partial s} = \frac{\partial I}{\partial r}\left(\frac{dr}{ds}\right) + \frac{\partial I}{\partial \mu}\left(\frac{d\mu}{ds}\right). \tag{2.43}$$

The geometry in this case is shown in Fig. 2.2. We examine the right triangle with sides dr, ds, and $-r d\theta$. The angle between dr and ds is $\theta + d\theta$ or, since $d\theta$ is a differential element, this angle can be taken as simply θ. Then

$$\cos \theta = \frac{dr}{ds}; \qquad \sin \theta = -r\frac{d\theta}{ds}. \tag{2.44}$$

The Equations of Radiation Hydrodynamics

In terms of $\mu \equiv \cos \theta$, Eq. (2.44) gives precisely what is needed, namely

$$\frac{dr}{ds} = \mu; \qquad \frac{d\mu}{ds} = \frac{1-\mu^2}{r}. \tag{2.45}$$

Thus, in spherical geometry with spherical symmetry,

$$\frac{\partial I}{\partial s} = \mu \frac{\partial I}{\partial r} + \frac{(1-\mu^2)}{r} \frac{\partial I}{\partial \mu}, \tag{2.46}$$

and the equation of transfer, Eq. (2.17), becomes

$$\frac{1}{c} \frac{\partial I(\nu, \mu)}{\partial t} + \mu \frac{\partial I(\nu, \mu)}{\partial r} + \frac{(1-\mu^2)}{r} \frac{\partial I(\nu, \mu)}{\partial \mu} + \sigma(\nu) I(\nu, \mu)$$

$$= S(\nu) + \sum_{n=0}^{\infty} \frac{2n+1}{2} P_n(\mu) \int_0^{\infty} d\nu' \frac{\nu}{\nu'} \sigma_{sn}(\nu' \to \nu) \int_{-1}^{1} d\mu' \, P_n(\mu') I(\nu', \mu'), \tag{2.47}$$

where $I \equiv I(r, \nu, \mu, t)$. As in the plane case, the scattering term has been simplified by use of the addition formula for the Legendre polynomials.

An interesting point here is that $d\mu/ds$ is not zero, but given by Eq. (2.45). On the other hand, in the general discussion of the equation of transfer in Sections 2 and 3 of this chapter, we set $d\mu/ds = 0$. The difference is that in this general discussion, μ was considered as the cosine of an angle defined with respect to a fixed (in space) axis; whereas, here the angle $\theta \equiv \cos^{-1}(\mu)$ is measured with respect to the radial coordinate which is *not* fixed in space (it varies with the photon coordinate r).

It is an instructive aside to return to the derivations in Sections 2 and 3 and, in the case of spherical symmetry, define the coordinates r and μ as appropriate in this case from the outset. If we do this, identifying x with r and deleting the extraneous variables y, z, and φ, we find that Eq. (2.13) or (2.27) reads, recognizing that $\dot{\nu} = 0$,

$$\frac{\partial f'(\nu, \mu)}{\partial t} + \frac{\partial(\dot{r}f')}{\partial r} + \frac{\partial(\dot{\mu}f')}{\partial \mu} = q'(\nu) - c\sigma_a(\nu) f'(\nu, \mu)$$

$$+ c \int_0^{\infty} d\nu' \int_{4\pi} d\Omega' [\sigma_s(\nu' \to \nu, \Omega' \cdot \Omega) f'(\nu', \mu') - \sigma_s(\nu \to \nu', \Omega \cdot \Omega') f'(\nu, \mu)], \tag{2.48}$$

where $f' \equiv f'(r, \nu, \mu, t)$. We have primed both f and q to distinguish these quantities as the special ones now under consideration. Introducing the specific intensity, simplifying the scattering term as we have done earlier, and recognizing that Eq. (2.45) is equivalent to

$$\dot{r} = c\mu; \qquad \dot{\mu} = \frac{c(1-\mu^2)}{r}, \tag{2.49}$$

we find that Eq. (2.48) gives the equation of transfer

$$\frac{1}{c}\frac{\partial I'(\nu,\mu)}{\partial t}+\mu\frac{\partial I'(\nu,\mu)}{\partial r}+\frac{1}{r}\frac{\partial}{\partial\mu}[(1-\mu^2)\,I'(\nu,\mu)]+\sigma(\nu)\,I'(\nu,\mu)$$

$$=S'(\nu)+\sum_{n=0}^{\infty}\frac{2n+1}{2}\,P_n(\mu)\int_0^{\infty}d\nu'\,\frac{\nu}{\nu'}\,\sigma_{sn}(\nu'\rightarrow\nu)\int_{-1}^{1}d\mu'P_n(\mu')\,I'(\nu',\mu'),\qquad(2.50)$$

where $I'\equiv I'(r,\nu,\mu,t)$.

Comparison of this result with Eq. (2.47) shows that these two equations of transfer, both presumably correct, differ in the $\partial/\partial\mu$ term. How does one reconcile the correctness of these two equations with the difference in their forms? The answer is that the difference results from a difference in the definitions of the distribution functions $I = ch\nu f$ and $I' = ch\nu f'$. The distribution function f is defined as a distribution per unit volume and solid angle, whereas f', from our identification of x with r and deletion of the variables y, z, and φ, is defined as a distribution function per unit radial coordinate r and cosine element μ. Hence we have the equality

$$f(4\pi r^2\,dr)\,d\nu(2\pi\,d\mu) = f'\,dr\,d\nu\,d\mu,\qquad(2.51)$$

since both sides of this equation represent the number of photons under consideration. This gives

$$f' = 8\pi^2r^2f;\qquad I' = 8\pi^2r^2I.\qquad(2.52)$$

Similarly for the functions S' and S we have

$$S' = 8\pi^2r^2S.\qquad(2.53)$$

Use of Eqs. (2.52) and (2.53) in Eq. (2.50) brings this equation of transfer into conformity with that given by Eq. (2.47). It is Eq. (2.47) which is the usual form of the equation of transfer for spherically symmetric systems, i.e., it is the form in which the specific intensity is defined on a per unit volume basis. Hopefully this side discussion has clarified, rather than confused, the situation regarding terms such as $d\mu/ds$.

If the system and boundary conditions have cylindrical symmetry and if the system is infinite in the z direction, the specific intensity again depends upon only one spatial variable, the perpendicular distance r from the axis of the cylinder (the z axis). However, in this case it requires two angles to specify Ω. These may be taken as θ, the angle between z and Ω, and a corresponding azimuthal angle φ, the angle between the projection of Ω in the $x-y$ plane and the cylindrical coordinate r, with φ being measured in the usual right hand screw sense. Then $I \equiv I(r,\theta,\varphi)$, and the directional derivative of I is given by

$$\frac{\partial I}{\partial s} = \frac{\partial I}{\partial r}\left(\frac{dr}{ds}\right)+\frac{\partial I}{\partial\theta}\left(\frac{d\theta}{ds}\right)+\frac{\partial I}{\partial\varphi}\left(\frac{d\varphi}{ds}\right).\qquad(2.54)$$

The changes of r, θ, and φ with path length follow from Fig. 2.3. In an attempt to make this figure more transparent, we have drawn it using an angle $\omega \equiv 2\pi-\varphi$ rather than φ. The angle ω is also an azimuthal angle about the z axis, but in the unconventional left hand

The Equations of Radiation Hydrodynamics

screw sense. We first note from this figure that

$$\frac{d\theta}{ds} = 0, \tag{2.55}$$

since θ is measured with respect to an axis fixed in space, the z axis. To evaluate the other two quantities needed to compute $\partial I/\partial s$, we examine the right triangle with sides dr, $ds\,\sin\theta$, and $-r\,d\omega$. The angle between dr and $ds\,\sin\theta$ is $\omega+d\omega$, or simply ω, since $d\omega$ is a differential element. Then, from the geometric interpretation of the sine and cosine functions,

$$\cos\omega = \frac{dr}{ds\,\sin\theta}\;;\qquad \sin\omega = \frac{-r\,d\omega}{\sin\theta\,ds}\,. \tag{2.56}$$

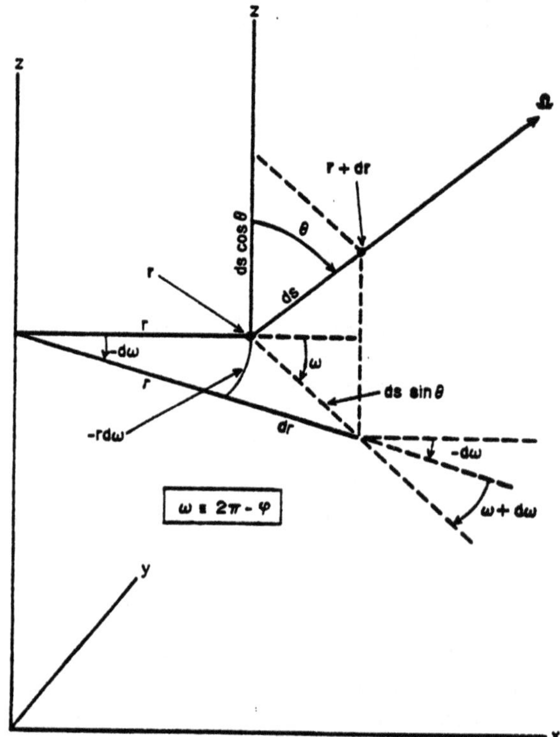

FIG. 2.3. Cylindrical geometry with cylindrical symmetry.

Introducing $\varphi \equiv 2\pi-\omega$ in Eq. (2.56) we find

$$\frac{dr}{ds} = \sin\theta\,\cos\varphi;\qquad \frac{d\varphi}{ds} = -\frac{1}{r}\sin\theta\,\sin\varphi. \tag{2.57}$$

Use of Eqs. (2.55) and (2.57) in Eq. (2.54) yields

$$\frac{\partial I}{\partial s} = \sin\theta\,\cos\varphi\,\frac{\partial I}{\partial r} - \frac{1}{r}\sin\theta\,\sin\varphi\,\frac{\partial I}{\partial\varphi}, \tag{2.58}$$

and the equation of transfer follows from Eq. (2.17) as

$$\frac{1}{c}\frac{\partial I(\nu,\theta,\varphi)}{\partial t}+\sin\theta\left[\cos\varphi\,\frac{\partial I(\nu,\theta,\varphi)}{\partial r}-\frac{1}{r}\sin\varphi\,\frac{\partial I(\nu,\theta,\varphi)}{\partial\varphi}\right]+\sigma(\nu)\,I(\nu,\theta,\varphi)$$

$$=S(\nu)+\int_0^\infty d\nu'\int_{4\pi} d\Omega'\,\frac{\nu}{\nu'}\,\sigma_s(\nu'\to\nu,\Omega'\cdot\Omega)\,I(\nu',\theta',\varphi'),\qquad(2.59)$$

where $I\equiv I(r,\nu,\theta,\varphi,t)$. No simplification is possible in the scattering term since the angular dependence of the specific intensity in this case is as general as in the original equation of transfer (i.e. two angles are required).

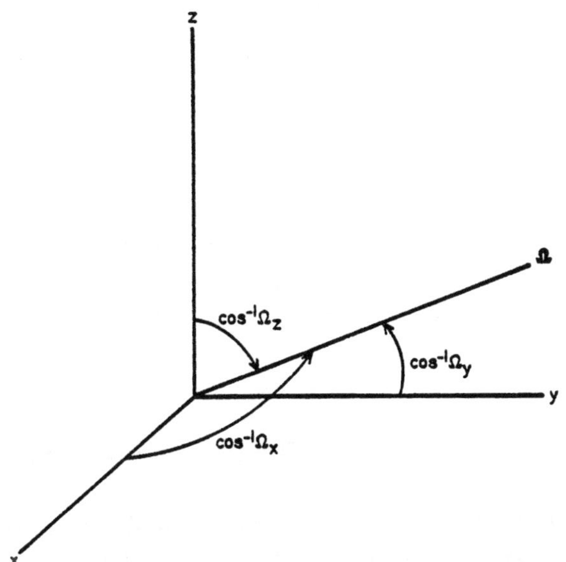

FIG. 2.4. The direction cosines in Cartesian geometry.

We end this section by giving, without derivation, the equation of transfer in three dimensional Cartesian, spherical, and cylindrical geometries. By three dimensional we mean that no assumptions are made concerning the symmetry of the systems. In Cartesian coordinates, Eq. (2.17) is obviously written

$$\frac{1}{c}\frac{\partial I(\nu,\Omega)}{\partial t}+\Omega_x\frac{\partial I(\nu,\Omega)}{\partial x}+\Omega_y\frac{\partial I(\nu,\Omega)}{\partial y}+\Omega_z\frac{\partial I(\nu,\Omega)}{\partial z}+\sigma(\nu)\,I(\nu,\Omega)$$

$$=S(\nu)+\int_0^\infty d\nu'\int_{4\pi} d\Omega'\,\frac{\nu}{\nu'}\,\sigma_s(\nu'\to\nu,\Omega'\cdot\Omega)\,I(\nu',\Omega'),\qquad(2.60)$$

where $I\equiv I(x,y,z,\nu,\Omega,t)$. Here $\Omega_x,\Omega_y,$ and Ω_z are the direction cosines with respect to the x, y, and z axes, respectively (Fig. 2.4). Any two of these three cosines can be taken as the two variables needed to define Ω since the third one follows from

$$\Omega_x^2+\Omega_y^2+\Omega_z^2 = 1. \tag{2.61}$$

In a general spherical system we take the spatial coordinates to be the standard ones, namely the radial distance r, the polar angle Θ between r and the z axis, and the azimuthal angle Φ between the x axis and the projection of r in the $x-y$ plane, with Φ measured in the conventional right hand screw sense. We describe Ω by the polar angle θ between r and Ω,

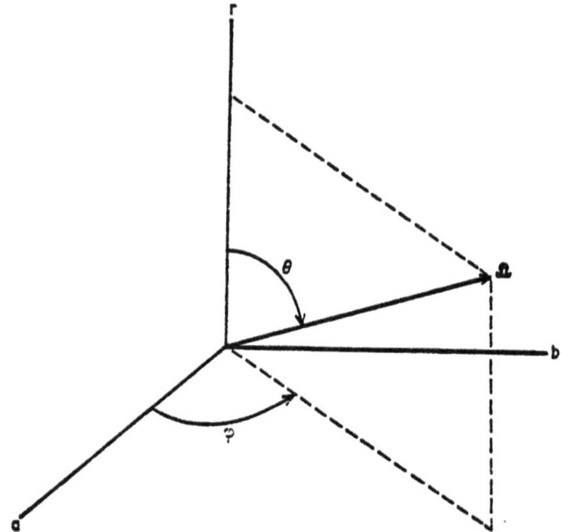

FIG. 2.5. The angles defining Ω in general spherical geometry.

and the azimuthal angle φ between the projection of Ω in the plane perpendicular to r and any reference axis in this plane, again defining this azimuthal angle in the right hand screw sense (Fig. 2.5). Then the equation of transfer can be written (Reed and Lathrop, 1970)

$$\frac{1}{c}\frac{\partial I(\nu, \mu, \varphi)}{\partial t}+\mu\,\frac{\partial I(\nu, \mu, \varphi)}{\partial r}+\frac{\eta}{r}\,\frac{\partial I(\nu, \mu, \varphi)}{\partial \Theta}+\frac{\xi}{r\sin\Theta}\,\frac{\partial I(\nu, \mu, \varphi)}{\partial \Phi}$$

$$+\frac{(1-\mu^2)}{r}\,\frac{\partial I(\nu, \mu, \varphi)}{\partial \mu}-\frac{\xi\cot\Theta}{r}\,\frac{\partial I(\nu, \mu, \varphi)}{\partial \varphi}+\sigma(\nu)\,I(\nu, \mu, \varphi)$$

$$= S(\nu)+\int_0^\infty dv' \int_{4\pi} d\Omega'\,\frac{\nu}{\nu'}\,\sigma_s(\nu' \to \nu, \Omega' \cdot \Omega)\,I(\nu', \mu', \varphi'), \tag{2.62}$$

where $I \equiv I(r, \Theta, \Phi, \nu, \mu, \varphi, t)$. Here we have defined

$$\left.\begin{aligned}\mu &= \cos\theta, \\ \eta &= \sin\theta\cos\varphi, \\ \xi &= \sin\theta\sin\varphi.\end{aligned}\right\} \tag{2.63}$$

These variables μ, η, and ξ have the geometric interpretation of direction cosines describing the vector $\boldsymbol{\Omega}$ (Fig. 2.6). From Eq. (2.63) we clearly have

$$\mu^2+\eta^2+\xi^2 = 1, \tag{2.64}$$

a relationship always satisfied by direction cosines.

Likewise, in a general cylindrical system, we take the spatial coordinates to be the standard ones, namely the perpendicular distance r from the z axis, the perpendicular distance z from the $x-y$ plane, and the angle Θ, defined in the right hand screw sense about z, between r and

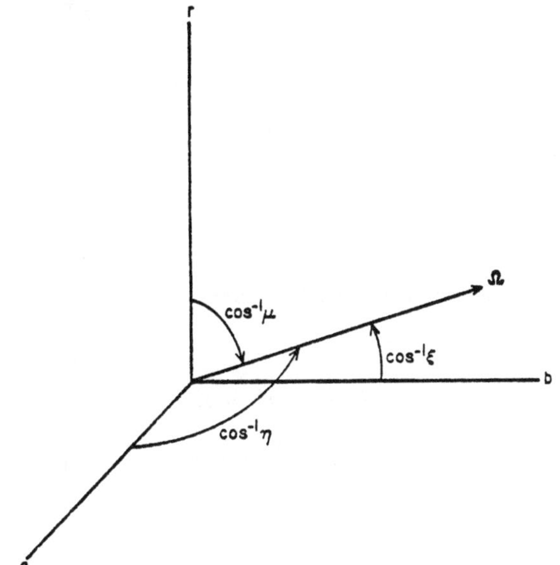

FIG. 2.6. The direction cosines in general spherical geometry.

the x axis. The direction $\boldsymbol{\Omega}$ can be defined in terms of the polar angle θ between the z axis and $\boldsymbol{\Omega}$, and the azimuthal angle φ between r and the projection of $\boldsymbol{\Omega}$ in the $x-y$ plane, again measuring φ in the right hand screw sense (Fig. 2.7). The equation of transfer then takes the form (Reed and Lathrop, 1970)

$$\frac{1}{c}\frac{\partial I(\nu, \theta, \varphi)}{\partial t} + \mu \frac{\partial I(\nu, \theta, \varphi)}{\partial r} + \frac{\eta}{r}\frac{\partial I(\nu, \theta, \varphi)}{\partial \Theta} + \xi \frac{\partial I(\nu, \theta, \varphi)}{\partial z}$$

$$-\frac{\eta}{r}\frac{\partial I(\nu, \theta, \varphi)}{\partial \varphi} + \sigma(\nu)\, I(\nu, \theta, \varphi) = S(\nu)$$

$$+ \int_0^\infty d\nu' \int_{4\pi} d\Omega' \frac{\nu}{\nu'}\, \sigma_s(\nu' \to \nu, \boldsymbol{\Omega}' \cdot \boldsymbol{\Omega})\, I(\nu', \theta', \varphi'), \tag{2.65}$$

where $I \equiv I(r, \Theta, z, \nu, \theta, \varphi, t)$. Here we have defined

$$\left.\begin{aligned} \xi &= \cos\theta, \\ \mu &= \sin\theta\cos\varphi, \\ \eta &= \sin\theta\sin\varphi. \end{aligned}\right\} \tag{2.66}$$

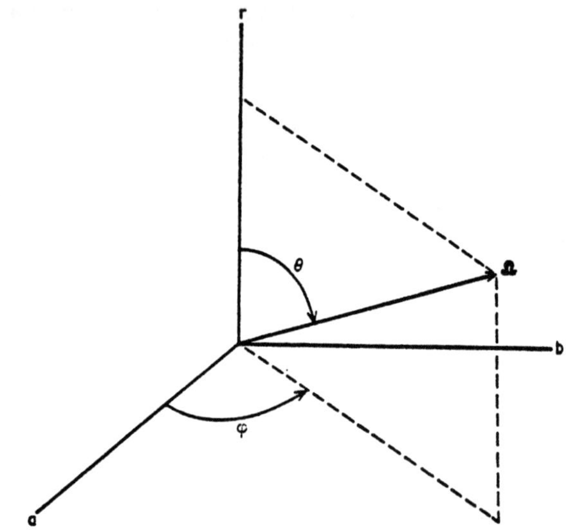

FIG. 2.7. The angles defining Ω in general cylindrical geometry.

Again we have

$$\xi^2 + \mu^2 + \eta^2 = 1, \tag{2.67}$$

and these three quantities have the geometric interpretation of direction cosines describing Ω (Fig. 2.8).

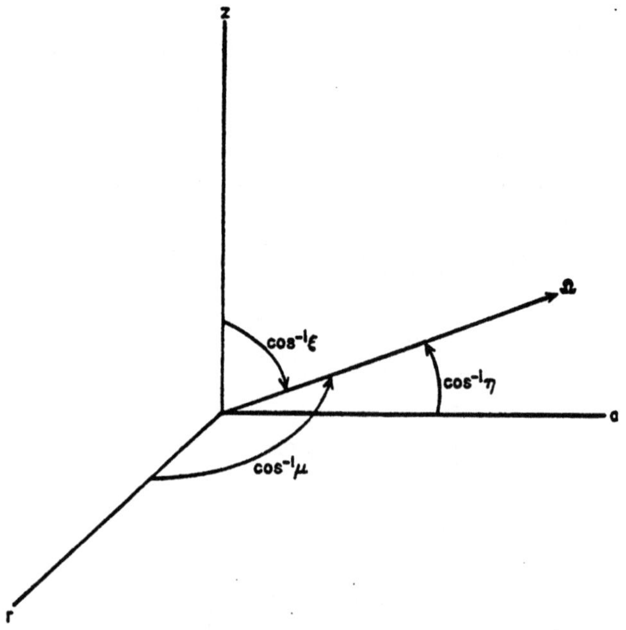

FIG. 2.8. The direction cosines in general cylindrical geometry.

6. The Equation of Transfer in Integral Form

Thus far we have concentrated exclusively on the integro-differential form of the equation of transfer. In this section we show that one can derive a purely integral transport equation.

We first consider the general time independent problem $I \equiv I(\mathbf{r}, v, \mathbf{\Omega})$ in which case the equation of transfer, Eq. (2.17), reads

$$\mathbf{\Omega} \cdot \nabla I(\mathbf{r}, v, \mathbf{\Omega}) + \sigma(\mathbf{r}, v)\, I(\mathbf{r}, v, \mathbf{\Omega}) = Q(\mathbf{r}, v, \mathbf{\Omega}). \tag{2.68}$$

Here we have defined $Q(\mathbf{r}, v, \mathbf{\Omega})$ as the total angular source resulting from both spontaneous emission and inscattering

$$Q(\mathbf{r}, v, \mathbf{\Omega}) = S(\mathbf{r}, v) + \int_0^\infty dv' \int_{4\pi} d\mathbf{\Omega}'\, \frac{v}{v'}\, \sigma_s(\mathbf{r}, v' \to v, \mathbf{\Omega}' \cdot \mathbf{\Omega})\, I(\mathbf{r}, v', \mathbf{\Omega}'). \tag{2.69}$$

The boundary condition on Eq. (2.68) is just Eq. (2.29) with the time argument deleted

$$I(\mathbf{r}_s, v, \mathbf{\Omega}) = \Gamma(\mathbf{r}_s, v, \mathbf{\Omega}), \quad \mathbf{n} \cdot \mathbf{\Omega} < 0, \tag{2.70}$$

where, as before, Γ is a specified function and $\mathbf{r}_s$ denotes a point on the surface of the system. Interpreting $\mathbf{\Omega} \cdot \nabla$ in Eq. (2.68) as a directional derivative in the direction $\mathbf{\Omega}$, and introducing the length s as the distance back along $\mathbf{\Omega}$ from the point $\mathbf{r}$, we rewrite Eq. (2.68) as

$$-\frac{\partial I(\mathbf{r} - s\mathbf{\Omega}, \mathbf{\Omega})}{\partial s} + \sigma(\mathbf{r} - s\mathbf{\Omega})\, I(\mathbf{r} - s\mathbf{\Omega}, \mathbf{\Omega}) = Q(\mathbf{r} - s\mathbf{\Omega}, \mathbf{\Omega}). \tag{2.71}$$

In writing Eq. (2.71) we have, for notational simplicity, dropped the frequency argument v everywhere since it only appears parametrically in Eqs. (2.68) and (2.70). However, it must be kept in mind that an equation such as Eq. (2.71) holds for each frequency v, and these equations are generally coupled since the scattering contribution to the total source Q couples the different frequencies.

Equation (2.71) is a first order differential equation whose solution is easily found through the standard use of an integrating factor. We rewrite Eq. (2.71) as

$$-\frac{\partial}{\partial s}\left\{ I(\mathbf{r} - s\mathbf{\Omega}, \mathbf{\Omega}) \exp\left[-\int_{s_0}^{s} ds''\sigma(\mathbf{r} - s''\mathbf{\Omega})\right]\right\} = Q(\mathbf{r} - s\mathbf{\Omega}, \mathbf{\Omega}) \exp\left[-\int_{s_0}^{s} ds''\sigma(\mathbf{r} - s''\mathbf{\Omega})\right],$$
$$\tag{2.72}$$

where s_0 is an arbitrary point along s. Integration of Eq. (2.72) from s to s_0 gives

$$I(\mathbf{r} - s\mathbf{\Omega}, \mathbf{\Omega}) \exp\left[-\int_{s_0}^{s} ds''\sigma(\mathbf{r} - s''\mathbf{\Omega})\right] - I(\mathbf{r} - s_0\mathbf{\Omega}, \mathbf{\Omega})$$
$$= \int_{s}^{s_0} ds'\, Q(\mathbf{r} - s'\mathbf{\Omega}, \mathbf{\Omega}) \exp\left[-\int_{s_0}^{s'} ds''\sigma(\mathbf{r} - s''\mathbf{\Omega})\right], \tag{2.73}$$

or, dividing through by the exponential factor on the left hand side of this equation,

$$I(\mathbf{r}-s\mathbf{\Omega}, \mathbf{\Omega}) = I(\mathbf{r}-s_0\mathbf{\Omega}, \mathbf{\Omega}) \exp\left[\int_{s_0}^{s} ds''\sigma(\mathbf{r}-s''\mathbf{\Omega})\right]$$
$$+ \int_{s}^{s_0} ds'Q(\mathbf{r}-s'\mathbf{\Omega}, \mathbf{\Omega}) \exp\left[\int_{s'}^{s} ds''\sigma(\mathbf{r}-s''\mathbf{\Omega})\right]. \tag{2.74}$$

Setting $s = 0$ in Eq. (2.74) we find that the specific intensity at the point $\mathbf{r}$ in the direction $\mathbf{\Omega}$ is given by

$$I(\mathbf{r}, \mathbf{\Omega}) = I(\mathbf{r}-s_0\mathbf{\Omega}, \mathbf{\Omega}) \exp\left[-\int_{0}^{s_0} ds''\sigma(\mathbf{r}-s''\mathbf{\Omega})\right]$$
$$+ \int_{0}^{s_0} ds'Q(\mathbf{r}-s'\mathbf{\Omega}, \mathbf{\Omega}) \exp\left[-\int_{0}^{s'} ds''\sigma(\mathbf{r}-s''\mathbf{\Omega})\right]. \tag{2.75}$$

We still need to evaluate the constant of integration $I(\mathbf{r}-s_0\mathbf{\Omega}, \mathbf{\Omega})$. To do this we use the boundary condition given by Eq. (2.70). Let s_0 be such that $\mathbf{r}-s_0\mathbf{\Omega} = \mathbf{r}_s$, a point on the boundary. This implies that $s_0 = |\mathbf{r}-\mathbf{r}_s|$ and hence, from Eq. (2.70),

$$I(\mathbf{r}-s_0\mathbf{\Omega}, \mathbf{\Omega})\Big|_{s_0=|\mathbf{r}-\mathbf{r}_s|} = \Gamma(\mathbf{r}_s, \mathbf{\Omega}). \tag{2.76}$$

Setting $s_0 = |\mathbf{r}-\mathbf{r}_s|$ in Eq. (2.75) and making use of Eq. (2.76), we obtain

$$I(\mathbf{r}, \mathbf{\Omega}) = \Gamma(\mathbf{r}_s, \mathbf{\Omega}) \exp\left[-\int_{0}^{|\mathbf{r}-\mathbf{r}_s|} ds''\sigma(\mathbf{r}-s''\mathbf{\Omega})\right]$$
$$+ \int_{0}^{|\mathbf{r}-\mathbf{r}_s|} ds'Q(\mathbf{r}-s'\mathbf{\Omega}, \mathbf{\Omega}) \exp\left[-\int_{0}^{s'} ds''\sigma(\mathbf{r}-s''\mathbf{\Omega})\right]. \tag{2.77}$$

Finally, recognizing that the source Q is zero outside the system, we can extend the upper integration limit in Eq. (2.77) to infinity, and our end result for the specific intensity is

$$I(\mathbf{r}, \mathbf{\Omega}) = \Gamma(\mathbf{r}_s, \mathbf{\Omega}) \exp\left[-\int_{0}^{|\mathbf{r}-\mathbf{r}_s|} ds''\sigma(\mathbf{r}-s''\mathbf{\Omega})\right]$$
$$+ \int_{0}^{\infty} ds'Q(\mathbf{r}-s'\mathbf{\Omega}, \mathbf{\Omega}) \exp\left[-\int_{0}^{s'} ds''\sigma(\mathbf{r}-s''\mathbf{\Omega})\right]. \tag{2.78}$$

It should be emphasized that Eq. (2.78) is not a solution to the equation of transfer since the source term Q depends upon the specific intensity through the scattering interaction. Rather, it is an integral equation to be solved for the specific intensity. However, in the special case of no scattering ($\sigma_s = 0$), Eq. (2.78) is the general solution to the time independent equation of transfer.

Equation (2.78) can also be interpreted physically (Fig. 2.9). The specific intensity at a point $\mathbf{r}$ in the direction $\boldsymbol{\Omega}$ is the sum of two terms. The first term is the intensity incident upon the

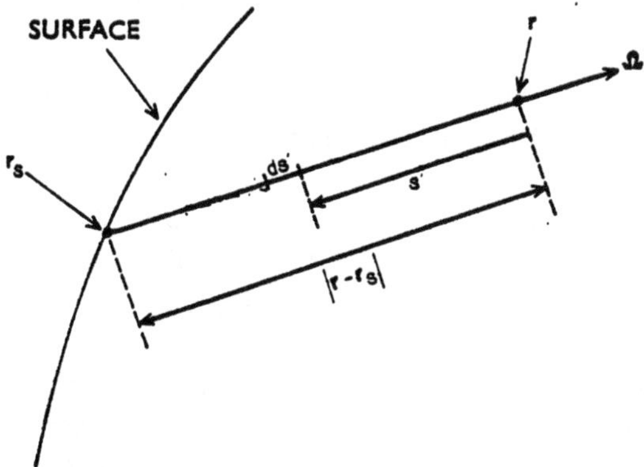

FIG. 2.9. The geometry associated with the integral equation of transfer.

surface of the system, exponentially attenuated by collisions over the distance between the surface point $\mathbf{r}_s$ and the point $\mathbf{r}$ under consideration. The second term gives the contribution to the specific intensity due to emission and scattering into the beam from each path length element ds' along $\boldsymbol{\Omega}$. The radiation originating in each element ds' must be exponentially attenuated to find its contribution to the specific intensity at the point $\mathbf{r}$. The quantity $\tau(\mathbf{r}, \mathbf{r}_s)$, defined as

$$\tau(\mathbf{r}, \mathbf{r}_s) \equiv \int_0^{|\mathbf{r}-\mathbf{r}_s|} ds'' \sigma(\mathbf{r}-s''\boldsymbol{\Omega}), \tag{2.79}$$

is referred to as the "optical distance" or "optical path length" between the points $\mathbf{r}$ and $\mathbf{r}_s$. As can be seen from Eq. (2.78), it is the optical distance between two points which is the relevant quantity in calculating the exponential attenuation of a beam of photons in traveling from one point to another.

We now consider the general time dependent situation allowing the collision coefficient σ as well as the specific intensity to depend upon time. We wish to obtain the integral equation, analogous to Eq. (2.78), in this case. We write the integro-differential equation as, suppressing the frequency dependence,

$$\frac{1}{c}\frac{\partial I(\mathbf{r}, \boldsymbol{\Omega}, t)}{\partial t} + \boldsymbol{\Omega}\cdot\nabla I(\mathbf{r}, \boldsymbol{\Omega}, t) + \sigma(\mathbf{r}, t)\, I(\mathbf{r}, \boldsymbol{\Omega}, t) = Q(\mathbf{r}, \boldsymbol{\Omega}, t), \tag{2.80}$$

where we have defined

$$Q(\mathbf{r}, \boldsymbol{\Omega}, t) \equiv Q(\mathbf{r}, \nu, \boldsymbol{\Omega}, t) = S(\mathbf{r}, \nu, t)$$

$$+ \int_0^\infty d\nu' \int_{4\pi} d\Omega' \frac{\nu}{\nu'} \sigma_s(\mathbf{r}, \nu' \to \nu, \boldsymbol{\Omega}'\cdot\boldsymbol{\Omega}, t)\, I(\mathbf{r}, \nu', \boldsymbol{\Omega}', t). \tag{2.81}$$

The boundary conditions on Eq. (2.80) in space and time are given by Eqs. (2.29) and (2.31). The key to an algebraic derivation of the integral equation from Eqs. (2.29), (2.31), and

(2.80) is the introduction of the total path length derivative, as discussed in Section 3 of this chapter, and the resulting identity

$$-\frac{d}{ds}I(\mathbf{r}-s\boldsymbol{\Omega},\,\boldsymbol{\Omega},\,t-s/c) = \left(\frac{1}{c}\,\frac{\partial}{\partial t}+\boldsymbol{\Omega}\cdot\nabla\right)I(\mathbf{r}-s\boldsymbol{\Omega},\,\boldsymbol{\Omega},\,t-s/c). \tag{2.82}$$

The derivation then proceeds along the lines just used in the time independent case. An alternate way to proceed, which we shall follow here, is to use physical arguments to construct the integral equation of transfer.

The integral equation simply follows from the fact that photons of direction $\boldsymbol{\Omega}$ which are at point $\mathbf{r}$ at time t, must have originated at some point $\mathbf{r}-s\boldsymbol{\Omega}$ at time $t-s/c$, due to the finite speed of light. One must also introduce the appropriate exponential, with the optical distance between $\mathbf{r}$ and $\mathbf{r}-s\boldsymbol{\Omega}$ as the exponent, which gives the probability that the photon does not suffer a collision on its flight from $\mathbf{r}-s\boldsymbol{\Omega}$ to $\mathbf{r}$. For the moment we consider a system with $\Gamma=0$ (no photons enter the system through the bounding surface) and $\Lambda=0$ (no photons are present in the system at $t=0$). Then the only source of radiation is the source function Q, and the physical arguments just given allow us to write

$$I(\mathbf{r},\,\boldsymbol{\Omega},\,t) = \int_0^\infty ds'\,Q(\mathbf{r}-s'\boldsymbol{\Omega},\boldsymbol{\Omega},\,t-s'/c)\,\exp\left[-\int_0^{s'} ds''\sigma(\mathbf{r}-s''\boldsymbol{\Omega},\,t-s''/c)\right]. \tag{2.83}$$

To obtain the integral equation in the general case ($\Gamma\neq0$; $\Lambda\neq0$) we note on physical grounds (it can also be shown mathematically from the integro-differential equation of transfer) that the incoming distribution $\Gamma(\mathbf{r}_s,\,\boldsymbol{\Omega},\,t)$ is equivalent to a surface source of photons, and the initial condition $\Lambda(\mathbf{r},\,\boldsymbol{\Omega})$ is equivalent to a volume source, at $t=0$, of photons. More specifically, interpretation of these spatial and temporal boundary conditions as sources implies that $Q(\mathbf{r},\,\boldsymbol{\Omega},\,t)$ in Eq. (2.83) should be replaced by

$$Q(\mathbf{r},\,\boldsymbol{\Omega},\,t) \rightarrow Q(\mathbf{r},\,\boldsymbol{\Omega},\,t)+\Gamma(\mathbf{r}_s,\,\boldsymbol{\Omega},\,t)\,\delta(s-s_0)H(t)+\frac{1}{c}\,\Lambda(\mathbf{r},\,\boldsymbol{\Omega})\,\delta(t)H(s_0-s). \tag{2.84}$$

The term involving Γ in Eq. (2.84) is the equivalent source due to photons entering the system through the bounding surface, and the term involving Λ is the equivalent source due to the initial condition. Here $\delta(\xi)$ is the Dirac delta function, s_0 again denotes (in the s variable) a point on the surface of the system ($s_0=|\mathbf{r}-\mathbf{r}_s|$), and $H(\xi)$ is the Heaviside or unit step function. The Heaviside functions in Eq. (2.84) indicate explicitly that Γ should be taken as zero for negative time, and Λ should be taken as zero for points outside the system. The source function Q is also, of course, zero outside the system. Use of Eq. (2.84) in Eq. (2.83) and integration over the two Dirac delta functions yields

$$I(\mathbf{r},\boldsymbol{\Omega},t) = \int_0^\infty ds'\,Q(\mathbf{r}-s'\boldsymbol{\Omega},\boldsymbol{\Omega},t-s'/c)\,\exp\left[-\int_0^{s'}ds''\sigma(\mathbf{r}-s''\boldsymbol{\Omega},t-s''/c)\right]$$

$$+\,\Gamma(\mathbf{r}_s,\boldsymbol{\Omega},t-|\mathbf{r}-\mathbf{r}_s|/c)\,\exp\left[-\int_0^{|\mathbf{r}-\mathbf{r}_s|}ds''\sigma(\mathbf{r}-s''\boldsymbol{\Omega},t-s''/c)\right]H(ct-|\mathbf{r}-\mathbf{r}_s|)$$

$$+\,\Lambda(\mathbf{r}-ct\boldsymbol{\Omega},\boldsymbol{\Omega})\,\exp\left[-\int_0^{ct}ds''\sigma(\mathbf{r}-s''\boldsymbol{\Omega},t-s''/c)\right]H(|\mathbf{r}-\mathbf{r}_s|-ct). \tag{2.85}$$

This result is the integral equation of transfer in the most general case. Although algebraically it appears very complex, the physical interpretation of this equation is quite simple as we have previously noted. It should be recalled that the functions $I, \Gamma, \Lambda, \sigma,$ and Q in Eq. (2.85) carry an implied frequency dependence, and the equations for various values of frequency are coupled through the scattering interaction contained in the source function Q. The presence of the scattering in Q also makes Eq. (2.85) an integral equation to be solved for the specific intensity, rather than a solution in itself.

7. Peierls' Equation

From the integral equation of transfer one can very easily obtain a result known as Peierls' equation. This equation, integral in character, describes time independent radiative transfer in a system which scatters photons isotropically and in which no photons enter the system through the bounding surface. The advantage of Peierls' equation over the equations of transfer already discussed is that it involves $\varrho(\mathbf{r}, \nu)$, defined as

$$\varrho(\mathbf{r}, \nu) \equiv \int_{4\pi} d\Omega I(\mathbf{r}, \nu, \Omega), \tag{2.86}$$

rather than the specific intensity. Hence the dependent variable in this case depends upon two less independent variables (the two angles describing Ω) than does the dependent variable in the equations of transfer heretofore discussed.

We take as our starting point the integral equation of transfer given by Eq. (2.78) with Γ set to zero (no photons enter the system through the bounding surface). Since the scattering is assumed isotropic and the emission source S is always isotropic, the total source Q is independent of direction, $Q \equiv Q(\mathbf{r}, \nu)$. Then Eq. (2.78) can be written

$$I(\mathbf{r}, \Omega) = \int_0^\infty ds' Q(\mathbf{r} - s'\Omega) \, \exp\left[-\int_0^{s'} ds'' \sigma(\mathbf{r} - s''\Omega)\right], \tag{2.87}$$

where, as before, we have dropped the frequency argument in all terms for notational simplicity. Integration of Eq. (2.87) over all solid angle gives, with the introduction of ϱ as defined by Eq. (2.86),

$$\varrho(\mathbf{r}) = \int_{4\pi} d\Omega \int_0^\infty ds' Q(\mathbf{r} - s'\Omega) \exp\left[-\int_0^{s'} ds'' \sigma(\mathbf{r} - s''\Omega)\right]. \tag{2.88}$$

Setting $\mathbf{r}' = \mathbf{r} - s'\Omega$ (hence $s' = |\mathbf{r} - \mathbf{r}'|$) we find

$$\varrho(\mathbf{r}) = \int_{4\pi} d\Omega \int_0^\infty d|\mathbf{r} - \mathbf{r}'| Q(\mathbf{r}') \exp\left[-\int_0^{|\mathbf{r} - \mathbf{r}'|} ds'' \sigma(\mathbf{r} - s''\Omega)\right]. \tag{2.89}$$

Recognizing the exponent as just the optical distance $\tau(\mathbf{r}, \mathbf{r}')$ defined by Eq. (2.79), and grouping terms, we rewrite Eq. (2.89) as

$$\varrho(\mathbf{r}) = \int_0^\infty |\mathbf{r} - \mathbf{r}'|^2 d|\mathbf{r} - \mathbf{r}'| \int_{4\pi} d\Omega \left[\frac{Q(\mathbf{r}')}{|\mathbf{r} - \mathbf{r}'|^2} e^{-\tau(\mathbf{r}, \mathbf{r}')}\right]. \tag{2.90}$$

The Equations of Radiation Hydrodynamics

The crucial point in the development is the recognition that $|\mathbf{r}-\mathbf{r}'|^2 d|\mathbf{r}-\mathbf{r}'| d\Omega$ is just a volume element, in spherical coordinates, centered around the point $\mathbf{r}$. One can rewrite this volume element more generally as simply $d\mathbf{r}'$ without reference to any particular coordinate system, and Eq. (2.90) can be written as

$$\varrho(\mathbf{r}) = \int_V d\mathbf{r}' \, \frac{Q(\mathbf{r}')}{|\mathbf{r}-\mathbf{r}'|^2} e^{-\tau(\mathbf{r},\,\mathbf{r}')}, \tag{2.91}$$

where the integration extends over the volume V of the system. To show that Eq. (2.91) is actually an integral equation for $\varrho(\mathbf{r})$, we need consider the dependence of $Q(\mathbf{r})$ upon $\varrho(\mathbf{r})$. From Eq. (2.69) we have

$$Q(\mathbf{r}) \equiv Q(\mathbf{r},\,\nu) = S(\mathbf{r},\,\nu) + \int_0^\infty d\nu' \frac{\nu}{\nu'} \sigma_s(\mathbf{r},\,\nu' \to \nu)\, \varrho(\mathbf{r},\,\nu'), \tag{2.92}$$

where we have made use of the fact that the differential scattering coefficient is assumed independent of angle (isotropic). Combining Eqs. (2.91) and (2.92) we obtain, with full display of the frequency variable,

$$\varrho(\mathbf{r},\,\nu) = \int_V d\mathbf{r}' \frac{4\pi \left[S(\mathbf{r}',\,\nu) + \displaystyle\int_0^\infty d\nu' \, \frac{\nu}{\nu'} \sigma_s(\mathbf{r}',\,\nu' \to \nu)\, \varrho(\mathbf{r}',\,\nu') \right]}{4\pi |\mathbf{r}-\mathbf{r}'|^2} e^{-\tau(\mathbf{r},\,\mathbf{r}')}. \tag{2.93}$$

Equation (2.93), clearly an integral equation for $\varrho(\mathbf{r},\,\nu)$, is the most general form of the result known as Peierls' equation. We have introduced the factors of 4π in both numerator and denominator to aid in the physical interpretation of this equation. The term $4\pi Q(\mathbf{r}')$ is just the angle integrated total source at a point $\mathbf{r}'$. To obtain the contribution of this source to $\varrho(\mathbf{r})$ one must first attenuate it by the proper exponential, namely $e^{-\tau(\mathbf{r},\,\mathbf{r}')}$, which is just the noncollision probability. One must also introduce the geometric attenuation due to spherical divergence, namely the area of the spherical shell at point $\mathbf{r}$ with center at $\mathbf{r}'$. This is just $4\pi|\mathbf{r}-\mathbf{r}'|^2$, the denominator in Eq. (2.93).

Let us consider the form Peierls' equation takes in the three standard one dimensional geometries. Considering first the plane case, we have

$$\varrho \equiv \varrho(z); \quad Q \equiv Q(z); \quad \sigma \equiv \sigma(z), \tag{2.94}$$

where z is one of the three Cartesian coordinates. Equation (2.91) becomes in this case

$$\varrho(z) = \int_0^{2\pi} d\theta' \int_0^\infty dr'r' \int_0^R dz' \, \frac{Q(z')}{|\mathbf{r}-\mathbf{r}'|^2} e^{-\tau(\mathbf{r},\,\mathbf{r}')}, \tag{2.95}$$

where we have envisioned a slab of thickness R in the z direction (occupying $0 < z < R$) and infinite in the x and y directions. The integration over V is performed in a cylindrical coordinate system $(r',\,\theta',\,z')$ with the origin chosen such that the vector $\mathbf{r}$ coincides with the

z' axis, and $z' = 0$ coincides with the bottom edge of the slab ($z = 0$). From Fig. 2.10 we see that

$$|\mathbf{r}-\mathbf{r}'| = \sqrt{r'^2+(z-z')^2}. \tag{2.96}$$

The optical distance between $\mathbf{r}$ and $\mathbf{r}'$ is given by

$$\tau(\mathbf{r}, \mathbf{r}') = \int_0^{|\mathbf{r}-\mathbf{r}'|} ds\,\sigma, \tag{2.97}$$

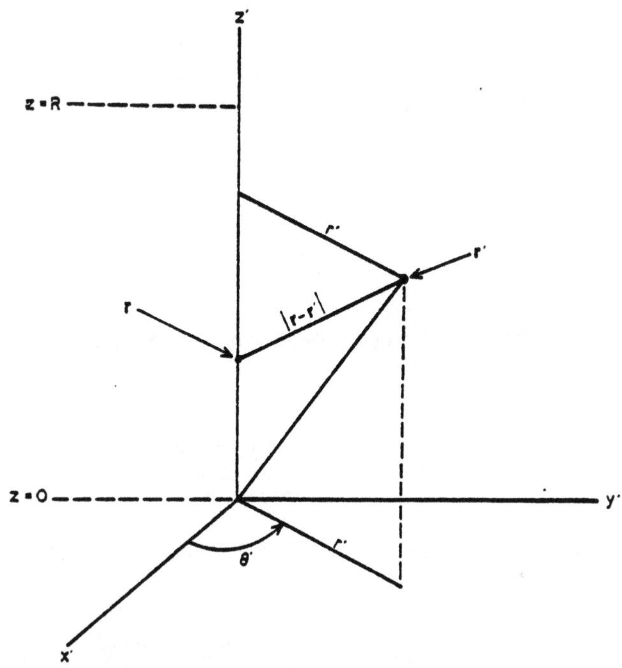

FIG. 2.10. Peierls' equation in plane geometry.

where σ is evaluated along the path length s between $\mathbf{r}$ and $\mathbf{r}'$. We introduce a new variable of integration in Eq. (2.97) (Fig. 2.11),

$$z'' = z+\frac{(z'-z)}{\sqrt{r'^2+(z-z')^2}}\,s. \tag{2.98}$$

Then

$$\tau(\mathbf{r}, \mathbf{r}') = \frac{\sqrt{r'^2+(z-z')^2}}{(z'-z)}\int_z^{z'} dz''\sigma(z''), \tag{2.99}$$

or, defining a slab optical depth $\tau(z)$,

$$\tau(z) \equiv \int_0^z dz''\sigma(z''), \tag{2.100}$$

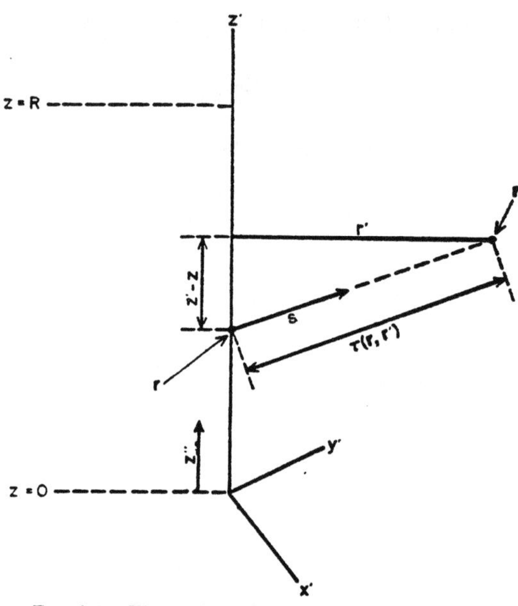

FIG. 2.11. The optical distance in plane geometry.

we can write

$$\tau(\mathbf{r}, \mathbf{r}') = \frac{\sqrt{r'^2 + (z - z')^2}}{(z' - z)} (\tau' - \tau),$$ (2.101)

where $\tau \equiv \tau(z)$ and $\tau' \equiv \tau(z')$. Using Eqs. (2.96) and (2.101) in Eq. (2.95) and performing the trivial integration over θ' we find

$$\varrho(z) = 2\pi \int_0^R dz' Q(z') \int_0^\infty dr' r' \frac{\exp\left[-\frac{\sqrt{r'^2 + (z - z')^2}\,(\tau' - \tau)}{(z' - z)}\right]}{r'^2 + (z - z')^2}.$$ (2.102)

Introducing a new variable of integration ξ in place of r'

$$\xi = \frac{\sqrt{r'^2 + (z - z')^2}\,(\tau' - \tau)}{(z' - z)},$$ (2.103)

we find that Eq. (2.102) takes the simpler form

$$\varrho(z) = 2\pi \int_0^R dz' Q(z') \int_{|\tau - \tau'|}^\infty d\xi \frac{e^{-\xi}}{\xi}.$$ (2.104)

(The bottom integration limit follows from the identity $|z - z'|(\tau' - \tau)/(z' - z) = |\tau - \tau'|$.) Finally, we can write Eq. (2.104) in the conventional form by introducing the exponential

34

integral, defined in nth order as

$$E_n(z) \equiv \int\limits_1^\infty dt\, \frac{e^{-zt}}{t^n} = \int\limits_0^1 dt\, t^{n-2}e^{-z/t}. \tag{2.105}$$

A discussion of this class of functions, as well as tables of values, can be found in Case *et al.* (1953a). In terms of the exponential integral of first order, Eq. (2.104) becomes

$$\varrho(z) = 2\pi \int\limits_0^R dz' E_1(|\tau - \tau'|)Q(z'). \tag{2.106}$$

In systems with spherical and cylindrical symmetry, one only obtains simple results, analogous to Eq. (2.106), if the collision coefficient σ is independent of position. Accordingly, we restrict ourselves to such systems. Considering first a spherical system occupying $0 < r < R$, we have

$$\varrho \equiv \varrho(r); \quad Q \equiv Q(r), \tag{2.107}$$

where r is the radial coordinate defining the vector **r**. We describe the vector **r'** by the standard spherical coordinates r', θ', and φ', and choose the z' axis to coincide with the vector **r** (Fig. 2.12). Then

$$|\mathbf{r} - \mathbf{r}'| = (r^2 + r'^2 - 2rr'\cos\theta')^{1/2}, \tag{2.108}$$

and since σ is assumed constant throughout the system

$$\tau(\mathbf{r}, \mathbf{r}') = \sigma|\mathbf{r} - \mathbf{r}'| = \sigma(r^2 + r'^2 - 2rr'\cos\theta')^{1/2}. \tag{2.109}$$

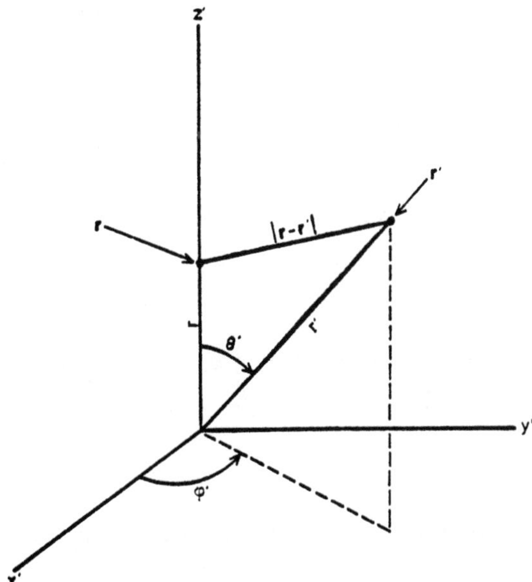

FIG. 2.12. Peierls' equation in spherical geometry.

Then Eq. (2.91) gives

$$\varrho(r) = 2\pi \int_0^R dr'\, r'^2 Q(r') \int_{-1}^1 d\mu'\, \frac{\exp\left[-\sigma(r^2+r'^2-2rr'\mu')^{1/2}\right]}{r^2+r'^2-2rr'\mu'}, \tag{2.110}$$

where we have set $\mu' = \cos\theta'$ and the 2π arises from the integral over φ'. Changing integration variables from μ' to ξ according to

$$\xi = \sigma(r^2+r'^2-2rr'\mu')^{1/2}, \tag{2.111}$$

we find

$$\varrho(r) = 2\pi \int_0^R dr'\, \frac{r'}{r}\, Q(r') \int_{\sigma|r-r'|}^{\sigma|r+r'|} d\xi\, \frac{e^{-\xi}}{\xi}. \tag{2.112}$$

Introducing the first order exponential integral as defined by Eq. (2.105) into Eq. (2.112), we obtain

$$\varrho(r) = 2\pi \int_0^R dr'\, \frac{r'}{r}\, Q(r')[E_1(\sigma|r-r'|)-E_1(\sigma|r+r'|)] \tag{2.113}$$

as Peierls' equation for a homogeneous, spherically symmetric system.

We now consider an infinite cylindrically symmetric system of radius R with σ again assumed independent of position. Then

$$\varrho \equiv \varrho(r); \quad Q \equiv Q(r), \tag{2.114}$$

where r here is the usual cylindrical coordinate describing the vector $\mathbf{r}$. We define the integration vector $\mathbf{r}'$ by the standard cylindrical coordinates r', θ', and z', choosing the x' axis to coincide with the vector $\mathbf{r}$ (Fig. 2.13). Then

$$|\mathbf{r}-\mathbf{r}'| = [z'^2+(r-r'\cos\theta')^2+(r'\sin\theta')^2]^{1/2} = (z'^2+r^2+r'^2-2rr'\cos\theta')^{1/2}, \tag{2.115}$$

and

$$\tau(\mathbf{r},\ \mathbf{r}') = \sigma(z'^2+r^2+r'^2-2rr'\cos\theta')^{1/2}, \tag{2.116}$$

since σ is assumed independent of position. Equation (2.91) gives in this case

$$\varrho(r) = \int_0^R dr'\, r' Q(r') \int_0^{2\pi} d\theta' \int_{-\infty}^{\infty} dz'\, \frac{e^{-\sigma\sqrt{z'^2+\alpha^2}}}{z'^2+\alpha^2}, \tag{2.117}$$

where we have defined

$$\alpha^2 \equiv r^2+r'^2-2rr'\cos\theta'. \tag{2.118}$$

Considering first the integral over z', we change integration variables from z' to ξ by setting

$$\alpha^2\xi^2 = z'^2+\alpha^2. \tag{2.119}$$

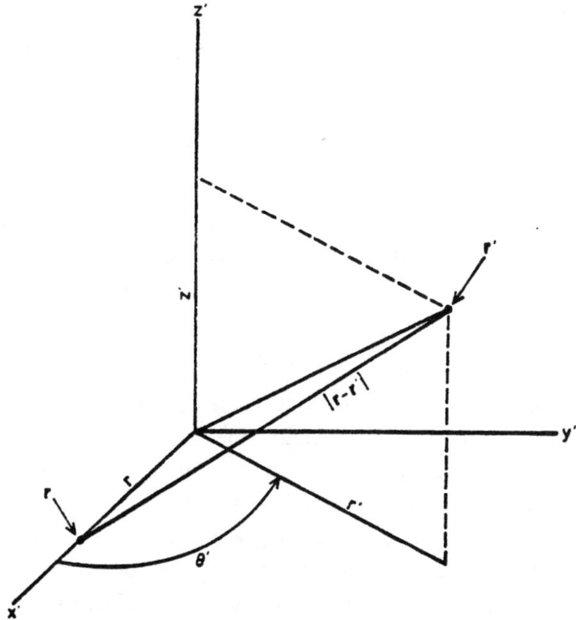

FIG. 2.13. Peierls' equation in cylindrical geometry.

This gives

$$\varrho(r) = 2 \int\limits_0^R dr' r' Q(r') \int\limits_0^{2\pi} d\theta' \int\limits_1^\infty d\xi \; \frac{e^{-\sigma\alpha\xi}}{\alpha\xi\sqrt{\xi^2-1}} \, . \tag{2.120}$$

We introduce another integral on the right hand side of Eq. (2.120) and write

$$\varrho(r) = 2\sigma \int\limits_0^R dr' r' Q(r') \int\limits_0^{2\pi} d\theta' \int\limits_1^\infty d\xi \int\limits_1^\infty dy \; \frac{e^{-\sigma\alpha\xi y}}{\sqrt{\xi^2-1}} \, . \tag{2.121}$$

By introducing this integration over y, the integral over ξ in Eq. (2.121) is now an integral representation of the zeroth order modified Bessel function of the third kind (Erdelyi *et al.*, 1953b) and we have

$$\varrho(r) = 2\sigma \int\limits_0^R dr' r' Q(r') \int\limits_1^\infty dy \int\limits_0^{2\pi} d\theta' K_0(\sigma\alpha y). \tag{2.122}$$

To proceed further, we require the Gegenbauer addition theorem for $K_0(w)$ which reads (Erdelyi *et al.*, 1953b)

$$K_0(w) = I_0(z)\, K_0(Z) + 2 \sum_{n=1}^\infty I_n(z) K_n(Z) \cos n\varphi, \tag{2.123}$$

where

$$w = (z^2 + Z^2 - 2zZ \cos \varphi)^{1/2}, \quad z < Z. \tag{2.124}$$

The Equations of Radiation Hydrodynamics

The argument $\sigma \alpha y$ of the Bessel function occurring in Eq. (2.122), with α given by Eq. (2.118), is exactly of this type. Making use of Eq. (2.123) in Eq. (2.122) we find, upon integrating over θ',

$$\varrho(r) = 4\pi\sigma \int_0^R dr'r'Q(r') \int_1^\infty dy I_0(\sigma r_< y)\, K_0(\sigma r_> y), \tag{2.125}$$

which is the final result for Peierls' equation in an infinite, homogeneous, cylindrically symmetric system. Here we have introduced the conventional notation

$$\left. \begin{aligned} r_< &= \min\,(r, r'), \\ r_> &= \max\,(r, r'). \end{aligned} \right\} \tag{2.126}$$

As we noted at the beginning of this section, Peierls' equation describes time independent radiative transfer in an isotropically scattering medium in which no photons enter the system through the bounding surface. One might inquire as to whether a more general equation of the Peierls' type exists. To this end we return to the integral equation describing radiative transfer in general, Eq. (2.85), and integrate over all solid angle. The terms involving Γ and Λ, the boundary and initial conditions, yield, when integrated over angle, a known function of space and time. The term involving the volume source Q can be handled as in Peierls' equation. Hence we find, if we define

$$\varrho(\mathbf{r}, t) \equiv \int_{4\pi} d\Omega I(\mathbf{r}, \Omega, t), \tag{2.127}$$

the result

$$\varrho(\mathbf{r}, t) = J(\mathbf{r}, t) + \int_V d\mathbf{r}' \, \frac{Q(\mathbf{r}', \Omega, t-|\mathbf{r}-\mathbf{r}'|/c)}{|\mathbf{r}-\mathbf{r}'|^2} \, e^{-\tau(\mathbf{r},\,\mathbf{r}',\,t)}, \tag{2.128}$$

where we have let τ denote an optical depth

$$\tau(\mathbf{r}, \mathbf{r}', t) = \int_0^{|\mathbf{r}-\mathbf{r}'|} ds''\sigma(\mathbf{r}-s''\Omega, t-s''/c), \tag{2.129}$$

and $J(\mathbf{r}, t)$ gives the contribution from the boundary and initial conditions

$$J(\mathbf{r}, t) = \int_{4\pi} d\Omega\, \Gamma(\mathbf{r}_s, \Omega, t-|\mathbf{r}-\mathbf{r}_s|/c) e^{-\tau(\mathbf{r},\,\mathbf{r}_s,\,t)}\, H(ct-|\mathbf{r}-\mathbf{r}_s|)$$

$$+ \int_{4\pi} d\Omega \Lambda(\mathbf{r}-ct\Omega, \Omega) \exp\left[-\int_0^{ct} ds''\sigma(\mathbf{r}-s''\Omega, t-s''/c)\right] H(|\mathbf{r}-\mathbf{r}_s|-ct). \tag{2.130}$$

Once again we have dropped the frequency argument in all terms for notational simplicity. Equation (2.128) is generally valid, but is only useful if the scattering is isotropic. In this case Eq. (2.81) gives

$$Q(\mathbf{r}, v, \Omega, t) = S(\mathbf{r}, v, t) + \int_0^\infty dv' \, \frac{v}{v'} \, \sigma_s\,(\mathbf{r}, v' \to v, t)\, \varrho(\mathbf{r}, v', t), \tag{2.131}$$

and Eq. (2.128) is an integral equation, in the space and frequency variables, for the quantity $\varrho(\mathbf{r}, \nu, t)$. In the general case of anisotropic scattering Q cannot be written in terms of $\varrho(\mathbf{r}, t)$ alone (Q depends upon the angular detail of the specific intensity) and hence for anisotropic scattering Eq. (2.128) represents an identity rather than an equation which can be solved for $\varrho(\mathbf{r}, t)$.

8. A Variational Characterization of the Equation of Transfer

In this section we give a variational characterization of the equation of radiative transfer. The usual procedure for obtaining such a characterization of any equation is to postulate a functional and then prove it has the desired properties. We shall show, however, that the appropriate functional can be derived using a Lagrange multiplier technique. The adjoint equation, always present in a variational description of a non-self-adjoint system, enters naturally in the course of the derivation. The resulting functional can be viewed in two distinct ways. In the first place, given first order approximations to the specific intensity and a certain adjoint function, it allows one to calculate a second order estimate of any functional of interest of the specific intensity. It is this viewpoint that allows us to derive the proper variational expression. Secondly, it can be used in the Lagrangian sense. That is, the demand that the functional be stationary under arbitrary variations is equivalent to solving the equation of transfer and the adjoint equation. In this sense the variational characterization is entirely equivalent to the equations of transfer already discussed. It can be considered on the same basis as the integro-differential and integral equations; i.e., it is a complete description (within the physics assumed thus far) of the transfer of energy by radiative processes. This Lagrangian viewpoint also allows a systematic procedure for deriving approximate descriptions of radiative transfer. One demands that the functional be stationary within a restricted class of admissible functions (trial functions).

We now proceed to develop a variational principle for the integro-differential equation of transfer, Eq. (2.17), together with the boundary and initial conditions, Eqs. (2.29) and (2.31). The same procedure could be used to construct variational principles for the general integral equation (2.85) and Peierls' equation (2.93). We assume a certain functional of the specific intensity $I(\mathbf{r}, \nu, \Omega, t)$, say $G[I]$, to be of particular interest, and consider the problem of obtaining a variational estimate of $G[I]$. The technique we use consists of viewing the equation of transfer and the boundary and initial conditions, Eqs. (2.17), (2.29), and (2.31), as constraints to be imposed in estimating $G[I]$ and using Lagrange multipliers to incorporate these constraints into the formalism. One merely multiplies all of the constraints by Lagrange multipliers and adds these to the functional of interest. Now, Eq. (2.17) holds for all $\mathbf{r}, \nu, \Omega$, and t; Eq. (2.29) holds for all $\mathbf{r}_s$ (surface points), ν, Ω such that $\mathbf{n} \cdot \Omega < 0$ ($\mathbf{n}$ is an outward normal vector at $\mathbf{r}_s$) and t; and Eq. (2.31) holds for all $\mathbf{r}, \nu,$ and Ω. Hence the sum of constraints takes the form of three multiple integrals, and we are led to the functional

$$F[i, j, k, l] = G[i] + \int_V d\mathbf{r} \int_0^\infty d\nu \int_{4\pi} d\Omega \int_0^T dt(Hi - S)j$$

$$+ \int_A ds \int_0^\infty d\nu \int_{\mathbf{n} \cdot \Omega < 0} d\Omega \int_0^T dt(i - \Gamma)k + \int_V d\mathbf{r} \int_0^\infty d\nu \int_{4\pi} d\Omega [i(0) - \Lambda]l, \qquad (2.132)$$

The Equations of Radiation Hydrodynamics

where H is the integro-differential operator of Eq. (2.17). In writing Eq. (2.132) we have assumed a system of volume V bounded by a surface of area A. Further, we have taken the time interval of interest to be $0 < t < T$. Here $i \equiv i(\mathbf{r}, v, \mathbf{\Omega}, t)$ is an approximation to $I(\mathbf{r}, v, \mathbf{\Omega}, t)$, considered to be in error by terms of first order, and $i(0) \equiv i(\mathbf{r}, v, \mathbf{\Omega}, 0)$. The functions $j \equiv j(\mathbf{r}, v, \mathbf{\Omega}, t)$, $k \equiv k(\mathbf{r}_s, v, \mathbf{\Omega}, t)$ for $\mathbf{n} \cdot \mathbf{\Omega} < 0$, and $l \equiv l(\mathbf{r}, v, \mathbf{\Omega})$ are the Lagrange multipliers. The functions S, Γ, and Λ are the source, boundary, and initial conditions occurring in the transport problem, Eqs. (2.17), (2.29), and (2.31). If the use of Lagrange multipliers is unfamiliar to the reader, Eq. (2.132) can be considered as the starting point in the development, with the particular form of this equation having been guessed. The only reason for introducing the Lagrange multiplier notion was to give a justification of Eq. (2.132) as the starting point.

We now consider the Lagrange multipliers $j(\mathbf{r}, v, \mathbf{\Omega}, t)$, $k(\mathbf{r}_s, v, \mathbf{\Omega}, t)$, and $l(\mathbf{r}, v, \mathbf{\Omega})$ to be first order approximations to functions $J(\mathbf{r}, v, \mathbf{\Omega}, t)$, $K(\mathbf{r}_s, v, \mathbf{\Omega}, t)$, and $L(\mathbf{r}, v, \mathbf{\Omega})$, respectively. The definition of these functions J, K, and L will be chosen such that the first order errors in i, j, k, and l will lead to a value of $F[i, j, k, l]$ which is a second order estimate of $G[I]$. Since by construction $F[i, j, k, l]$ gives $G[I]$ when evaluated with $i = I$, we need only demand that the first variation of $F[i, j, k, l]$ vanish when evaluated with the exact solutions I, J, K, and L. This requirement yields the defining equations for the functions J, K, and L.

Taking the first variation of Eq. (2.132), we obtain

$$\delta F[i, j, k, l] = \int_V d\mathbf{r} \int_0^\infty dv \int_{4\pi} d\Omega \int_0^T dt \{ G'[i]\delta i + (Hi - S)\delta j + jH\delta i \}$$

$$+ \int_A ds \int_0^\infty dv \int_{\mathbf{n} \cdot \mathbf{\Omega} < 0} d\Omega \int_0^T dt [(i - \Gamma)\delta k + k\delta i]$$

$$+ \int_V d\mathbf{r} \int_0^\infty dv \int_{4\pi} d\Omega \{ [i(0) - \Lambda]\delta l + l\delta i(0) \}, \tag{2.133}$$

where $G'[i] \equiv G'[i; \mathbf{r}, v, \mathbf{\Omega}, t]$ is the first functional derivative of $G[i(\mathbf{r}, v, \mathbf{\Omega}, t)]$. This functional derivative is defined such that

$$\delta G[i(\xi)] \equiv G[i(\xi) + \delta i(\xi)] - G[i(\xi)] \equiv \int dy\, G'[i(\xi); y]\, \delta i(y) \tag{2.134}$$

in the limit as $\delta i(\xi)$ approaches zero. Here ξ and y stand for the seven variables $\mathbf{r}, v, \mathbf{\Omega}$, and t. Setting

$$\delta i(y) = \varepsilon\delta(y - x); \quad \delta i(\xi) = \varepsilon\delta(\xi - x), \tag{2.135}$$

we find that Eq. (2.134) yields

$$G'[i(\xi); x] = \lim_{\varepsilon \to 0} \left\{ \frac{G[i(\xi) + \varepsilon\delta(\xi - x)] - G[i(\xi)]}{\varepsilon} \right\}, \tag{2.136}$$

where $\delta(z)$ in Eq. (2.136) and on the right hand sides of Eq. (2.135) is the Dirac delta function. Equation (2.136) is a formal definition of the first functional derivative of $G[i(\xi)]$. A complete discussion of functionals and their derivatives is given by Volterra (1959). To proceed with the variational characterization of the equation of transfer, we consider the term $jH\delta i$ in Eq. (2.133) which reads

$$jH\delta i = j\left[\frac{1}{c}\frac{\partial(\delta i)}{\partial t}+\boldsymbol{\Omega}\boldsymbol{\cdot}\nabla(\delta i) + \dots\right],\qquad(2.137)$$

where a sevenfold integration, as in Eq. (2.133), over these terms is understood. We integrate the $\partial/\partial t$ term by parts (carrying, of course, the endpoint terms), recognize that $j\boldsymbol{\Omega}\boldsymbol{\cdot}\nabla(\delta i)=\nabla\boldsymbol{\cdot}(\boldsymbol{\Omega}j\delta i)-\delta i\boldsymbol{\Omega}\boldsymbol{\cdot}\nabla j$, and use Gauss' theorem on the divergence term to convert the volume integral to a surface integral. In addition we interchange the integration variables v and v' in the integrand of the inscattering term contained in $jH\delta i$ (we also interchange $\boldsymbol{\Omega}$ and $\boldsymbol{\Omega}'$ in this term but this introduces no change since these angles appear only as $\boldsymbol{\Omega}\boldsymbol{\cdot}\boldsymbol{\Omega}'$). We then find that Eq. (2.133) can be rewritten, grouping terms,

$$\delta F[i, j, k, l] = \int\limits_{V} d\mathbf{r} \int\limits_{0}^{\infty} dv \int\limits_{4\pi} d\Omega \int\limits_{0}^{T} dt\{(Hi-S)\delta j+(H^*j+G'[i])\delta i\}$$

$$+\int\limits_{A} ds \int\limits_{0}^{\infty} dv \int\limits_{\mathbf{n}\boldsymbol{\cdot}\boldsymbol{\Omega}<0} d\Omega \int\limits_{0}^{T} dt\{(i-\Gamma)\delta k+[k+(\mathbf{n}\boldsymbol{\cdot}\boldsymbol{\Omega})j]\delta i\}$$

$$+\int\limits_{A} ds \int\limits_{0}^{\infty} dv \int\limits_{\mathbf{n}\boldsymbol{\cdot}\boldsymbol{\Omega}>0} d\Omega \int\limits_{0}^{T} dt(\mathbf{n}\boldsymbol{\cdot}\boldsymbol{\Omega})j\delta i$$

$$+\int\limits_{V} d\mathbf{r} \int\limits_{0}^{\infty} dv \int\limits_{4\pi} d\Omega\{[i(0)-\Lambda]\delta l+[l-j(0)/c]\delta i(0)+[j(T)/c]\delta i(T)\}.$$

$$(2.138)$$

The operator H^* in Eq. (2.138) is called the operator "adjoint" to H and is given by

$$H^*(\ ) = -\frac{1}{c}\frac{\partial(\ )}{\partial t} -\boldsymbol{\Omega}\boldsymbol{\cdot}\nabla(\ )+\sigma(\mathbf{r}, v, t)(\ )$$

$$+\int\limits_{0}^{\infty} dv' \int\limits_{4\pi} d\Omega'\frac{v'}{v}\sigma_s(\mathbf{r}, v \rightarrow v',\boldsymbol{\Omega}\boldsymbol{\cdot}\boldsymbol{\Omega}', t)(\ ).\qquad(2.139)$$

This operator differs from H in the signs of the time derivative and gradient terms, and the interchange of v and v' in the inscattering operator.

The requirement that $\delta F[I, J, K, L]$ vanish for independent and arbitrary variations of δj, δk, and δl yields the equation of transfer and its boundary and initial conditions, Eqs. (2.17), (2.29), and (2.31). The requirement that $\delta F[I, J, K, L]$ vanish for independent and arbitrary variations of $\delta i(\mathbf{r}, \boldsymbol{\Omega})$, $\delta i(\mathbf{r}_s, \mathbf{n}\boldsymbol{\cdot}\boldsymbol{\Omega} < 0)$, $\delta i(\mathbf{r}_s, \mathbf{n}\boldsymbol{\cdot}\boldsymbol{\Omega} > 0)$, $\delta i(0)$ and $\delta i(T)$ yields the five equations

$$H^*J(\mathbf{r}, v, \boldsymbol{\Omega}, t) = -G'[I],\qquad(2.140)$$

$$J(\mathbf{r}_s, v, \boldsymbol{\Omega}, t) = 0, \mathbf{n}\boldsymbol{\cdot}\boldsymbol{\Omega} > 0,\qquad(2.141)$$

$$J(\mathbf{r}, v, \boldsymbol{\Omega}, T) = 0,\qquad(2.142)$$

and

$$K(\mathbf{r}_s, v, \boldsymbol{\Omega}, t)+(\mathbf{n}\boldsymbol{\cdot}\boldsymbol{\Omega}) J(\mathbf{r}_s, v, \boldsymbol{\Omega}, t) = 0, \mathbf{n}\boldsymbol{\cdot}\boldsymbol{\Omega} < 0,\qquad(2.143)$$

$$L(\mathbf{r}, v, \boldsymbol{\Omega})-J(\mathbf{r}, v, \boldsymbol{\Omega}, 0)/c = 0.\qquad(2.144)$$

Equation (2.140) is the adjoint equation of transfer with Eqs. (2.141) and (2.142) acting as the spatial and temporal boundary conditions on this equation. Equations (2.143) and (2.144) define the functions K and L in terms of the solution to the adjoint problem. In view of these two equations, we use

$$k(\mathbf{r}_s, v, \mathbf{\Omega}, t) = -(\mathbf{n}\cdot\mathbf{\Omega})j(\mathbf{r}_s, v, \mathbf{\Omega}, t), \quad \mathbf{n}\cdot\mathbf{\Omega} < 0, \tag{2.145}$$

$$l(\mathbf{r}, v, \mathbf{\Omega}) = j(\mathbf{r}, v, \mathbf{\Omega}, 0)/c, \tag{2.146}$$

as trial functions for K and L in Eq. (2.132). We then find, dropping the arguments k and l in $F[i, j, k, l]$ since they no longer appear explicitly in the functional,

$$F[i, j] = G[i] + \int_V d\mathbf{r} \int_0^\infty dv \int_{4\pi} d\mathbf{\Omega} \int_0^T dt (Hi - S)j$$

$$+ \int_A ds \int_0^\infty dv \int_{\mathbf{n}\cdot\mathbf{\Omega}<0} d\mathbf{\Omega} \int_0^T dt (\mathbf{n}\cdot\mathbf{\Omega})(\Gamma - i)j$$

$$+ \int_V d\mathbf{r} \int_0^\infty dv \int_{4\pi} d\mathbf{\Omega} [i(0) - \Lambda] j(0). \tag{2.147}$$

Equation (2.147) is the variational characterization of the equation of transfer. It has the property that if $i(\mathbf{r}, v, \mathbf{\Omega}, t)$ is a first order estimate of $I(\mathbf{r}, v, \mathbf{\Omega}, t)$ defined by Eqs. (2.17), (2.29), and (2.31), and $j(\mathbf{r}, v, \mathbf{\Omega}, t)$ is a first order estimate of $J(\mathbf{r}, v, \mathbf{\Omega}, t)$ defined by Eqs. (2.140) through (2.142), then $F[i, j]$ is a second order estimate of the functional of interest $G[I]$. The functions i and j to be used in Eq. (2.147) need not satisfy the spatial and temporal boundary conditions on the transport and adjoint transport problems, Eqs. (2.29), (2.31), (2.141), and (2.142).

An alternate way of viewing Eq. (2.147) is in the Lagrangian sense. Demanding that $F[i, j]$ be stationary, at the point $i = I$ and $j = J$, under independent and arbitrary variations of i and j is equivalent to demanding that I is a solution of the transport problem defined by Eqs. (2.17), (2.29), and (2.31) and J is a solution of the adjoint problem defined by Eqs. (2.140) through (2.142). This point of view allows one to systematically derive approximate descriptions of the transport and adjoint transport problems. One assumes functional forms for $i(\mathbf{r}, v, \mathbf{\Omega}, t)$ and $j(\mathbf{r}, v, \mathbf{\Omega}, t)$ and demands that $F[i, j]$ be stationary within this restricted class of trial functions. The resulting Euler–Lagrange equations constitute approximate descriptions of the radiative transfer and adjoint transport problems. In this use of the variational method one is not interested in any particular functional of the intensity but rather in an approximation to the intensity itself. Consequently the adjoint equation is of no direct importance and $G[I]$ is more or less arbitrary. An appropriate choice for $G[I]$ is one which simplifies the Lagrangian use of the variational principle. We shall now show that, for a restricted class of transport problems, a proper choice of the functional $G[I]$ leads to a simple relationship between $I(\mathbf{r}, v, \mathbf{\Omega}, t)$ and $J(\mathbf{r}, v, \mathbf{\Omega}, t)$. Use of this relationship allows us to effectively eliminate the adjoint function from the variational characterization of the equation of transfer. Then only a single trial function, $i(\mathbf{r}, v, \mathbf{\Omega}, t)$, is needed to derive an approximate description of radiative transfer.

Consider a class of transport problems in which

$$\sigma(\mathbf{r}, \nu, t) = \sigma(\mathbf{r}, \nu, T-t), \tag{2.148}$$

$$\frac{\nu}{\nu'} \sigma_s(\mathbf{r}, \nu' \to \nu, \boldsymbol{\Omega}' \cdot \boldsymbol{\Omega}, t) = \frac{\nu'}{\nu} \sigma_s(\mathbf{r}, \nu \to \nu', \boldsymbol{\Omega} \cdot \boldsymbol{\Omega}', T-t). \tag{2.149}$$

A physically meaningful example from within this class is one in which both σ and σ_s are independent of time, and the scattering interaction gives no change in frequency (the frequency dependence of the differential scattering coefficient then contains a Dirac delta function with argument $\nu - \nu'$). The significance of Eqs. (2.148) and (2.149) is that, for this class of problems, the operators H and H^* are simply related according to

$$H^*(\mathbf{r}, \nu, \boldsymbol{\Omega}, t) = H(\mathbf{r}, \nu, -\boldsymbol{\Omega}, T-t). \tag{2.150}$$

Let us now choose the functional $G[I]$ as

$$
\begin{aligned}
G[I] = &- \int_V d\mathbf{r} \int_0^\infty d\nu \int_{4\pi} d\boldsymbol{\Omega} \int_0^T dt\, S(\mathbf{r}, \nu, T-t) I(\mathbf{r}, \nu, \boldsymbol{\Omega}, t) \\
&- \int_A ds \int_0^\infty d\nu \int_{\mathbf{n} \cdot \boldsymbol{\Omega} > 0} d\boldsymbol{\Omega} \int_0^T dt\, \Gamma(\mathbf{r}_s, \nu, -\boldsymbol{\Omega}, T-t)(\mathbf{n} \cdot \boldsymbol{\Omega}) I(\mathbf{r}_s, \nu, \boldsymbol{\Omega}, t) \\
&- \frac{1}{c} \int_V d\mathbf{r} \int_0^\infty d\nu \int_{4\pi} d\boldsymbol{\Omega}\, \Lambda(\mathbf{r}, \nu, -\boldsymbol{\Omega}) I(\mathbf{r}, \nu, \boldsymbol{\Omega}, t).
\end{aligned}
\tag{2.151}
$$

The first functional derivative of $G[I]$ as given by Eq. (2.151) is

$$
\begin{aligned}
G'[I] = &-S(\mathbf{r}, \nu, T-t) - \Gamma(\mathbf{r}_s, \nu, -\boldsymbol{\Omega}, T-t)(\mathbf{n} \cdot \boldsymbol{\Omega})\, U(\mathbf{n} \cdot \boldsymbol{\Omega})\, \delta(\mathbf{n} \cdot \mathbf{r} - \mathbf{n} \cdot \mathbf{r}_s) \\
&- \frac{1}{c} \Lambda(\mathbf{r}, \nu, -\boldsymbol{\Omega})\, \delta(t-T),
\end{aligned}
\tag{2.152}
$$

where $U(z)$ is the unit step (Heaviside) function and $\delta(z)$ is the Dirac delta function. The adjoint equation then becomes, from Eqs. (2.140) and (2.152),

$$
\begin{aligned}
H^*(\mathbf{r}, \nu, \boldsymbol{\Omega}, t) J(\mathbf{r}, \nu, \boldsymbol{\Omega}, t) = &\; S(\mathbf{r}, \nu, T-t) + \Gamma(\mathbf{r}_s, \nu, -\boldsymbol{\Omega}, T-t)\, (\mathbf{n} \cdot \boldsymbol{\Omega})\, U(\mathbf{n} \cdot \boldsymbol{\Omega})\, \delta(\mathbf{n} \cdot \mathbf{r} - \mathbf{n} \cdot \mathbf{r}_s) \\
&+ \frac{1}{c} \Lambda(\mathbf{r}, \nu, -\boldsymbol{\Omega})\, \delta(t-T).
\end{aligned}
\tag{2.153}
$$

The delta functions in Eq. (2.153) represent a surface source and a pulse source, at $t = T$, for the adjoint equation. It can be shown quite easily that these two sources can be replaced with appropriate boundary conditions on the space and time variables. Omitting the mathematical details, Eq. (2.153), with space and time boundary conditions given by Eqs. (2.141) and (2.142), is equivalent to the adjoint problem

$$
\left.
\begin{aligned}
H^*(\mathbf{r}, \nu, \boldsymbol{\Omega}, t) J(\mathbf{r}, \nu, \boldsymbol{\Omega}, t) &= S(\mathbf{r}, \nu, T-t), \\
J(\mathbf{r}_s, \nu, \boldsymbol{\Omega}, t) &= \Gamma(\mathbf{r}_s, \nu, -\boldsymbol{\Omega}, T-t), \qquad \mathbf{n} \cdot \boldsymbol{\Omega} > 0, \\
J(\mathbf{r}, \nu, \boldsymbol{\Omega}, T) &= \Lambda(\mathbf{r}, \nu, -\boldsymbol{\Omega}).
\end{aligned}
\right\}
\tag{2.154}
$$

The Equations of Radiation Hydrodynamics

It is to be recalled that the equation of transfer reads

$$H(\mathbf{r}, \nu, \mathbf{\Omega}, t)\, I(\mathbf{r}, \nu, \mathbf{\Omega}, t) = S(\mathbf{r}, \nu, t),$$
$$I(\mathbf{r}_s, \nu, \mathbf{\Omega}, t) = \Gamma(\mathbf{r}_s, \nu, \mathbf{\Omega}, t), \qquad \mathbf{n}\cdot\mathbf{\Omega} < 0, \tag{2.155}$$
$$I(\mathbf{r}, \nu, \mathbf{\Omega}, 0) = \Lambda(\mathbf{r}, \nu, \mathbf{\Omega}).$$

For the class of problems under consideration, i.e., for those which satisfy Eq. (2.150), a comparison of Eqs. (2.154) and (2.155) establishes the identity

$$J(\mathbf{r}, \nu, \mathbf{\Omega}, t) = I(\mathbf{r}, \nu, -\mathbf{\Omega}, T-t). \tag{2.156}$$

In view of Eq. (2.156), we set

$$j(\mathbf{r}, \nu, \mathbf{\Omega}, t) = i(\mathbf{r}, \nu, -\mathbf{\Omega}, T-t) \tag{2.157}$$

in the functional $F[i, j]$ as given by Eq. (2.147). The resulting functional, $F[i]$, is given by

$$
\begin{aligned}
F[i] = G[i] &+ \int_V d\mathbf{r} \int_0^\infty d\nu \int_{4\pi} d\mathbf{\Omega} \int_0^T dt (Hi-S)\, i(\mathbf{r}, \nu, -\mathbf{\Omega}, T-t) \\
&+ \int_A ds \int_0^\infty d\nu \int_{\mathbf{n}\cdot\mathbf{\Omega}<0} d\mathbf{\Omega} \int_0^T dt (\mathbf{n}\cdot\mathbf{\Omega})(\Gamma-i)\, i(\mathbf{r}, \nu, -\mathbf{\Omega}, T-t) \\
&+ \int_V d\mathbf{r} \int_0^\infty d\nu \int_{4\pi} d\mathbf{\Omega} [i(0)-\Lambda]\, i(\mathbf{r}, \nu, -\mathbf{\Omega}, T),
\end{aligned}
\tag{2.158}
$$

where the arguments not fully displayed are the appropriate ones chosen from $\mathbf{r}, \nu, \mathbf{\Omega}$, and t. For example, $i(0)-\Lambda \equiv i(\mathbf{r}, \nu, \mathbf{\Omega}, 0)-\Lambda(\mathbf{r}, \nu, \mathbf{\Omega})$, and similarly for the terms $Hi-S$ and $\Gamma-i$. Hence for the class of problems defined by Eqs. (2.148) and (2.149) one has a variational characterization of the equation of transfer which involves no adjoint function. In the general case, however, such a characterization as given by Eq. (2.147) does involve an adjoint equation of transfer.

9. Induced Processes and Local Thermodynamic Equilibrium

The equation of transfer discussed thus far may properly be termed the classical equation of transfer since its derivation was based solely on classical physics concepts. We have considered this equation in great detail since it, or a simplification of it, is the description of energy transport used in the vast majority of radiation hydrodynamic calculations. Nevertheless, it neglects some interesting physics and we consider one such item here.

Specifically, we consider the manifestation in the equation of transfer of the quantum statistics obeyed by photons. Since photons are bosons, both the processes of emission and scattering are enhanced by the number of photons already in the final state following the interaction. This enhancement is generally referred to as resulting from "induced processes". The quantitative statement of this enhancement is simply stated as: If P represents the basic probability of a photon event (emission or scattering) then, due to induced effects, the actual probability P' is given by (Feynman *et al.*, 1966a)

$$P' = P(1+n), \tag{2.159}$$

where n is the number of photons in the final state of the transition. In the present context, the final state corresponds to the basic, or unit cell, of phase space. In terms of the distribution function $f(\mathbf{r}, \nu, \Omega, t)$ introduced in Chapter I, the number of photons at time t in a unit cell of $\mathbf{r}, \nu, \Omega$ space is given by

$$n = \int d\mathbf{r} \int_{\Delta} d\nu \int d\Omega f(\mathbf{r}, \nu, \Omega, t), \qquad (2.160)$$

where Δ denotes the unit cell of phase space. It is convenient to transform from ν, Ω space to $\mathbf{p}$ (momentum) space. We have, representing the momentum differential in spherical coordinates,

$$d\mathbf{p} = p^2 dp\, d\Omega, \qquad (2.161)$$

or, since $p = h\nu/c$ for photons,

$$d\mathbf{p} = (h/c)^3 \nu^2\, d\nu\, d\Omega. \qquad (2.162)$$

Introducing the specific intensity $I \equiv ch\nu f$ in Eq. (2.160) and making use of Eq. (2.162), we find

$$n = \frac{c^2}{h^4} \int d\mathbf{r} \int_{\Delta} d\mathbf{p} I(\mathbf{r}, \nu, \Omega, t)/\nu^3, \qquad (2.163)$$

where Δ now denotes the basic cell in $\mathbf{r}, \mathbf{p}$ space. This basic element of phase space is just $\Delta \equiv \Delta\mathbf{p}\,\Delta\mathbf{r} = h^3$ (Landau and Lifshitz, 1965). However, since photons occur in two states of polarization, each h^3 of phase space can accommodate two photons. Hence the element of phase space appropriate in Eq. (2.163) is

$$\Delta = h^3/2. \qquad (2.164)$$

This leads to

$$n = \frac{c^2}{h^4\nu^3} I(\mathbf{r}, \nu, \Omega, t)\, \Delta = \frac{c^2}{2h\nu^3} I(\mathbf{r}, \nu, \Omega, t), \qquad (2.165)$$

and thus

$$P' = P[1 + c^2 I/2h\nu^3]. \qquad (2.166)$$

Equation (2.166) implies that the probabilities of emission and scattering in the equation of transfer should be increased by the factor $1 + c^2 I/2h\nu^3$, where the frequency and angle arguments of the specific intensity correspond to the state of the photon after the emission or scattering process has occurred. With this change, the classical equation of transfer, Eq. (2.28), becomes

$$\frac{1}{c}\frac{\partial I(\nu, \Omega)}{\partial t} + \Omega \cdot \nabla I(\nu, \Omega) = S(\nu)[1 + c^2 I(\nu, \Omega)/2h\nu^3] - \sigma_a(\nu)\, I(\nu, \Omega)$$

$$+ \int_0^\infty d\nu' \int_{4\pi} d\Omega' \frac{\nu}{\nu'} \sigma_s(\nu' \to \nu, \Omega' \cdot \Omega)\, I(\nu', \Omega')[1 + c^2 I(\nu, \Omega)/2h\nu^3]$$

$$- \int_0^\infty d\nu' \int_{4\pi} d\Omega' \sigma_s(\nu \to \nu', \Omega \cdot \Omega')\, I(\nu, \Omega)[1 + c^2 I(\nu', \Omega')/2h\nu'^3]. \qquad (2.167)$$

The Equations of Radiation Hydrodynamics

Equation (2.167) is the equation of radiative transfer including the effects of induced processes. It can be seen that induced scattering severely complicates the equation of transfer in that it leads to nonlinear terms, quadratic in the specific intensity. It might also be noted that in the limiting case of no frequency change upon scattering [such that $\sigma_s(\nu' \to \nu, \Omega' \cdot \Omega)$ contains a Dirac delta function with argument $\nu - \nu'$] the induced inscattering and outscattering terms identically cancel one another, and the equation of transfer again becomes linear. The induced contribution to the emission remains, however, but this term is always linear in character. As we shall discuss in more detail in Chapter VIII, these induced scattering terms are necessary in the equation of transfer for the scattering operator to give the correct equilibrium distribution, namely the Planck function. A final note of interest concerning induced processes is that they result from a physical principle closely connected with the Pauli exclusion principle. The Pauli principle reduces the probability of fermion events by the factor $1 - n$, and hence if the specific intensity described fermions the appropriate factors in Eq. (2.167) would be $1 - c^2 I/2h\nu^3$. As is clear from this short discussion, the Planck function and the Pauli principle are closely connected. The underlying physics of this connection is beyond the scope of this book, but we will be able to demonstrate very simply in Chapter VIII that the neglect of induced processes leads to the Wien, rather than the Planck, function as the equilibrium distribution for the specific intensity.

Another item of interest to consider here is the concept of local thermodynamic equilibrium (LTE). With reference to Eq. (2.167), the source term S represents the source of photons due to spontaneous emission from atoms, and the functions σ_a and σ_s determine the interaction of photons with the matter. In general these three quantities depend upon the microscopic description of the atoms which compose the matter, i.e., the population of the various states of the atoms, and there is no simple relationship between the three quantities. A simplifying assumption in this regard often invoked in radiative transfer work is the LTE assumption. It is assumed that the properties of the matter are dominated by atomic collisions which establish thermodynamic equilibrium locally at position $\mathbf{r}$ and time t, and that the radiation field, even if it deviates substantially from the equilibrium Planck distribution, does not affect this equilibrium. That is, at a given instant of time and point in space it suffices to specify, in addition to the atomic composition, two thermodynamic quantities such as temperature and density in order to compute the source term S, absorption coefficient σ_a, and scattering coefficient σ_s. Equilibrium statistical mechanics, together with quantum mechanics, can then in principle be used to compute S, σ_a, and σ_s. In particular, the Saha and Boltzmann laws, appropriate to thermodynamic equilibrium (Cox and Giuli, 1968), can be used to determine the relative abundance of the ionic species and the population of the states within a given ionic species. The LTE assumption also leads to a simple relationship between S and σ_a, as we now show.

As it stands, Eq. (2.167) is not restricted to LTE situations but describes a more general class of problems. To see the effect of the LTE assumption on the equation of transfer, it is convenient to eliminate S and σ_a in Eq. (2.167) in favor of B and σ_a', defined by the relationships

$$S = \sigma_a' B, \tag{2.168}$$

$$\sigma_a = \sigma_a'(1 + c^2 B/2h\nu^3). \tag{2.169}$$

At this point B is *not* to be interpreted as the Planck function, but is merely a new variable defined in terms of σ_a and S according to Eqs. (2.168) and (2.169). In terms of B and σ_a', Eq. (2.167) can be written

$$\frac{1}{c}\frac{\partial I(\nu, \Omega)}{\partial t} + \Omega \cdot \nabla I = \sigma_a'(\nu)[B(\nu) - I(\nu, \Omega)]$$

$$+ \int_0^\infty d\nu' \int_{4\pi} d\Omega' \frac{\nu}{\nu'} \sigma_s(\nu' \rightarrow \nu, \Omega' \cdot \Omega)\, I(\nu', \Omega')[1 + c^2 I(\nu, \Omega)/2h\nu^3]$$

$$- \int_0^\infty d\nu' \int_{4\pi} d\Omega' \sigma_s(\nu \rightarrow \nu', \Omega \cdot \Omega')\, I(\nu, \Omega)[1 + c^2 I(\nu', \Omega')/2h\nu'^3]. \qquad (2.170)$$

For the moment, let us take σ_s in Eq. (2.170) to be zero, although the argument we are about to make can also be made with scattering included, using the detailed balance relationship to be discussed in Chapter VIII. In complete thermodynamic equilibrium, the radiation field is independent of space and time and hence, in this situation with the neglect of scattering, Eq. (2.170) reads

$$\sigma_a'(\nu)[B(\nu) - I(\nu, \Omega)] = 0. \qquad (2.171)$$

It is well known that in complete thermodynamic equilibrium the equation of transfer must give the Planck black body distribution for the specific intensity I. For this to be the case, it is clear from Eq. (2.171) that B must be the Planck function as well. Since the LTE assumption states that the radiation field does not affect the properties of the matter, in particular the source function B, we conclude that under the LTE simplification B is the Planck function no matter what the radiation field is. That is, under LTE we have

$$B(\nu) = \frac{2h\nu^3}{c^2}\,(e^{h\nu/kT} - 1)^{-1}, \qquad \qquad (2.172)$$

where T is the local temperature of the matter. Further, use of Eq. (2.172) in Eq. (2.169) shows that the function σ_a' is given in the LTE approximation by

$$\sigma_a'(\nu) = \sigma_a(\nu)(1 - e^{-h\nu/kT}). \qquad (2.173)$$

Here $\sigma_a(\nu)$ is the absorption coefficient appropriate to thermodynamic equilibrium and the exponential factor is the effective decrease in the absorption coefficient due to stimulated emission.

The form of Eq. (2.170), involving emission and absorption in the form $\sigma_a'(B - I)$, is the conventional way of writing the transport equation in radiative transfer, even if the LTE assumption is not invoked (in which case B is not the Planck function). However, the LTE assumption is generally made in radiation hydrodynamic work because of the vast simplification it introduces; namely, thermodynamics can be used to describe the matter.

10. The Validity of the Equation of Transfer

It is natural to inquire as to the validity of Eq. (2.170) as the description of energy transport by radiative processes. In particular, what approximations in the underlying physics are contained in this equation? These approximations fall into two classes—those which are inherent in any transfer equation description of radiation energy transport, and those which could be incorporated into Eq. (2.170) at the expense of simplicity. It should be emphasized at the outset of this discussion that the entire question of the inherent validity of transport equations is by no means settled, but is still being actively researched. However, some comments of a qualitative nature can be made.

Most of the approximations contained in Eq. (2.170) stem from the fact that this equation basically treats photons as classical, point particles. It is known, however, that photons exhibit wave behavior; i.e., a photon is in reality a wave packet. To be able to treat the wave packet as a point particle, it is necessary that the spread of the packet, in both physical and momentum space, be small. By this we mean that the spread should be small compared to the resolution of interest in the space (r) and momentum (represented by v and Ω) variables. This is clearly required since we assume in writing the specific intensity I as a function of the variables r, v, and Ω that it is sufficient to specify the phase space coordinates of the center of the wave packet, and that any information concerning the distribution about this center is irrelevant. Because of the uncertainty principle, which states that the spreads in physical and momentum space cannot both be made arbitrarily small at the same time, these considerations impose a minimum on the spatial and momentum resolution which is possible. Secondly, Eq. (2.170) deals with intensities rather than wave amplitudes, and hence the possibility of interference between photons is nonexistent in Eq. (2.170). This implies that a condition on Eq. (2.170) is that the density of photons is low, i.e., low enough so that the overlap in the tails of the wave packets of two photons is negligibly small. This restriction is somewhat too strong since, given a time resolution of interest, photons of sufficiently different frequencies do not interfere even if they coincide spatially (Slater and Frank, 1947). This fact is needed to justify the assumption that the source photons emitted at the same space point do not interfere with each other; i.e., the source photons in Eq. (2.170) are incoherent. Also, Eq. (2.170) assumes that all collision and emission processes occur instantaneously. That is, the loss or gain of photons due to these processes is characterized by σ_a', σ_s, and B at a given instant of time rather than being dependent upon some sort of time average of these quantities over the collision or emission time. This essentially imposes a minimum on the time resolution which the equation of transfer can supply.

Finally, any transport equation for photons cannot describe the strong wave behavior manifested in diffraction and reflection. These phenomena depend upon interference among the waves arising from different scattering centers which scatter the same photon. For interference of this type to take place, two conditions must be satisfied. First, the scattering centers must be correlated as in a crystal, and, secondly, the spatial extent of the wave packet must be such that several scattering centers are encompassed by a photon. A qualitative discussion of this type of interference due to scattering is given by Slater and Frank (1947), and a simple quantitative example is given by Morse and Feshbach (1961). Both of these references employ an infinite plane wave in their discussion, i.e., a photon of infinite spatial extent. Such a wave is generally the basis of any discussion or calculation of such effects. Hence Eq. (2.170), which treats scattering as independent and isolated events, is only a proper description if the scattering centers are randomly distributed as in a gas, or if the wave packets are small compared to the distance between scattering centers. In either event, the material has no preferential direction as far as a photon is concerned, and hence the absorption coefficient σ_a' and the source B are independent of the photon propagation direction Ω, and the scattering kernel σ_s only depends upon the scattering angle, i.e., upon $\Omega \cdot \Omega'$. These dependences are indicated in Eq. (2.170).

The approximations just discussed are inherent in any transport equation. However, there is a second class of assumptions contained in Eq. (2.170) which can be relaxed and the corresponding physical effects incorporated into an equation of transfer. The first item we mention is that Eq. (2.170) does not account for the two states of polarization of photons. As we shall discuss in Chapter IV, four parameters are required to specify the state of polarization of a beam of light. Further, the state of polarization and hence these four parameters in general change when a photon undergoes a scattering event. Accordingly, a proper description of photon transport including polarization effects involves four equations of transfer, which in general are coupled. Equation (2.170) can be considered as the result of averaging this set of four equations over polarization states, assuming the light to be unpolarized. This averaging process, however, introduces an error and this constitutes an approximation in Eq. (2.170).

Another feature of Eq. (2.170) is that it neglects effects due to refraction and dispersion. It is known that a photon will move at less than the vacuum speed of light in matter with a refractive index other than unity. Also, if the refractive index is a function of position, the photon will not stream in straight lines between collisions but will undergo (continuous) refraction. In addition, if the refractive index is time dependent, a photon will (continuously) change its frequency as it streams between collisions. The origin of these effects is found in the scattering of photons, and in particular in an interference phenomenon. In this case the interference is between the scattered wave in the near forward direction and the incident wave. This type of interference is discussed, for example, by Feynman *et al.* (1966b). Again, this discussion is based upon an infinite plane wave. The assumption in Eq. (2.170) is that the matter has a refractive index of unity. However, refractive and dispersive effects can be incorporated into the streaming terms in Eq. (2.170), and Chapter V treats this topic.

The final approximation contained in Eq. (2.170) which we shall discuss is the angular dependences of σ_a', σ_s, and B. We have previously argued that for large wave packets the equation of transfer is only valid if the interaction centers of the matter are randomly distributed. This implies no inherent preferred direction in the matter, and hence σ_a' and B should be independent of Ω, and σ_s should depend only upon $\Omega \cdot \Omega'$. However, the fact that in radiation hydrodynamic problems the material is in general in motion changes the situation. As seen by the inertial frame observer, this motion does introduce a preferred direction in the matter, namely the direction of motion of the fluid. This, in turn, in the relativistic limit, introduces an Ω dependence into σ_a' and B, and separate Ω and Ω' dependences into σ_s. It should be emphasized that these angular dependences are not inherent properties of the material, but arise only from the relative motion between the fluid and the observer. Hence these angular dependences can be computed from the special theory of relativity. This subject is treated in Chapter VI and the Appendix. These angular effects turn out to be, as one might expect, of the order of u/c, where u is the macroscopic speed of the fluid and c is the speed of light.

III

Approximate Descriptions
of Radiative Transfer

---

1. Introduction

The equation of transfer derived in the last chapter is obviously very complicated; the specific intensity, which is the dependent variable in this equation, depends in general upon seven independent variables $(\mathbf{r}, \nu, \Omega, t)$. Even in the simplest physically interesting situation of time and frequency independent transport in plane geometry (then only two independent variables z and μ are involved), one can obtain analytic solutions in only a very small number of limiting cases. Hence in general one must approximate the equation of transfer, either analytically or numerically, in order to obtain a solution.

Most approximate descriptions of radiative transfer are based upon the integro-differential equation (2.170) rather than the integral equation. The frequency and angle dependences of the specific intensity, which give rise to the integral terms in this equation, are generally approximated analytically. This leads to a finite (and hopefully small) number of coupled differential equations in the space and time variables. These equations are then conventionally solved numerically via finite difference techniques.

It is our purpose in this chapter to discuss a limited number of the analytic approximations employed in frequency and angle. The methods we shall consider certainly do not represent the totality of all methods which can, and have, been used in radiation hydrodynamic calculations. They are, however, the techniques most commonly used in practice. The finite difference methods used in space and time will not be considered. Such techniques, especially with the advent of high speed computers, can be very sophisticated and represent a discipline within themselves.

The first part of this chapter deals with low order angular approximations to the equation of transfer, namely diffusion-like equations which encompass the so-called Eddington approximation as a special case. This leads naturally to a discussion of the spherical harmonic and discrete ordinate methods, which are capable of giving any desired accuracy. The diffusion descriptions of radiative transfer are related to (and in some cases, identical with) the lowest order spherical harmonic and discrete ordinate approximations. The final part of the chapter deals with the so-called multigroup method of representing the frequency variable in the equation of transfer. A special case of this is the one-group, or gray, approximation involving special average absorption coefficients, called the Planck and Rosseland means.

50

Sandwiched between our treatments of the various approximations in angle and frequency is a discussion of equilibrium diffusion theory. This corresponds to a very low order approximation in both angle and frequency and leads to an exceedingly simple expression for the rate of energy transport by radiative processes. Although this description is very crude, it is a widely used approximation in radiation hydrodynamic calculations because of its simplicity. This is particularly true in multidimensional problems since multidimensional hydrodynamic calculations are in themselves quite complex numerically, and one hopes to be able to get by with a simple treatment of the radiation flow.

Aside from the discussion of equilibrium diffusion theory, we will not explicitly consider techniques which involve approximations in both frequency and angle simultaneously. In practice, the general procedure is to combine the multigroup method with any one of the angular approximations discussed in this chapter.

2. The Eddington or Diffusion Approximation

The basic assumption underlying the classical diffusion, or Eddington, description of radiative transfer is that the angular dependence of the specific intensity can be represented by the first two terms in a spherical harmonic expansion. That is, it is assumed that

$$I(\mathbf{r}, v, \mathbf{\Omega}, t) = \frac{1}{4\pi} I_0(\mathbf{r}, v, t) + \frac{3}{4\pi} \mathbf{\Omega} \cdot \mathbf{I}_1(\mathbf{r}, v, t), \tag{3.1}$$

where I_0 and $\mathbf{I}_1$ are the expansion coefficients to be determined. The term involving $\mathbf{I}_1$ in Eq. (3.1) represents a first order anisotropy correction to the (presumed) dominant I_0 term, and hence this representation would be expected to be most accurate in situations where the specific intensity is almost isotropic. The coefficients I_0 and $\mathbf{I}_1$ have a physical interpretation. Integration of Eq. (3.1) over all solid angle establishes

$$I_0(\mathbf{r}, v, t) = \int_{4\pi} d\Omega I(\mathbf{r}, v, \mathbf{\Omega}, t), \tag{3.2}$$

and multiplication of Eq. (3.1) by $\mathbf{\Omega}$ prior to a similar integration yields

$$\mathbf{I}_1(\mathbf{r}, v, t) = \int_{4\pi} d\Omega \mathbf{\Omega} I(\mathbf{r}, v, \mathbf{\Omega}, t). \tag{3.3}$$

Hence, in analogy to Eqs. (1.9) and (1.16), $I_0(\mathbf{r}, v; t)$ is, to within a factor of the speed of light, just the energy density per unit frequency, and $\mathbf{I}_1(\mathbf{r}, v, t)$ is the radiative flux per unit frequency.

To obtain the desired equations for I_0 and $\mathbf{I}_1$, we use the assumed representation of the specific intensity, Eq. (3.1), in the integro-differential equation of transfer, Eq. (2.170), and form the first two angular moments. That is, integration of Eq. (2.170), with $I(\mathbf{r}, v, \mathbf{\Omega}, t)$ given by Eq. (3.1), over all solid angle yields, after a bit of straightforward algebra,

The Equations of Radiation Hydrodynamics

$$\frac{1}{c}\frac{\partial I_0(\nu)}{\partial t} + \nabla \cdot \mathbf{I}_1(\nu) = \sigma_a'(\nu)[4\pi B(\nu) - I_0(\nu)] - \sigma_s(\nu)\, I_0(\nu)$$

$$+ \int_0^\infty d\nu' \frac{\nu}{\nu'} \sigma_{s0}(\nu' \to \nu)\, I_0(\nu')$$

$$+ \frac{c^2}{8\pi h} I_0(\nu) \int_0^\infty d\nu' \left[\frac{1}{\nu^2\nu'}\sigma_{s0}(\nu' \to \nu) - \frac{1}{\nu'^3}\sigma_{s0}(\nu \to \nu') \right] I_0(\nu')$$

$$+ \frac{3c^2}{8\pi h} \mathbf{I}_1(\nu) \cdot \int_0^\infty d\nu' \left[\frac{1}{\nu^2\nu'}\sigma_{s1}(\nu' \to \nu) - \frac{1}{\nu'^3}\sigma_{s1}(\nu \to \nu') \right] \mathbf{I}_1(\nu'). \tag{3.4}$$

Similarly, multiplication of Eq. (2.170) by Ω prior to an integration over all angle yields

$$\frac{1}{c}\frac{\partial \mathbf{I}_1(\nu)}{\partial t} + \frac{1}{3}\nabla I_0(\nu) = -[\sigma_a'(\nu) + \sigma_s(\nu)]\,\mathbf{I}_1(\nu) + \int_0^\infty d\nu' \frac{\nu}{\nu'} \sigma_{s1}(\nu' \to \nu)\, \mathbf{I}_1(\nu')$$

$$+ \frac{c^2}{8\pi h} I_0(\nu) \int_0^\infty d\nu' \left[\frac{1}{\nu^2\nu'}\sigma_{s1}(\nu' \to \nu) - \frac{1}{\nu'^3}\sigma_{s1}(\nu \to \nu') \right] \mathbf{I}_1(\nu')$$

$$+ \frac{c^2}{8\pi h} \mathbf{I}_1(\nu) \int_0^\infty d\nu' \left[\frac{1}{\nu^2\nu'}\sigma_{s0}(\nu' \to \nu) - \frac{1}{\nu'^3}\sigma_{s0}(\nu \to \nu') \right] I_0(\nu'). \tag{3.5}$$

Here we have defined

$$\sigma_{sn}(\nu' \to \nu) = 2\pi \int_{-1}^{1} d\mu_0 \sigma_s(\nu' \to \nu, \mu_0)\, P_n(\mu_0), \tag{3.6}$$

where $P_n(\mu_0)$ is the Legendre polynomial of degree n, and $\sigma_s(\nu)$ is the total scattering coefficient as defined by Eq. (1.32), i.e.,

$$\sigma_s(\nu) = \int_0^\infty d\nu' \sigma_{s0}(\nu \to \nu'). \tag{3.7}$$

Equations (3.4) and (3.5) form a closed set of equations for $I_0(\mathbf{r}, \nu, t)$ and $\mathbf{I}_1(\mathbf{r}, \nu, t)$ if appropriate initial and boundary conditions are specified. The initial conditions follow directly from the initial condition on the integro-differential transport equation. Referring to Eq. (2.31), we have

$$I_0(\mathbf{r}, \nu, 0) = \int_{4\pi} d\Omega \Lambda(\mathbf{r}, \nu, \Omega), \tag{3.8}$$

$$\mathbf{I}_1(\mathbf{r}, \nu, 0) = \int_{4\pi} d\Omega \Omega \Lambda(\mathbf{r}, \nu, \Omega), \tag{3.9}$$

where $\Lambda(\mathbf{r}, \nu, \Omega)$ is a specified function representing the initial condition for Eq. (2.170).

The boundary conditions on Eqs. (3.4) and (3.5) are not as straightforward to write down. The structures of these equations require a single condition between $I_0(\mathbf{r}, \nu, t)$ and $\mathbf{I}_1(\mathbf{r}, \nu, t)$

at each boundary point $\mathbf{r}_s$. It is clear that, because of its simple angular dependence, the Eddington representation of the specific intensity, Eq. (3.1), cannot satisfy the integro-differential boundary condition, Eq. (2.29), for an arbitrary incoming distribution $\Gamma(\mathbf{r}_s,\ \nu,\ \Omega,\ t)$. The best one can do is demand that Eq. (2.29) be satisfied in an integral sense. That is, we use Eq. (3.1) in Eq. (2.29), multiply the result by a weight function $w(\Omega)$, and integrate over all incoming directions. This gives

$$\int_{\mathbf{n}\cdot\Omega<0} d\Omega\, w(\Omega)\left[\frac{1}{4\pi}I_0(\mathbf{r}_s,\ \nu,\ t)+\frac{3}{4\pi}\Omega\cdot\mathbf{I}_1(\mathbf{r}_s,\ \nu,\ t)-\Gamma(\mathbf{r}_s,\ \nu,\ \Omega,\ t)\right]=0, \qquad (3.10)$$

where $\mathbf{n}$ is a unit outward normal vector at the surface point $\mathbf{r}_s$. Equation (3.10), once $w(\Omega)$ has been specified, is the required boundary condition on Eqs. (3.4) and (3.5).

The two boundary conditions used in practice are referred to as the Marshak and Mark conditions. The Marshak condition corresponds to the choice

$$w(\Omega) = \mathbf{n}\cdot\Omega, \qquad (3.11)$$

and Eq. (3.10) then yields

$$\tfrac{1}{4}I_0(\mathbf{r}_s,\ \nu,\ t)-\tfrac{1}{2}\mathbf{n}\cdot\mathbf{I}_1(\mathbf{r}_s,\ \nu,\ t) = \int_{\mathbf{n}\cdot\Omega<0} d\Omega\,|\mathbf{n}\cdot\Omega|\,\Gamma(\mathbf{r}_s,\ \nu,\ \Omega,\ t). \qquad (3.12)$$

The Marshak boundary condition has the physical interpretation that the normal component of the flux per unit frequency is the integral quantity conserved in passing from the exact condition, Eq. (2.29), to the integral condition, Eq. (3.10). To obtain the Mark condition, we represent Ω by a polar angle $\theta \equiv \cos^{-1}(\mu)$, measured with respect to the normal $\mathbf{n}$, and a corresponding azimuthal angle φ. Then Eq. (3.10) can be written

$$\int_0^{2\pi} d\varphi \int_{-1}^0 d\mu\, w(\mu,\varphi)\left[\frac{1}{4\pi}I_0(\mathbf{r}_s,\ \nu,\ t)+\frac{3}{4\pi}\Omega\cdot\mathbf{I}_1(\mathbf{r}_s,\ \nu,\ t)-\Gamma(\mathbf{r}_s,\ \nu,\ \mu,\ \varphi,\ t)\right]=0. \qquad (3.13)$$

We now set

$$w(\mu,\ \varphi) = \delta(\mu-\mu_0), \qquad (3.14)$$

where $\delta(z)$ is the Dirac delta function and μ_0 is in the range $-1 < \mu_0 < 0$. The Mark boundary condition follows from the choice $\mu_0 = -1/\sqrt{3}$, and Eq. (3.13) becomes

$$\frac{1}{2}I_0(\mathbf{r}_s,\ \nu,\ t)-\frac{\sqrt{3}}{2}\mathbf{n}\cdot\mathbf{I}_1(\mathbf{r}_s,\ \nu,\ t) = \int_0^{2\pi} d\varphi\,\Gamma(\mathbf{r}_s,\ \nu,\ -1/\sqrt{3},\ \varphi,\ t). \qquad (3.15)$$

Equation (3.15) has the interpretation that the exact boundary condition is satisfied at a single polar angle point, $\mu = -1/\sqrt{3}$. This particular angle is chosen because of considerations such as the following. Consider time independent transport in a homogeneous, purely absorbing, planar system. According to the equation of transfer, Eq. (2.42), photons incident on the surface will be absorbed such that, at a depth z from the surface, $\exp(-\sigma_a z/\mu)$ represents the probability of survival for photons of polar angle $\theta \equiv \cos^{-1}(\mu)$. On the other hand, Eqs. (3.4) and (3.5) give $\exp-(\sqrt{3}\sigma_a z)$ as the survival probability for all photons. Hence $|\mu| = 1/\sqrt{3}$ can be considered as the average angle associated with the Eddington approximation. Experience indicates that the Marshak condition, Eq. (3.12), is generally more accurate than the Mark condition, Eq. (3.15). In problems with one dimensional symmetry (slabs, spheres, cylinders) one can develop a boundary condition using a variational technique (Federighi, 1964; Pomraning, 1964) which is more accurate than the Marshak condition, but this technique has not been extended to a general, three dimensional, situation.

Equations (3.4) and (3.5) are generally referred to as the $P-1$ approximation. (The origin of this term will become apparent in Section 5 of this chapter.) For these equations to reduce to a diffusion-like description of radiative transfer, we must demand, to be consistent with normal usage of the term diffusion, that Eq. (3.5) reduce to a Fick's law of diffusion, i.e., $\mathbf{I}_1 = -D\nabla I_0$, where $D \equiv D(\mathbf{r}, \nu, t)$ is the local diffusion coefficient at frequency ν. This requires that we neglect the $(1/c)\,\partial\mathbf{I}_1/\partial t$ term in Eq. (3.5) which, if carried, would give rise to a wave rather than a pure diffusion character. We also need assume, in Eq. (3.5) only, that the scattering kernel is diagonal

$$\sigma_{sn}(\nu \to \nu') = A_n(\nu)\,\delta(\nu-\nu'), \qquad n = 0 \quad \text{and} \quad 1, \tag{3.16}$$

where $\delta(z)$ is the Dirac delta function. Then the induced scattering terms cancel identically and all other terms can be combined to give

$$\mathbf{I}_1(\mathbf{r}, \nu, t) = -\frac{1}{3\sigma_{\mathrm{tr}}}\,\nabla I_0(\mathbf{r}, \nu, t), \tag{3.17}$$

where the so-called transport coefficient, $\sigma_{\mathrm{tr}} \equiv \sigma_{\mathrm{tr}}(\mathbf{r}, \nu, t)$, is defined as

$$\sigma_{\mathrm{tr}} = \sigma_a' + \sigma_s - A_1. \tag{3.18}$$

To complete the development, we need specify A_1. This is accomplished by demanding that Eq. (3.16) for $n = 1$ be exact when integrated over all ν', i.e.,

$$A_1(\nu) = \int_0^\infty d\nu'\sigma_{s1}(\nu \to \nu') = 2\pi \int_0^\infty d\nu' \int_{-1}^1 d\mu_0\mu_0\sigma_s(\nu \to \nu', \mu_0). \tag{3.19}$$

With this determination of $A_1(\nu)$, our definition of $\sigma_{\mathrm{tr}}(\mathbf{r}, \nu, t)$ is identical to that introduced by Frank-Kamenetskii (1962) in his discussion of the linear equation of transfer. Hence, as far as Fick's law of diffusion is concerned, the nonlinear terms due to induced scattering have no effect.

However, these induced terms are still present in Eq. (3.4), although they can be simplified as follows. Since the truncated spherical harmonic representation, Eq. (3.1), is only strictly valid if $|\mathbf{I}_1| \ll I_0$, it is a consistent approximation to neglect the $\mathbf{I}_1(\nu)\cdot\mathbf{I}_1(\nu')$ term in Eq. (3.4). Making this simplification and using Eq. (3.17) in Eq. (3.4), we find the diffusion equation

$$\frac{1}{c}\frac{\partial I_0(\nu)}{\partial t} - \nabla\cdot D(\nu)\nabla I_0(\nu) = \sigma_a'(\nu)[4\pi B(\nu)-I_0(\nu)] - \sigma_s(\nu)\,I_0(\nu)$$

$$+ \int_0^\infty d\nu'\,\frac{\nu}{\nu'}\,\sigma_{s0}(\nu' \to \nu)\,I_0(\nu')$$

$$+ \frac{c^2}{8\pi h}\,I_0(\nu) \int_0^\infty d\nu'\left[\frac{1}{\nu^2\nu'}\,\sigma_{s0}(\nu' \to \nu) - \frac{1}{\nu'^3}\,\sigma_{s0}(\nu \to \nu')\right] I_0(\nu'), \tag{3.20}$$

where we have defined the diffusion coefficient $D \equiv D(\mathbf{r}, \nu, t)$ as

$$D = \frac{1}{3\sigma_{\mathrm{tr}}}. \tag{3.21}$$

The initial condition on Eq. (3.20) is just Eq. (3.8), and the Marshak and Mark boundary conditions become

$$\tfrac{1}{4}I_0(\mathbf{r}_s, \nu, t) + \tfrac{1}{2}D\mathbf{n}\cdot\nabla I_0(\mathbf{r}_s, \nu, t) = \int\limits_{\mathbf{n}\cdot\mathbf{\Omega}<0} d\Omega\,|\,\mathbf{n}\cdot\mathbf{\Omega}\,|\,\Gamma(\mathbf{r}_s, \nu, \mathbf{\Omega}, t), \tag{3.22}$$

$$\frac{1}{2}I_0(\mathbf{r}_s, \nu, t) + \frac{\sqrt{3}}{2}D\mathbf{n}\cdot\nabla I_0(\mathbf{r}_s, \nu, t) = \int\limits_{0}^{2\pi} d\varphi\,\Gamma(\mathbf{r}_s, \nu, -1/\sqrt{3}, \varphi, t). \tag{3.23}$$

Equations (3.8), (3.20), and (3.22) or (3.23) represent the general form of the classical diffusion or Eddington approximation for an arbitrary scattering process. In radiation hydrodynamic problems, the quantities of interest are the radiative energy, flux, and pressure tensor as defined by Eqs. (1.9), (1.16), and (1.19). In the Eddington approximation these are given by

$$u(\mathbf{r}, t) = \frac{1}{c}\int\limits_{0}^{\infty} d\nu I_0(\mathbf{r}, \nu, t), \tag{3.24}$$

$$\mathbf{F}(\mathbf{r}, t) = - \int\limits_{0}^{\infty} d\nu D(\mathbf{r}, \nu, t)\,\nabla I_0(\mathbf{r}, \nu, t), \tag{3.25}$$

$$p(\mathbf{r}, t) = \frac{1}{3c}\,\mathsf{I}\int\limits_{0}^{\infty} d\nu I_0(\mathbf{r}, \nu, t) = \frac{1}{3}\,\mathsf{I}u(\mathbf{r}, t), \tag{3.26}$$

where I is the unit dyad (a 3×3 tensor with ones on the diagonal and zeros elsewhere). Although the classical diffusion or Eddington approximation is much simpler than the transport description from which it was derived, it should describe the energy flow due to radiative processes in a semi-quantitative sense. This description should be particularly accurate if the specific intensity of radiation is almost isotropic. Of course, the angular detail of the specific intensity has been lost since the essence of the Eddington approximation is the simple angular dependence assumed in Eq. (3.1).

3. Asymptotic Diffusion Theory

The critical assumption in reducing the equation of transfer, Eq. (2.170), to the diffusion description, Eq. (3.20), is that the specific intensity of radiation is almost isotropic, as expressed quantitatively by Eq. (3.1). This leads to a diffusion coefficient defined by Eq. (3.21). One would find other diffusion coefficients if other angular distributions were assumed. We consider the effect of one such distribution here, namely the asymptotic angular distribution of the equation of transfer. This time independent distribution is that found deep within (a few mean free paths from all boundaries) a source free $(B = 0)$ homogeneous medium in

The Equations of Radiation Hydrodynamics

which photons of different frequencies diffuse independently (the scattering kernel contains a Dirac delta function with argument $v-v'$). For simplicity we also consider the scattering kernel to be isotropic. Hence we have

$$\sigma_s(v' \rightarrow v, \boldsymbol{\Omega}' \cdot \boldsymbol{\Omega}) = \frac{1}{4\pi}\,\sigma_s(v')\,\delta(v-v').$$

(3.27)

The necessary analysis for treating the case of anisotropic scattering has been given by Mika (1961) and McInerney (1964), as we shall point out more explicitly toward the end of this section.

Let us first consider the problem in planar geometry. For the conditions just discussed, the transport equation (2.170) becomes

$$\mu\,\frac{\partial I(z,\,\mu)}{\partial z} + \sigma I(z,\,\mu) = \frac{\sigma_s}{2}\int_{-1}^{1} d\mu'\,I(z,\,\mu'),$$

(3.28)

where z is the spatial coordinate and μ is the cosine of the angle between the z axis and the direction of travel of the photon. Equation (3.28) is understood to hold at each frequency v, although for simplicity in notation we have dropped the argument v in σ_s, σ, and $I(z,\,\mu)$. The coefficient $\sigma = \sigma_a' + \sigma_s$ is just the total interaction coefficient. Disregarding the boundary conditions on Eq. (3.28) (we seek an asymptotic solution only valid far from the boundaries), we look for a solution of the separable form

$$I(z,\,\mu) = \psi(\mu)\,e^{\alpha\sigma z},$$

(3.29)

where α and $\psi(\mu)$ are to be determined. Use of Eq. (3.29) in Eq. (3.28) yields, upon cancellation of the common exponential factor,

$$\psi(\mu) = \frac{\sigma_s}{2\sigma(1+\alpha\mu)}\int_{-1}^{1} d\mu'\,\psi(\mu').$$

(3.30)

Integration of Eq. (3.30) over all values of μ gives, upon cancellation of the integral term,

$$\frac{2\alpha}{\tilde{\omega}} = \ln\left(\frac{1+\alpha}{1-\alpha}\right),$$

(3.31)

as the characteristic equation for α. Here we have defined

$$\tilde{\omega} = \sigma_s/\sigma = \sigma_s/(\sigma_a'+\sigma_s).$$

(3.32)

It is easily shown that for a given value of $\tilde{\omega}$ in the range $0 \le \tilde{\omega} \le 1$ there are two real allowable values of α, one being the negative of the other. For $\tilde{\omega} = 1$ (pure scattering), $\alpha^2 = 0$, and for $\tilde{\omega} = 0$ (pure absorption), $\alpha^2 = 1$. Table 3.1 gives values of α^2 for intermediate values of $\tilde{\omega}$. Since there are two permissible values of α, we find from Eqs. (3.29) and (3.30) that the general asymptotic solution is

$$I(z,\,\mu) = \frac{Ae^{\alpha\sigma z}}{1+\alpha\mu} + \frac{Be^{-\alpha\sigma z}}{1-\alpha\mu},$$

(3.33)

where A and B are arbitrary constants.

TABLE 3.1. *Asymptotic Diffusion Theory*

$\bar{\omega}$	α	α^2	σD
0.0	1.000000	1.000000	1.000000
0.1	1.000000	1.000000	0.900000
0.2	0.999909	0.999818	0.800145
0.3	0.997414	0.994834	0.703635
0.4	0.985624	0.971454	0.617631
0.5	0.957504	0.916814	0.545367
0.6	0.907332	0.823252	0.485878
0.7	0.828635	0.686636	0.436913
0.8	0.710412	0.504685	0.396287
0.9	0.525430	0.276076	0.362219
0.92	0.474002	0.224678	0.356065
0.94	0.413976	0.171376	0.350108
0.96	0.340829	0.116165	0.344339
0.98	0.242983	0.059041	0.338750
0.99	0.172511	0.029760	0.336021
1.00	0.000000	0.000000	0.333333

Let us now compute in plane geometry the quantities I_0 and I_1, defined in general geometry by Eqs. (3.2) and (3.3). From Eq. (3.33) we find

$$I_0(z) \equiv 2\pi \int_{-1}^{1} d\mu I(z, \mu) = \frac{2\pi}{\alpha} \ln\left(\frac{1+\alpha}{1-\alpha}\right)(Ae^{\alpha\sigma z} + Be^{-\alpha\sigma z}), \qquad (3.34)$$

$$I_1(z) \equiv 2\pi \int_{-1}^{1} d\mu\mu I(z, \mu) = \frac{2\pi}{\alpha}\left[2 - \frac{1}{\alpha}\ln\left(\frac{1+\alpha}{1-\alpha}\right)\right](Ae^{\alpha\sigma z} - Be^{-\alpha\sigma z}). \qquad (3.35)$$

These results can be rewritten in a simpler form using Eq. (3.31) as

$$I_0(z) = \frac{4\pi}{\bar{\omega}}(Ae^{\alpha\sigma z} + Be^{-\alpha\sigma z}), \qquad (3.36)$$

$$I_1(z) = -\frac{4\pi}{\alpha\bar{\omega}}(1-\bar{\omega})(Ae^{\alpha\sigma z} - Be^{-\alpha\sigma z}). \qquad (3.37)$$

By direct computation, one can easily verify, for arbitrary values of A and B, that

$$I_1(z) = -D\frac{dI_0(z)}{dz}, \qquad (3.38)$$

where D, the diffusion coefficient, is given by

$$D = \frac{(1-\bar{\omega})}{\sigma\alpha^2}. \qquad (3.39)$$

For $\bar{\omega} = 1$ (pure scattering) $\sigma D = 1/3$, whereas for $\bar{\omega} = 0$ (pure absorption) $\sigma D = 1$. Table 3.1 gives values of σD for intermediate values of $\bar{\omega}$. Equation (3.38) is the Fick's law of diffusion in planar geometry corresponding to an asymptotic angular distribution for the specific intensity.

We now show that such a Fick's law of diffusion holds for a general geometry. The pertinent equation of transfer, analogous to Eq. (3.28), is

$$\mathbf{\Omega} \cdot \nabla I(\mathbf{r}, \mathbf{\Omega}) + \sigma I(\mathbf{r}, \mathbf{\Omega}) = \frac{\sigma_s}{4\pi} \int_{4\pi} d\mathbf{\Omega}' I(\mathbf{r}, \mathbf{\Omega}'). \tag{3.40}$$

Defining

$$I_0(\mathbf{r}) \equiv \int_{4\pi} d\mathbf{\Omega} I(\mathbf{r}, \mathbf{\Omega}), \tag{3.41}$$

we have

$$I_0(\mathbf{r}) = \tilde{\omega}\sigma \int_V d\mathbf{r}' \frac{e^{-\sigma|\mathbf{r}-\mathbf{r}'|}}{4\pi |\mathbf{r}-\mathbf{r}'|^2} I_0(\mathbf{r}'), \tag{3.42}$$

as the Peierls' integral equation corresponding to Eq. (3.40) (see Section 7 of Chapter II). Since we seek only an asymptotic solution, the volume V in Eq. (3.42) can be taken as infinite. For this reason, we are also allowed to disregard boundary contributions to Eq. (3.42), as we have. We look for solutions to Eq. (3.42) of the form

$$I_0(\mathbf{r}) = e^{\boldsymbol{\alpha} \cdot \mathbf{r}}, \tag{3.43}$$

where $\boldsymbol{\alpha} = \alpha\mathbf{u}$, with $\mathbf{u}$ a unit vector. Inserting Eq. (3.43) into Eq. (3.42) and carrying out the integration over all space, we obtain

$$e^{\boldsymbol{\alpha} \cdot \mathbf{r}} = \frac{\tilde{\omega}}{2\alpha} \ln\left(\frac{1+\alpha}{1-\alpha}\right) e^{\boldsymbol{\alpha} \cdot \mathbf{r}}. \tag{3.44}$$

Hence Eq. (3.43) is a solution if α again satisfies the characteristic relationship, Eq. (3.31). Thus

$$I_0(\mathbf{r}) = e^{\sigma\alpha\mathbf{u} \cdot \mathbf{r}} \tag{3.45}$$

is a solution for arbitrary $\mathbf{u}$. More generally, we can superimpose solutions of this type and write

$$I_0(\mathbf{r}) = \int d\mathbf{u} f(\mathbf{u}) e^{\sigma\alpha\mathbf{u} \cdot \mathbf{r}}, \tag{3.46}$$

where $f(\mathbf{u})$ is an arbitrary function.

Equation (3.46) is the general asymptotic solution for $I_0(\mathbf{r})$. To obtain Fick's law of diffusion we first use Eq. (3.46) to compute the specific intensity of radiation. We use the integral equation (2.83) which in the restricted transport problem we are discussing reads

$$I(\mathbf{r}, \mathbf{\Omega}) = \frac{\tilde{\omega}\sigma}{4\pi} \int_0^\infty ds\, e^{-\sigma s} I_0(\mathbf{r}-s\mathbf{\Omega}). \tag{3.47}$$

Use of Eq. (3.46) for $I_0(\mathbf{r})$ in Eq. (3.47) yields

$$I(\mathbf{r}, \mathbf{\Omega}) = \frac{\tilde{\omega}\sigma}{4\pi} \int d\mathbf{u} f(\mathbf{u}) e^{\sigma\alpha\mathbf{u} \cdot \mathbf{r}} \int_0^\infty ds\, e^{-\sigma(1+\alpha\mathbf{u} \cdot \mathbf{\Omega})s}, \tag{3.48}$$

or, carrying out the integral over s,

$$I(\mathbf{r}, \boldsymbol{\Omega}) = \frac{\tilde{\omega}}{4\pi} \int du f(\mathbf{u}) \frac{e^{\sigma\alpha\mathbf{u}\cdot\mathbf{r}}}{1+\alpha\mathbf{u}\cdot\boldsymbol{\Omega}} . \tag{3.49}$$

We now define the ith component of the radiative flux per unit frequency as

$$I_1^{(i)}(\mathbf{r}) \equiv \int_{4\pi} d\Omega\,\Omega^i I(\mathbf{r}, \boldsymbol{\Omega}), \tag{3.50}$$

where Ω^i is the projection of the vector $\boldsymbol{\Omega}$ along the ith direction. Use of Eq. (3.49) in Eq. (3.50) gives

$$I_1^{(i)}(\mathbf{r}) = \frac{\tilde{\omega}}{4\pi} \int du f(\mathbf{u})\, e^{\sigma\alpha\mathbf{u}\cdot\mathbf{r}} \int_{4\pi} d\Omega \frac{\Omega^i}{1+\alpha\mathbf{u}\cdot\boldsymbol{\Omega}} . \tag{3.51}$$

To evaluate the integral over $\boldsymbol{\Omega}$, we choose a coordinate system such that the vector $\mathbf{u}$ coincides with the z axis and a unit vector $\mathbf{i}$ along the ith direction is in the $y-z$ plane (Fig. 3.1).

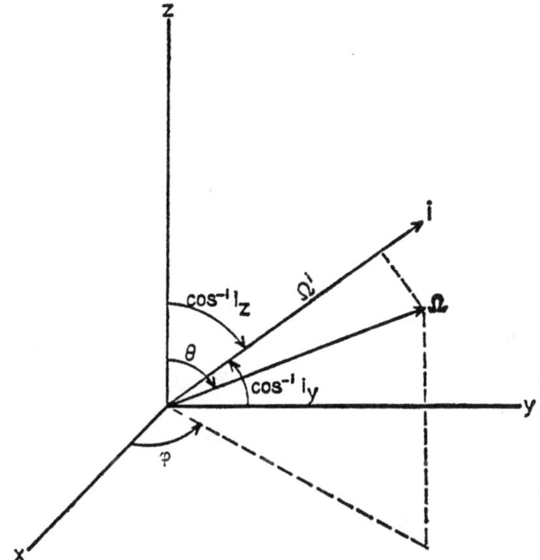

FIG. 3.1. The flux in asymptotic diffusion theory.

We then define $\boldsymbol{\Omega}$ by $\mu \equiv \cos\theta$, where θ is the polar angle, and φ, the corresponding azimuthal angle. We define the integral under consideration as J, and hence

$$J \equiv \int_{4\pi} d\Omega \frac{\Omega^i}{1+\alpha\mathbf{u}\cdot\boldsymbol{\Omega}} = \int_0^{2\pi} d\varphi \int_{-1}^1 d\mu \frac{\Omega^i}{1+\alpha\mathbf{u}\cdot\boldsymbol{\Omega}} . \tag{3.52}$$

Now, since both $\mathbf{u}$ and $\boldsymbol{\Omega}$ are unit vectors, we have

$$\mathbf{u}\cdot\boldsymbol{\Omega} = \cos\theta = \mu. \tag{3.53}$$

Further,

$$\Omega^i = \mathbf{i}\cdot\mathbf{\Omega} = i_y \sin\theta \sin\varphi + i_z \cos\theta, \tag{3.54}$$

where i_y and i_z are the direction cosines defining the unit vector $\mathbf{i}$. (The cosine $i_x = 0$ since $\mathbf{i}$ lies in the $y-z$ plane.) Use of Eqs. (3.53) and (3.54) in Eq. (3.52) yields

$$J = \int_0^{2\pi} d\varphi \int_{-1}^{1} d\mu \, \frac{i_y \sin\theta \sin\varphi + i_z \cos\theta}{1+\alpha\mu} = 2\pi i_z \int_{-1}^{1} d\mu \, \frac{\mu}{1+\alpha\mu}. \tag{3.55}$$

Carrying out the integral over μ, making use of Eq. (3.31) to simplify the result, we find

$$J = -\frac{4\pi i_z}{\alpha\tilde\omega} (1-\tilde\omega). \tag{3.56}$$

Thus Eq. (3.51) becomes

$$I_1^{(i)}(\mathbf{r}) = -\frac{(1-\tilde\omega)}{\alpha} \int d\mathbf{u} f(\mathbf{u}) \, e^{\sigma\alpha\mathbf{u}\cdot\mathbf{r}} \, i_z . \tag{3.57}$$

Now, the projection of the vector $\mathbf{u}$ on the direction i is given by

$$u^i = |\mathbf{u}| \, i_z, \tag{3.58}$$

or, since $\mathbf{u}$ is a unit vector,

$$i_z = u^i. \tag{3.59}$$

Hence Eq. (3.57) can be written

$$I_1^{(i)}(\mathbf{r}) = \frac{-(1-\tilde\omega)}{\alpha} \int d\mathbf{u} f(\mathbf{u}) \, e^{\sigma\alpha\mathbf{u}\cdot\mathbf{r}} \, u^i = -\frac{(1-\tilde\omega)}{\sigma\alpha^2} \frac{\partial}{\partial x_i} \int d\mathbf{u} f(\mathbf{u}) \, e^{\sigma\alpha\mathbf{u}\cdot\mathbf{r}}. \tag{3.60}$$

Equation (3.60) is just the ith component of the vector relationship

$$\mathbf{I}_1(\mathbf{r}) = -D\nabla I_0(\mathbf{r}), \tag{3.61}$$

where the diffusion coefficient D is given by Eq. (3.39). This result is just Fick's law of diffusion in general geometry. Clearly in the plane case, Eq. (3.61) reduces to our previous result, Eq. (3.38).

We now use these results to formulate a diffusion-like approximation to the general equation of transfer, Eq. (2.170). Integration of Eq. (2.170) over all solid angle yields

$$\frac{1}{c} \frac{\partial I_0(\nu)}{\partial t} + \nabla\cdot\mathbf{I}_1(\nu) = \sigma_a'(\nu)[4\pi B(\nu) - I_0(\nu)] - \sigma_s(\nu) I_0(\nu)$$

$$+ \int_0^{\infty} d\nu' \, \frac{\nu}{\nu'} \, \sigma_{s0}(\nu' \to \nu) \, I_0(\nu')$$

$$+ \frac{c^2}{8\pi h} I_0(\nu) \int_0^{\infty} d\nu' \left[\frac{1}{\nu^2\nu'} \sigma_{s0}(\nu' \to \nu) - \frac{1}{\nu'^3} \sigma_{s0}(\nu \to \nu') \right] I_0(\nu'), \tag{3.62}$$

where we have suppressed the spatial and temporal arguments of all quantities. In writing Eq. (3.62) we have assumed, for the purpose of treating the induced scattering terms only, that the specific intensity can be taken as isotropic. The error associated with this assumption should be small compared to the overall error involved in describing radiative transfer by a diffusion equation. Aside from this, Eq. (3.62) is exact. For it to be a useful result, however, one needs an additional relationship giving $I_1(v)$ in terms of $I_0(v)$. The assumption in asymptotic diffusion theory is that Eq. (3.61), derived under quite restrictive circumstances, is generally valid. That is, we assume, with full display of the arguments, that

$$\mathbf{I}_1(\mathbf{r},\, v,\, t) = -D(\mathbf{r},\, v,\, t)\, \nabla I_0(\mathbf{r},\, v,\, t). \tag{3.63}$$

Use of Eq. (3.63) in Eq. (3.62) yields

$$\frac{1}{c}\frac{\partial I_0(v)}{\partial t} - \nabla \cdot D(v)\, \nabla I_0(v) = \sigma_a'(v)\,[4\pi B(v) - I_0(v)] - \sigma_s(v)\, I_0(v)$$

$$+ \int_0^\infty dv'\, \frac{v}{v'}\, \sigma_{s0}(v' \to v)\, I_0(v')$$

$$+ \frac{c^2}{8\pi h}\, I_0(v) \int_0^\infty dv'\, \left[\frac{1}{v^2 v'}\, \sigma_{s0}(v' \to v) - \frac{1}{v'^3}\, \sigma_{s0}(v \to v') \right] I_0(v'), \tag{3.64}$$

where the diffusion coefficient $D \equiv D(\mathbf{r}, v, t)$ is given by Eq. (3.39).

Although Eq. (3.64) was derived under the assumption of isotropic scattering, the same form of the equation holds for a general anisotropic scattering law. Equation (3.39) still defines the diffusion coefficient, but the value of α is no longer given by Eq. (3.31). The analysis which gives α as a functional of the scattering law in the general case has been developed by Mika (1961) and McInerney (1964).

To complete the specification of asymptotic diffusion theory, we need discuss the initial and boundary conditions. As in the Eddington, or classical diffusion, description discussed in the last section, Eq. (3.8) remains the appropriate initial condition. An analogue of the Marshak boundary condition can be derived as follows. We first consider the left hand boundary of a planar system, say z_l, and recognize that Eq. (3.33) can be rewritten as

$$I(z_l,\, \mu) = \frac{E}{1 - \alpha^2 \mu^2} + \frac{F\mu}{1 - \alpha^2 \mu^2}, \tag{3.65}$$

where E and F are constants (which depend upon z_l). From the definitions of I_0 and I_1, Eqs. (3.2) and (3.3), we can write, making use of Eq. (3.31) to simplify the results,

$$I_0(z_l) = 2\pi \int_{-1}^{1} d\mu \left[\frac{E + F\mu}{1 - \alpha^2 \mu^2} \right] = \frac{4\pi E}{\bar{\omega}}, \tag{3.66}$$

$$I_1(z_l) = 2\pi \int_{-1}^{1} d\mu\, \mu \left[\frac{E + F\mu}{1 - \alpha^2 \mu^2} \right] = \frac{4\pi (1 - \bar{\omega}) F}{\alpha^2 \bar{\omega}}. \tag{3.67}$$

The Equations of Radiation Hydrodynamics

Hence Eq. (3.65) can be rewritten

$$I(z_l, \mu) = \frac{\tilde{\omega}}{4\pi} I_0(z_l) \frac{1}{1-\alpha^2\mu^2} + \frac{\alpha^2\tilde{\omega}}{4\pi(1-\tilde{\omega})} I_1(z_l) \frac{\mu}{1-\alpha^2\mu^2} . \tag{3.68}$$

Referring to the transport boundary condition, Eq. (2.29), we demand that Eq. (3.68) preserve the incoming radiative flux, which is the philosophy underlying the Marshak boundary condition. That is, we write

$$2\pi \int_0^1 d\mu\mu \left[\frac{\tilde{\omega}}{4\pi} I_0(z_l) \frac{1}{1-\alpha^2\mu^2} + \frac{\alpha^2\tilde{\omega}}{4\pi(1-\tilde{\omega})} I_1(z_l) \frac{\mu}{1-\alpha^2\mu^2} - \Gamma(z_l, \mu) \right] = 0, \tag{3.69}$$

or, carrying out the integrals and explicitly indicating all of the variables,

$$\frac{\tilde{\omega}}{4\alpha^2} \ln\left(\frac{1}{1-\alpha^2}\right) I_0(z_l, v, t) + \frac{1}{2} I_1(z_l, v, t) = 2\pi \int_0^1 d\mu\mu\Gamma(z_l, v, \mu, t). \tag{3.70}$$

Use of Fick's law of diffusion, Eq. (3.38), in Eq. (3.70) yields as the boundary condition

$$\frac{\tilde{\omega}}{4\alpha^2} \ln\left(\frac{1}{1-\alpha^2}\right) I_0(z_l, v, t) - \frac{D}{2} \frac{d}{dz} I_0(z_l, v, t) = 2\pi \int_0^1 d\mu\mu\Gamma(z_l, v, \mu, t). \tag{3.71}$$

A similar result can be derived for the right hand boundary of a planar system. Finally, in analogy to the Marshak condition for the Eddington approximation, Eq. (3.22), we generalize Eq. (3.71) to

$$\frac{\tilde{\omega}}{4\alpha^2} \ln\left(\frac{1}{1-\alpha^2}\right) I_0(\mathbf{r}_s, v, t) + \frac{1}{2} D\mathbf{n}\cdot\nabla I_0(\mathbf{r}_s, v, t) = \int_{\mathbf{n}\cdot\mathbf{\Omega}<0} d\Omega\, |\mathbf{n}\cdot\mathbf{\Omega}|\, \Gamma(\mathbf{r}_s, v, \mathbf{\Omega}, t) \tag{3.72}$$

as the boundary condition in general geometry. Here $\mathbf{r}_s$ denotes a surface point and $\mathbf{n}$ is a unit outward normal vector at the point $\mathbf{r}_s$.

As $\tilde{\omega}$ approaches unity (pure scattering) all aspects of asymptotic diffusion theory agree with those of the Eddington approximation. In particular, σD goes to a value of one-third, and the asymptotic boundary condition, Eq. (3.72), goes to the Eddington condition, Eq. (3.22). For this reason, it is often said that the Eddington approximation is only valid for almost pure scattering problems. This statement is only true as far as asymptotic solutions in a source free medium are concerned. The accurate statement concerning the Eddington approximation is that it is only strictly valid when the specific intensity of radiation is almost isotropic without regard to the amount of absorption present. Asymptotic diffusion theory is a strictly proper description of radiative transfer when the specific intensity is in a nearly asymptotic state, as is clear from its derivation.

Finally, we give the energy density, radiative flux, and pressure tensor corresponding to asymptotic diffusion theory. The first two quantities are obviously given by

$$u(\mathbf{r},\, t) = \frac{1}{c} \int_0^\infty dv I_0(\mathbf{r},\, v,\, t),$$

(3.73)

$$\mathbf{F}(\mathbf{r},\, t) = -\int_0^\infty dv D(\mathbf{r},\, v,\, t) \nabla I_0(\mathbf{r},\, v,\, t),$$

(3.74)

with D in Eq. (3.74) representing the asymptotic diffusion coefficient given by Eq. (3.39). The pressure tensor is more difficult to obtain, involving analysis similar to but more complex than that which led to Eq. (3.61) for the radiative flux per unit frequency. We content ourselves with quoting the result given by Case *et al.* (1953b). According to Eq. (1.18), the *ij* component of the pressure tensor is given by

$$p_{ij}(\mathbf{r},\, t) = \frac{1}{c} \int_0^\infty dv \int_{4\pi} d\Omega \Omega_i \Omega_j I(\mathbf{r},\, v,\, \mathbf{\Omega},\, t).$$

(3.75)

The reference just cited gives, for an asymptotic distribution,

$$\int_{4\pi} d\Omega \Omega_i \Omega_j I = \frac{1}{\sigma^2 \alpha^2} \frac{(3\sigma D - 1)}{2} \frac{\partial^2 I_0}{\partial x_i \, \partial x_j} + \frac{(1 - \sigma D)}{2} I_0 \delta_{ij},$$

(3.76)

where δ_{ij} is the Kronecker delta function (unity if $i = j$, zero otherwise).

One point of interest is that it is easily shown by direct computation from Eq. (3.76) that, if σ, α, and D are slowly varying in $\mathbf{r}$,

$$\nabla \cdot \int d\Omega \mathbf{\Omega}\mathbf{\Omega} I = \sigma D \nabla I_0.$$

(3.77)

To derive Eq. (3.77) one needs the fact that I_0 satisfies the Helmholtz equation

$$\nabla^2 I_0(\mathbf{r}) = \alpha^2 \sigma^2 I_0(\mathbf{r}),$$

(3.78)

as can be verified by applying the ∇^2 (Laplacian) operator to Eq. (3.46). We shall make use of Eq. (3.77) in the next section.

As in the case of the Eddington approximation, asymptotic diffusion theory gives a semi-quantitative description of radiative transfer at the expense of angular detail. For any given problem it is generally not *a priori* clear which of these two diffusion descriptions is more accurate. Historically, the Eddington approximation has been more popular than the asymptotic theory. Experience indicates that for general usage the two approximations are of comparable accuracy, and one cannot categorically point to one as being superior to the other.

4. A New Diffusion Theory

Both the classical (Eddington) and asymptotic diffusion theories would be expected to be quite accurate in the limiting situations of nearly isotropic and nearly asymptotic distributions, respectively, for the specific intensity. In this section we formulate a diffusion theory which contains both of these correct limits, and effectively interpolates between these limiting situations in the general case.

The basis of this diffusion theory is again the zeroth angular moment of the equation of transfer, as given by Eq. (3.62). We repeat this equation here for convenience, i.e.,

$$\frac{1}{c}\frac{\partial I_0(\nu)}{\partial t} + \nabla \cdot \mathbf{I}_1(\nu) = \sigma_a'(\nu)[4\pi B(\nu) - I_0(\nu)] - \sigma_s(\nu)I_0(\nu)$$

$$+ \int_0^\infty d\nu' \frac{\nu}{\nu'} \sigma_{s0}(\nu' \to \nu) I_0(\nu')$$

$$+ \frac{c^2}{8\pi h} I_0(\nu) \int_0^\infty d\nu' \left[\frac{1}{\nu^2\nu'} \sigma_{s0}(\nu' \to \nu) - \frac{1}{\nu'^3} \sigma_{s0}(\nu \to \nu') \right] I_0(\nu'). \quad (3.79)$$

As in all diffusion theories, we require in addition to Eq. (3.79) a Fick's law of diffusion of a form similar to that encountered in the last two sections. Actually, the Fick's law we shall derive here is of the form

$$\sigma \mathbf{I}_1 = -\nabla(\sigma D I_0). \quad (3.80)$$

Nevertheless, the main distinguishing feature between various diffusion theories is the expression for the diffusion coefficient D. The diffusion coefficient we shall obtain in this section limits to the Eddington and asymptotic results in the appropriate circumstances, but differs from these limiting values in more general situations.

The analysis leading to Eq. (3.80) is again based upon the monochromatic, time independent equation of transfer similar to Eq. (3.40) of the last section. However, in the present analysis we allow a nonzero temperature (hence $B \neq 0$), and further allow the scattering coefficient σ_s, the absorption coefficient σ_a', and the total interaction coefficient σ to be spatially dependent. Under these conditions the equation of transfer, Eq. (2.170), becomes

$$\mathbf{\Omega} \cdot \nabla I(\mathbf{r}, \mathbf{\Omega}) + \sigma(\mathbf{r}) I(\mathbf{r}, \mathbf{\Omega}) = \sigma_a'(\mathbf{r}) B(T) + \frac{\sigma_s(\mathbf{r})}{4\pi} \int_{4\pi} d\Omega' I(\mathbf{r}, \mathbf{\Omega}'), \quad (3.81)$$

where once again we have suppressed the frequency variable. In writing Eq. (3.81) we have assumed for simplicity of presentation that the scattering is isotropic. The extension of the analysis of this section to the case of an anisotropic scattering kernel has been given by Pomraning (1969a).

Initially we consider systems with plane symmetry, and Eq. (3.81) becomes

$$\mu \frac{\partial I(z, \mu)}{\partial z} + \sigma(z) I(z, \mu) = \sigma_a'(z) B(T) + \frac{\sigma_s(z)}{2} \int_{-1}^{1} d\mu' I(z, \mu'). \quad (3.82)$$

We write the solution to this equation of transfer as the sum of an even and an odd function of μ, i.e.,

$$I(z, \mu) = I_0(z)\, e(z, \mu) + I_1(z)\, o(z, \mu), \tag{3.83}$$

where $e(z, \mu)$ is even in μ and normalized according to

$$2\pi \int_{-1}^{1} d\mu e(z, \mu) = 1, \tag{3.84}$$

and $o(z, \mu)$ is odd in μ with the normalization

$$2\pi \int_{-1}^{1} d\mu \mu o(z, \mu) = 1. \tag{3.85}$$

Because of the particular normalizations we have imposed, the quantities I_0 and I_1 of Eq. (3.83) are identical to those introduced earlier and defined by Eqs. (3.2) and (3.3) in general geometry. We now form the first two angular moments of Eq. (3.82). Integration over all solid angle gives

$$\frac{dI_1(z)}{dz} = \sigma_a'(z)[4\pi B(T) - I_0(z)], \tag{3.86}$$

which is just the continuity equation expressing conservation of photons. Of more interest is the result of multiplication of Eq. (3.82) by μ prior to integration over all solid angle. This gives

$$\frac{d}{dz}[\sigma(z)\, D(z)\, I_0(z)] + \sigma(z)\, I_1(z) = 0, \tag{3.87}$$

where we have defined $\sigma(z)D(z)$ as the weighted average of μ^2 with respect to $e(z, \mu)$, i.e.,

$$\sigma(z)\, D(z) = 2\pi \int_{-1}^{1} d\mu \mu^2 e(z, \mu). \tag{3.88}$$

It should be noted that no approximations have been introduced at this point; Eqs. (3.86) and (3.87) follow exactly from Eq. (3.82).

Equation (3.87) is the Fick's law of diffusion we seek; it remains to derive an expression for the diffusion coefficient $D(z)$. The Eddington approximation corresponds to an assumption of isotropy for $e(z, \mu)$, i.e.,

$$e(z, \mu) = \frac{1}{4\pi}. \tag{3.89}$$

Equation (3.88) then gives

$$D(z) = \frac{1}{3\sigma(z)}, \tag{3.90}$$

which is the Eddington result, Eq. (3.21). In asymptotic diffusion theory one chooses $e(z, \mu)$ as the asymptotic distribution [see Eq. (3.68)]

$$e(z, \mu) = \frac{\tilde{\omega}}{4\pi} \frac{1}{1 - \alpha^2 \mu^2}, \tag{3.91}$$

The Equations of Radiation Hydrodynamics

which leads to

$$D(z) = \frac{1-\tilde{\omega}}{\sigma\alpha^2} , \tag{3.92}$$

with α given by Eq. (3.31). Equation (3.92) coincides with our result of the last section, Eq. (3.39). In our present analysis we derive an expression for $e(z, \mu)$ under the sole assumption that this function is slowly varying in space. Substitution of Eq. (3.83) into Eq. (3.82) yields, with the neglect of the spatial derivatives of the angular functions,

$$\frac{\mu e(z, \mu)}{\sigma(z) \, D(z)} \frac{d}{dz} [\sigma(z) \, D(z) \, I_0(z)] + \mu o(z, \mu) \frac{dI_1(z)}{dz}$$

$$+ \sigma(z) \, [I_0(z) \, e(z, \mu) + I_1(z) \, o(z, \mu)]$$

$$= \sigma_a'(z) \, B(T) + \frac{\sigma_s(z)}{4\pi} \, I_0(z). \tag{3.93}$$

Separating even and odd (in μ) parts of Eq. (3.93) and using Eqs. (3.86) and (3.87) to eliminate the spatial derivatives in the result, we find the two equations

$$o(z, \mu) = \frac{\mu e(z, \mu)}{\sigma(z) \, D(z)} , \tag{3.94}$$

$$\mu[4\pi B(T) - I_0(z)] \, \sigma_a'(z) \, o(z, \mu) + \sigma(z) \, I_0(z) \, e(z, \mu)$$

$$= \frac{1}{4\pi} [4\pi\sigma_a'(z) \, B(T) + \sigma_s(z) \, I_0(z)]. \tag{3.95}$$

Elimination of $o(z, \mu)$ between Eqs. (3.94) and (3.95) gives the required expression for $e(z, \mu)$ as

$$e(z, \mu) = \frac{\sigma(z) \, D(z) \, \gamma(z)}{4\pi\{\sigma(z) \, D(z) - \mu^2[1 - \gamma(z)]\}} , \tag{3.96}$$

where we have defined

$$\gamma(z) = \frac{\sigma_s(z) \, I_0(z) + 4\pi\sigma_a'(z) \, B(T)}{\sigma(z) \, I_0(z)} . \tag{3.97}$$

Multiplication of Eq. (3.96) by μ^2 and integration over all solid angle gives an equation to be solved for $D(z)$ [see Eq. (3.88)]. We find

$$D(z) = \frac{1 - \gamma(z)}{\sigma(z) \, \alpha^2(z)} , \tag{3.98}$$

where $\alpha(z)$ is defined as the solution of the transcendental equation

$$\frac{2\alpha(z)}{\gamma(z)} = \ln \left(\frac{1 + \alpha(z)}{1 - \alpha(z)}\right). \tag{3.99}$$

Equations (3.98) and (3.99) define the diffusion coefficient $D(z)$. That is, given $\gamma(z)$, Eq. (3.99) can be solved for $\alpha(z)$ and Eq. (3.98) then gives $D(z)$. However, $\gamma(z)$ depends upon $I_0(z)$ according to Eq. (3.97), and $I_0(z)$, is, of course, unknown until the diffusion equation has been solved. But to solve the diffusion equation one needs know $D(z)$. Hence the diffusion equation and the expressions defining the diffusion coefficient must be solved simultaneously. Since this set of equations is nonlinear, this would appear to limit the usefulness of this type of diffusion theory. In general, such is not the case, and we shall return to this point at the end of the section.

We note that Eqs. (3.98) and (3.99) are very reminiscent of the asymptotic expressions of the last section [see Eqs. (3.31) and (3.39)]. In fact, Eqs. (3.98) and (3.99) are identical to the asymptotic results with the parameter γ playing the role of $\tilde{\omega} = \sigma_s/\sigma$. Hence for a problem with an external source $(B \neq 0)$, γ can be considered as an effective value of $\tilde{\omega}$ if the problem had no external source (i.e., if the temperature was sufficiently low such that the source of photons due to spontaneous emission could be neglected). That is, in the absence of an external source we have

$$\tilde{\omega} \equiv \frac{\sigma_s}{\sigma} = \frac{\sigma_s I_0}{\sigma I_0} = \frac{\text{source of photons}}{\text{interaction of photons}}, \tag{3.100}$$

since the only source of photons is those which emerge from a scattering interaction. In a problem with spontaneous emission $(B \neq 0)$, the total source is due to this emission plus the scattering interaction. Hence one could logically define an effective $\tilde{\omega}$ in this case as

$$\tilde{\omega}_{\text{eff}} = \frac{\text{source of photons}}{\text{interaction of photons}} = \frac{\sigma_s I_0 + 4\pi \sigma_a' B}{\sigma I_0}, \tag{3.101}$$

and as an approximation, treat the problem as one with no external source according to the asymptotic methods of the last section. The result of this procedure would be precisely Eqs. (3.98) and (3.99), with γ denoting the effective value of $\tilde{\omega}$ to account for the external source. It is quite interesting that this line of reasoning gives the same results as the analysis undertaken here. We shall use this effective $\tilde{\omega}$ approach to generalize our results to arbitrary geometry.

Before doing this, however, let us show that the present formulation reduces to the Eddington and asymptotic results in the appropriate limiting situations. Consider first a spatial region far from boundaries in an optically thick medium. If the temperature T is slowly varying over several mean free paths, then the specific intensity is known to be almost isotropic and the Eddington approximation $\sigma D = 1/3$ will describe the radiative transfer. In this physical situation, $I_1(z) \approx 0$ (there is very little flux); hence Eq. (3.86) gives

$$I_0(z) \approx 4\pi B(T), \tag{3.102}$$

and Eq. (3.97) yields

$$\gamma(z) \approx 1. \tag{3.103}$$

We then find from Eq. (3.99)

$$\alpha^2(z) \approx 3[1 - \gamma(z)], \tag{3.104}$$

and finally Eq. (3.98) gives

$$\sigma(z)\, D(z) = 1/3, \tag{3.105}$$

which is just the Eddington result. Consider now a spatial region far from boundaries in an optically thick, homogeneous, source free $(B = 0)$ medium. This is just an asymptotic situation, and the correct diffusion coefficient is the asymptotic one of the last section. In this situation, Eq. (3.97) gives

$$\gamma = \frac{\sigma_s I_0}{\sigma I_0} = \frac{\sigma_s}{\sigma} = \tilde{\omega}, \tag{3.106}$$

and, as we have pointed out earlier, the present formulation reduces to asymptotic diffusion theory. Thus the diffusion coefficient derived in this section reduces to the Eddington and asymptotic values in the proper limiting circumstances.

As we pointed out earlier, the similarity of our plane geometry results to asymptotic diffusion theory (with γ playing the role of an effective $\tilde{\omega}$) provides a method of generalization to arbitrary geometry. In this case Eq. (3.81) is the equation of transfer we need deal with. If we define

$$I_0(\mathbf{r}) \equiv \int_{4\pi} d\Omega I(\mathbf{r}, \Omega), \tag{3.107}$$

and

$$\gamma(\mathbf{r}) = \frac{\sigma_s(\mathbf{r})I_0(\mathbf{r})+4\pi\sigma_a'(\mathbf{r})B(T)}{\sigma(\mathbf{r})I_0(\mathbf{r})}, \tag{3.108}$$

then Eq. (3.81) can be formally rewritten

$$\Omega \cdot \nabla I(\mathbf{r})+\sigma(\mathbf{r})I(\mathbf{r}, \Omega) = \frac{\sigma(\mathbf{r})\,\gamma(\mathbf{r})}{4\pi} \int_{4\pi} d\Omega' I(\mathbf{r}, \Omega'). \tag{3.109}$$

Forming the first two angular moments of Eq. (3.109), we find

$$\nabla \cdot \mathbf{I}_1(\mathbf{r})+\sigma(\mathbf{r})[1-\gamma(\mathbf{r})]\,I_0(\mathbf{r}) = 0, \tag{3.110}$$

$$\nabla \cdot \int_{4\pi} d\Omega\Omega\Omega I(\mathbf{r}, \Omega)+\sigma(\mathbf{r})\mathbf{I}_1(\mathbf{r}) = 0, \tag{3.111}$$

where $\mathbf{I}_1(\mathbf{r})$ is given by

$$\mathbf{I}_1(\mathbf{r}) \equiv \int_{4\pi} d\Omega\Omega\, I(\mathbf{r}, \Omega). \tag{3.112}$$

Use of the definition of $\gamma(\mathbf{r})$, Eq. (3.108), in Eq. (3.110) gives the usual conservation equation

$$\nabla \cdot \mathbf{I}_1(\mathbf{r}) = \sigma_a'(\mathbf{r})[4\pi B(T)-I_0(\mathbf{r})]. \tag{3.113}$$

To proceed further, we need to introduce an approximation. In our plane geometry formulation, we assumed the angular functions $e(z, \mu)$ and $o(z, \mu)$ were slowly varying functions of space. This is equivalent to the assumptions that $\sigma(z)$ and $\gamma(z)$ are slowly varying functions of space and that one is far, in optical distance, from the boundaries of the system under consideration. These same assumptions here allow Eq. (3.109) to be treated as an infinite medium equation (far from boundaries) with spatially constant cross sections [the functions $\sigma(\mathbf{r})$ and $\gamma(\mathbf{r})$ are slowly varying]. Under these circumstances, the solution to Eq. (3.109) is an asymptotic one, and in particular Eq. (3.77) of the last section holds, i.e.,

$$\nabla \cdot \int_{4\pi} d\Omega\Omega\Omega I(\mathbf{r}, \Omega) = \nabla [\sigma(\mathbf{r})\,D(\mathbf{r})\,I_0(\mathbf{r})], \tag{3.114}$$

with the diffusion coefficient $D(\mathbf{r})$ given by

$$D(\mathbf{r}) = \frac{1-\gamma(\mathbf{r})}{\sigma(\mathbf{r})\alpha^2(\mathbf{r})}, \tag{3.115}$$

where $\alpha(\mathbf{r})$ is a solution of

$$\frac{2\alpha(\mathbf{r})}{\gamma(\mathbf{r})} = \ln\left(\frac{1+\alpha(\mathbf{r})}{1-\alpha(\mathbf{r})}\right). \tag{3.116}$$

Use of Eq. (3.114) in Eq. (3.111) gives Fick's law of diffusion as

$$\mathbf{I}_1(\mathbf{r}) = -\frac{1}{\sigma(\mathbf{r})}\,\nabla\left[\sigma(\mathbf{r})\,D(\mathbf{r})\,I_0(\mathbf{r})\right]. \tag{3.117}$$

This result clearly reduces to our previous one, Eq. (3.87), in the case of plane symmetry.

It should be noted that since Eq. (3.116) is based upon the assumption that σ and γ (and hence D) are sensibly independent of position, one could just as well have written Eq. (3.117) with σD taken outside the gradient operator. We chose to use the form given in Eq. (3.117) since experience indicates that it is generally more accurate (we shall return to this point shortly). Further, Eq. (3.117) as written is consistent with our plane geometry result, Eq. (3.87). [Actually, in Eq. (3.87) one could also take σD outside the derivative operation since the assumption that $e(z, \mu)$ is sensibly independent of z, used to obtain the planar approximation for σD, is tantamount to assuming that σD is independent of z.]

To obtain a general diffusion description of radiative transfer, we assume that Eq. (3.117) is generally valid for all radiative transfer problems, and write, with full display of all the variables,

$$\mathbf{I}_1(\mathbf{r}, \nu, t) = -\frac{1}{\sigma(\mathbf{r}, \nu, t)}\,\nabla\left[\sigma(\mathbf{r}, \nu, t)\,D(\mathbf{r}, \nu, t)\,I_0(\mathbf{r}, \nu, t)\right]. \tag{3.118}$$

Use of Eq. (3.118) in Eq. (3.79) gives the final result for the present diffusion description as

$$\frac{1}{c}\frac{\partial I_0(\nu)}{\partial t} - \nabla\cdot\frac{1}{\sigma(\nu)}\,\nabla\left[\sigma(\nu)\,D(\nu)\,I_0(\nu)\right] = \sigma_a'(\nu)[4\pi B(\nu) - I_0(\nu)]$$

$$+ \int_0^\infty d\nu'\,\frac{\nu}{\nu'}\,\sigma_{s0}(\nu' \to \nu)\,I_0(\nu') - \sigma_s(\nu)\,I_0(\nu)$$

$$+ \frac{c^2}{8\pi h}\,I_0(\nu)\int_0^\infty d\nu'\left[\frac{1}{\nu^2\nu'}\sigma_{s0}(\nu' \to \nu) - \frac{1}{\nu'^3}\sigma_{s0}(\nu \to \nu')\right]I_0(\nu'), \tag{3.119}$$

with $D \equiv D(\mathbf{r}, \nu, t)$ given by Eq. (3.115). The initial and boundary conditions on this equation can be logically taken as the same as those encountered in asymptotic diffusion theory, Eqs. (3.8) and (3.72), with $\bar{\omega}$ replaced by γ and the value of α corresponding to Eq. (3.116). At the risk of being repetitive, we rewrite these conditions here for completeness. We have as the initial and boundary conditions on Eq. (3.119),

$$I_0(\mathbf{r}, \nu, 0) = \int_{4\pi} d\Omega A(\mathbf{r}, \nu, \Omega), \tag{3.120}$$

$$\frac{\gamma}{4\alpha^2}\ln\left(\frac{1}{1-\alpha^2}\right)I_0(\mathbf{r}_s, \nu, t) + \frac{1}{2}D\mathbf{n}\cdot\nabla I_0(\mathbf{r}_s, \nu, t)$$

$$= \int_{\mathbf{n}\cdot\mathbf{\Omega}<0} d\Omega\,|\mathbf{n}\cdot\mathbf{\Omega}|\,\Gamma(\mathbf{r}_s, \nu, \Omega, t), \tag{3.121}$$

where $\mathbf{r}_s$ denotes a surface point and $\mathbf{n}$ is a unit outward normal vector.

The Equations of Radiation Hydrodynamics

To solve Eqs. (3.119) through (3.121), one must know $\gamma(\mathbf{r}, \nu, t)$, $\alpha(\mathbf{r}, \nu, t)$, and $D(\mathbf{r}, \nu, t)$. These quantities in turn depend upon $I_0(\mathbf{r}, \nu, t)$ in a very nonlinear manner. However, in all but the simplest physical situations one would solve, even assuming γ, α, and D known, Eqs. (3.119) through (3.121) via finite difference methods on digital computers. The additional complexity of γ, α, and D depending upon I_0 could very easily be incorporated into such finite difference methods. At a given iterate or time step, one could use the values of γ, α, and D as calculated from I_0 according to the last iterate or time step. Thus the nonlinear character of the present diffusion approximation would seem to be a minor complication from the point of view of numerical calculations.

An examination of Eq. (3.119) shows that the quantities

$$\mathbf{Q}_1 \equiv \frac{1}{\sigma} \nabla (\sigma D I_0) \tag{3.122}$$

and

$$Q_2 \equiv \sigma D I_0 \tag{3.123}$$

are continuous at all points $\mathbf{r}$. Equation (3.118) identifies $\mathbf{Q}_1$ as the negative of the radiative flux per unit frequency, and hence the present diffusion approximation has the characteristic that the flux is everywhere continuous. This is a desirable property in any approximate description of radiative transfer since it implies conservation of photons. However, the continuity of Q_2 implies that I_0 is discontinuous wherever σD is discontinuous, and this occurs at discontinuities in the source $\sigma'_a B$ and the material properties σ_s and σ. The fact that the present approximation leads to discontinuities in I_0 is not surprising. For example, in plane geometry the basic assumption is that the angular functions $e(z, \mu)$ and $o(z, \mu)$ are slowly varying functions of space, and this is not the case near discontinuities in $\sigma'_a B$, σ_s, and σ. The resulting discontinuity in I_0 is an attempt by the formalism to somehow account for the rapid variation of the angular functions near a discontinuity. That is, by allowing such a behavior for I_0 the solution can be made more accurate far from the discontinuity. This behavior has been noted in other approximate descriptions of transport theory (Davison, 1957a; Selengut, 1962; Pomraning and Clark, 1963). It should be emphasized that a discontinuous I_0 (which implies a discontinuous energy density) does not make the present diffusion theory any less physical than the Eddington or asymptotic theories which do maintain continuity of I_0. The simplicity of all three formulations of a diffusion equation precludes an accurate description near discontinuities in $\sigma'_a B$, σ_s, and σ in any event.

Numerical examples given by Winslow (1968) and Pomraning (1969b) indicate that the diffusion theory formulated in this section is a remarkably accurate description of transport theory.

As a final item, we note that the energy density $u(\mathbf{r}, t)$, radiative flux $\mathbf{F}(\mathbf{r}, t)$, and pressure tensor $p(\mathbf{r}, t)$ for the diffusion description under discussion are given by

$$u(\mathbf{r}, t) = \frac{1}{c} \int_0^\infty d\nu I_0(\mathbf{r}, \nu, t), \tag{3.124}$$

$$\mathbf{F}(\mathbf{r}, t) = -\int_0^\infty d\nu \, \frac{1}{\sigma(\mathbf{r}, \nu, t)} \, \nabla [\sigma(\mathbf{r}, \nu, t) \, D(\mathbf{r}, \nu, t) \, I_0(\mathbf{r}, \nu, t)], \tag{3.125}$$

$$p_{ij}(\mathbf{r}, t) = \frac{1}{c} \int_0^\infty dv \int_{4\pi} d\Omega \, \Omega_i \Omega_j I(\mathbf{r}, v, \mathbf{\Omega}, t), \qquad (3.126)$$

where

$$\int_{4\pi} d\Omega \, \Omega_i \Omega_j I = \frac{1}{\sigma^2 \alpha^2} \frac{(3\sigma D - 1)}{2} \frac{\partial^2 I_0}{\partial x_i \, \partial x_j} + \frac{(1 - \sigma D)}{2} I_0 \delta_{ij}, \qquad (3.127)$$

with δ_{ij} denoting the Kronecker delta function. In writing Eq. (3.127) one has the freedom, consistent with the assumptions utilized to derive the diffusion theory, to commute the $\partial/\partial x_i$ operator with α, σ, and D. There are no results indicating the most accurate ordering of these terms. [The same comment, incidentally, applies to Eq. (3.76) in regard to asymptotic diffusion theory.]

5. The P–N (Spherical Harmonic) Approximation

The diffusion approximations to the equation of transfer discussed in the last three sections have one overriding characteristic in common: they are all of limited accuracy. If for a given problem their error is unacceptable, there is no way, within the framework of the approximations, to systematically improve their accuracy.

We discuss in this section an approximation to the equation of transfer, or more precisely a set of approximations, which is capable of estimating the solution to the equation of transfer to within any desired accuracy criteria. This is the so-called spherical harmonic, (or P–N) method, with N denoting the order of the approximation. For N infinite, the P–N solution is the exact solution to the equation of transfer. The lowest order P–N solution corresponds to $N = 1$, and is very closely related to the Eddington approximation discussed in Section 2 of this chapter. In formulating the P–N method we shall neglect the nonlinear induced scattering terms in the equation of transfer. These terms have not been treated in the literature within the P–N framework, although conceptually they would seem to offer little difficulty. Algebraically, however, they would undoubtedly lead to quite complex results, and for this reason it seems preferable to ignore them in this introduction to the P–N method.

Initially we consider systems with plane symmetry. The relevant equation of transfer is then Eq. (2.42). If we set $S = \sigma_a' B$ according to Eq. (2.168), we have, with $\sigma = \sigma_s + \sigma_a'$,

$$\frac{1}{c} \frac{\partial I(v, \mu)}{\partial t} + \mu \frac{\partial I(v, \mu)}{\partial z} + \sigma(v) \, I(v, \mu) = \sigma_a'(v) \, B(v)$$

$$+ \sum_{n=0}^\infty \frac{2n+1}{2} P_n(\mu) \int_0^\infty dv' \, \frac{v}{v'} \, \sigma_{sn}(v' \to v) \int_{-1}^1 d\mu' P_n(\mu') \, I(v', \mu'). \qquad (3.128)$$

We expand the specific intensity in Legendre polynomials according to

$$I(z, v, \mu, t) = \sum_{n=0}^\infty \frac{2n+1}{4\pi} I_n(z, v, t) \, P_n(\mu), \qquad (3.129)$$

The Equations of Radiation Hydrodynamics

where, due to the orthogonality of the $P_n(\mu)$, the expansion coefficients are given by

$$I_n(z, v, t) = 2\pi \int_{-1}^{1} d\mu P_n(\mu)\, I(z, v, \mu, t). \tag{3.130}$$

Insertion of Eq. (3.129) into Eq. (3.128) yields, upon equating coefficients of $P_n(\mu)$, the infinite set of equations for the expansion coefficients

$$\frac{1}{c}\frac{\partial I_0(v)}{\partial t} + \frac{\partial I_1(v)}{\partial z} + \sigma(v)I_0(v) = 4\pi\sigma_a'(v)\,B(v) + \int_0^\infty dv'\,\frac{v}{v'}\,\sigma_{s0}(v' \to v)I_0(v'), \tag{3.131}$$

$$\frac{2n+1}{.c}\frac{\partial I_n(v)}{\partial t} + n\frac{\partial I_{n-1}(v)}{\partial z} + (2n+1)\sigma(v)I_n(v) + (n+1)\frac{\partial I_{n+1}(v)}{\partial z}$$

$$= (2n+1)\int_0^\infty dv'\,\frac{v}{v'}\,\sigma_{sn}(v' \to v)I_n(v'), \quad n \geq 1. \tag{3.132}$$

Let us consider Eq. (3.131) and the first N equations of the infinite set given by Eq. (3.132). This amounts to $N+1$ equations in $N+2$ unknowns, namely $I_0, I_1, \ldots, I_{N+1}$. Hence it is necessary in order to close this set to somehow reduce the number of unknowns by one. Since Eq. (3.129) is assumed to be a convergent expansion, the I_n must decrease with increasing n and the natural and simplest truncation procedure is to make the approximation

$$I_{N+1}(z, v, t) = 0. \tag{3.133}$$

We then have as the last equation in the P–N set

$$\frac{2N+1}{c}\frac{\partial I_N(v)}{\partial t} + N\frac{\partial I_{N-1}(v)}{\partial z} + (2N+1)\sigma(v)I_N(v) = (2N+1)\int_0^\infty dv'\,\frac{v}{v'}\,\sigma_{sN}(v' \to v)I_N(v'). \tag{3.134}$$

Equation (3.131), the first $N-1$ equations of Eq. (3.132), and Eq. (3.134) constitute $N+1$ equations in $N+1$ unknowns $I_0, I_1, \ldots, I_N$, and are the equations of the Nth order P–N approximation in plane geometry.

As N becomes infinite, the solution of the P–N equations approaches the solution of the equation of transfer. However, experience shows that even in very low order the P–N method is quite accurate. For example, with $N = 1$ we have as the P–N equations

$$\frac{1}{c}\frac{\partial I_0(v)}{\partial t} + \frac{\partial I_1(v)}{\partial z} + \sigma(v)I_0(v) = 4\pi\sigma_a'(v)\,B(v) + \int_0^\infty dv'\,\frac{v}{v'}\,\sigma_{s0}(v' \to v)\,I_0(v'), \tag{3.135}$$

and

$$\frac{3}{c}\frac{\partial I_1(v)}{\partial t} + \frac{\partial I_0(v)}{\partial z} + 3\sigma(v)\,I_1(v) = 3\int_0^\infty dv'\,\frac{v}{v'}\,\sigma_{s1}(v' \to v)\,I_1(v'), \tag{3.136}$$

which are very reminiscent of the Eddington equations discussed in Section 2 of this chapter.

Davison (1957b) has pointed out that the even order P–N approximations (i.e., $N = 2, 4, \ldots$) have difficulties which are not suffered in odd order, and that the succeeding even order approximation is less accurate than the one lower odd order approximation. For these reasons, only odd order approximations are used in practice, and for the remainder of our discussion we shall restrict our attention to N odd.

The initial conditions for the P–N approximation follow immediately from the initial condition on the equation of transfer, Eq. (2.31). We have

$$I_n(z, v, 0) = 2\pi \int_{-1}^{1} d\mu P_n(\mu) \Lambda(z, v, \mu), \quad n = 0, 1, \ldots, N. \tag{3.137}$$

Concerning boundary conditions, we require $(N+1)/2$ conditions on each face of the planar system. If we consider the left hand surface, say $z = z_l$, these conditions can be taken as $(N+1)/2$ weighted averages of the exact boundary condition, Eq. (2.29). That is, we write

$$2\pi \int_{0}^{1} d\mu w_m(\mu)[I(z_l, v, \mu, t) - \Gamma(z_l, v, \mu, t)] = 0; \quad m = 1, 2, \ldots, (N+1)/2, \tag{3.138}$$

where the $w_m(\mu)$, arbitrary linearly independent functions, are the weight functions. Since the P–N approximation consists of setting $I_n(z, v, t) = 0$ for $n > N$, we use a truncated version of Eq. (3.129) in Eq. (3.138) and write

$$\sum_{n=0}^{N} \frac{2n+1}{2} I_n(z_l, v, t) \int_{0}^{1} d\mu w_m(\mu) P_n(\mu) = 2\pi \int_{0}^{1} d\mu w_m(\mu) \Gamma(z_l, v, \mu, t), \quad m = 1, 2, \ldots, (N+1)/2. \tag{3.139}$$

Once the $w_m(\mu)$ have been specified, Eq. (3.139) is the required $(N+1)/2$ relationships among the $I_n(z, v, t)$ at $z = z_l$. The so-called Marshak conditions consist of the choice

$$w_m(\mu) = P_{2m-1}(\mu), \quad m = 1, 2, \ldots, (N+1)/2, \tag{3.140}$$

or equivalently,

$$w_m(\mu) = \mu^{2m-1}, \quad m = 1, 2, \ldots, (N+1)/2. \tag{3.141}$$

The Marshak condition for $m = 1$ has the physical interpretation of preserving the incoming flux per unit frequency. The Mark boundary conditions correspond to choosing the weight functions as Dirac delta functions

$$w_m(\mu) = \delta(\mu - \mu_m), \quad m = 1, 2, \ldots, (N+1)/2, \tag{3.142}$$

where the μ_m are the positive roots of the $(N+1)$th Legendre polynomial, i.e.,

$$P_{N+1}(\mu_m) = 0, \quad \mu_m > 0, m = 1, 2, \ldots, (N+1)/2. \tag{3.143}$$

It can be shown (Davison, 1957c) that in the special case of no incoming flux ($\Gamma = 0$), the Mark conditions are equivalent to surrounding the system with a source free, pure absorber and carrying out the P–N calculation over all space, assuming $I_n(z, v, t)$ to vanish at $z = -\infty$. Similar considerations give the Marshak and Mark boundary conditions at the right hand surface of a planar system. The only difference is that the integrals in Eq. (3.139) cover the range $-1 < \mu < 0$ rather than $0 < \mu < 1$, and one uses the negative, rather than the positive, roots from Eq. (3.143). In practice, the Marshak conditions generally prove to be more accurate than the Mark conditions. Boundary conditions derived from variational considerations (Federighi, 1964; Pomraning, 1964) appear to be more accurate than either those of Marshak or Mark, but have the disadvantage that they have not been generalized to multidimensional situations.

It should also be remarked that other truncation schemes, rather than setting $I_{N+1}(z, v, t) = 0$, have been proposed within the context of the P–N method (Carlson and Lathrop, 1965; Pomraning, 1965a). What is required in general to truncate the infinite set of equations is a method of relating $I_{N+1}(z, v, t)$ to the lower order expansion coefficients. This can be done with some generality in the following way. Suppose that one expects the angular distribution to be approximately described by a specified function $f(z, v, \mu, t)$. For the purposes of truncation, we then assume that the specific intensity of radiation can be represented by

$$I(z, v, \mu, t) = \sum_{n=0}^{N-2} c_n(z, v, t)P_n(\mu)$$

$$+ c_d(z, v, t) f_o(z, v, \mu, t) + c_e(z, v, t) f_e(z, v, \mu, t), \tag{3.144}$$

where f_o and f_e are the odd and even parts (in μ) respectively of $f(z, v, \mu, t)$, and the c_n, c_d, and c_e are expansion coefficients. Multiplying Eq. (3.144) by $P_{N-1}(\mu)$ and integrating over all solid angle, we find (we assume N odd)

$$I_{N-1}(z, v, t) = 2\pi c_e(z, v, t) \int_{-1}^{1} d\mu \, P_{N-1}(\mu)f_e(z, v, \mu, t). \tag{3.145}$$

Similarly, one can obtain an expression for $I_{N+1}(z, v, t)$. Taking the ratio of these two results, we obtain

$$I_{N+1}(z, v, t) = \frac{\int_{-1}^{1} d\mu P_{N+1}(\mu) f_e(z, v, \mu, t)}{\int_{-1}^{1} d\mu P_{N-1}(\mu) f_e(z, v, \mu, t)} I_{N-1}(z, v, t) \tag{3.146}$$

as the truncation condition. Hence Eq. (3.132) for $n = N$ becomes

$$\frac{2N+1}{c} \frac{\partial I_N(v)}{\partial t} + \frac{\partial}{\partial z} \{[N + (N+1) R_N(v)] I_{N-1}(v)\}$$

$$+ (2N+1)\sigma(v) I_N(v) = (2N+1) \int_{0}^{\infty} dv' \, \frac{v}{v'} \sigma_{sN}(v' \to v) I_N(v'), \tag{3.147}$$

where $R_N \equiv R_N(z, v, t)$ is the ratio of integrals appearing in Eq. (3.146). This modified P–N approximation then consists of Eq. (3.131), the first $N-1$ equations of Eq. (3.132), and Eq. (3.147) as the truncating equation. The initial conditions on this set of equations are still given by Eq. (3.137). The extension of the Marshak boundary conditions is algebraically messy and will not be given here. These conditions have been discussed by Pomraning (1965a).

A particularly interesting example of this type of truncation is that corresponding to the choice of $f(z, v, \mu, t)$ as the asymptotic distribution. For example, for isotropic scattering we have [see Eq. (3.65)]

$$f_e(z, v, \mu, t) = \frac{1}{1 - \alpha^2 \mu^2} \, . \tag{3.148}$$

From Eq. (3.146) we then find

$$R_N(z, v, t) \equiv \frac{I_{N+1}(z, v, t)}{I_{N-1}(z, v, t)} = \frac{Q_{N+1}(1/\alpha)}{Q_{N-1}(1/\alpha)} , \tag{3.149}$$

where

$$Q_n(z) \equiv \frac{1}{2} \int_{-1}^{1} d\xi \, \frac{P_n(\xi)}{z - \xi} \tag{3.150}$$

is the nth order Legendre function of the second kind. One can show that this "asymptotic" P–N approximation gives the exact asymptotic behavior for the I_n in the appropriate situation (time-independent, monochromatic transport in a source-free homogeneous infinite medium). Further, for $N > 1$ this approximation will, in all orders, accurately describe the almost isotropic transport problem (for which the Eddington approximation is known to be valid). Calculations on simple test problems indicate that, for a given N, this modified P–N method is more accurate than conventional P–N theory. In summary, this modified P–N approximation appears to be quite accurate in low order, properly displays both the asymptotic and Eddington limits, and is capable of giving an arbitrarily small error (by choosing N great enough) for any transport problem of interest.

The energy density, radiative flux, and pressure tensor in the P–N or modified P–N method are given by

$$u(z, t) = \frac{1}{c} \int_{0}^{\infty} dv I_0(z, v, t), \tag{3.151}$$

$$F_z(z, t) = \int_{0}^{\infty} dv I_1(z, v, t), \tag{3.152}$$

and, for $N > 1$,

$$p_{ij}(z, t) = 0, \quad i \neq j, \tag{3.153}$$

$$p_{zz}(z, t) = \frac{1}{c} \int_{0}^{\infty} dv \left[\frac{1}{3} I_0(z, v, t) + \frac{2}{3} I_2(z, v, t) \right], \tag{3.154}$$

$$p_{xx}(z, t) = p_{yy}(z, t) = \frac{1}{c} \int_{0}^{\infty} dv \left[\frac{1}{3} I_0(z, v, t) - \frac{1}{3} I_2(z, v, t) \right]. \tag{3.155}$$

We note that

$$p_{xx} + p_{yy} + p_{zz} = u, \tag{3.156}$$

a property true in general as discussed in Chapter I [see Eq. (1.21)].

The Equations of Radiation Hydrodynamics

With this introduction to the *P–N* approximation for systems with plane symmetry, we now extend the method to arbitrary, three dimensional geometry. For simplicity, we initially assume the scattering to be isotropic. We shall indicate the extension to a general scattering kernel shortly. The relevant equation of transfer is Eq. (2.17) which reads, for isotropic scattering and with $S = \sigma'_a B$ and $\sigma = \sigma_s + \sigma'_a$,

$$\frac{1}{c}\frac{\partial I(v,\Omega)}{\partial t} + \Omega \cdot \nabla I(v,\Omega) + \sigma(v)I(v,\Omega) = \sigma'_a(v)B(v)$$

$$+\frac{1}{4\pi}\int_0^\infty dv'\,\frac{v}{v'}\,\sigma_{s0}(v' \to v)\int_{4\pi} d\Omega'\,I(v',\Omega'). \tag{3.157}$$

To facilitate the expansion of $I(\mathbf{r}, v, \Omega, t)$ in a full set of spherical harmonics, we define a vector $\mathbf{U}$ of arbitrary magnitude U in the direction Ω, i.e.,

$$\mathbf{U} = U\Omega. \tag{3.158}$$

We expand the specific intensity in surface harmonics according to

$$I(\mathbf{r}, v, \Omega, t) = \frac{1}{4\pi}\sum_{n=0}^\infty\sum_{m=-n}^n A_n^m I_n^m(\mathbf{r}, v, t)\,Y_n^m(\Omega). \tag{3.159}$$

The surface harmonics, which are complete on the unit sphere, are defined as

$$Y_n^m(\Omega) = P_n^{|m|}(\cos\theta)\,e^{im\varphi}, \tag{3.160}$$

where θ is a polar angle, φ is the corresponding azimuthal angle, and the $P_n^{|m|}(x)$ are the usual associated Legendre functions (see, for example, Abramowitz and Stegun, 1964). The surface harmonics and their complex conjugates are orthogonal on the unit sphere according to

$$\int_{4\pi} d\Omega\, Y_n^m(\Omega)\, Y_l^{j*}(\Omega) = 4\pi\delta_{jm}\delta_{in}/A_n^m, \tag{3.161}$$

where the normalization coefficient A_n^m is given by

$$A_n^m = 4\pi\left[\int_{4\pi} d\Omega\, Y_n^m(\Omega)\, Y_n^{m*}(\Omega)\right]^{-1} = \frac{(2n+1)(n-|m|)!}{(n+|m|)!}. \tag{3.162}$$

If we define

$$\psi_n \equiv \frac{U^n}{2n+1}\sum_{m=-n}^n A_n^m I_n^m(\mathbf{r}, v, t)\,Y_n^m(\Omega), \tag{3.163}$$

Eq. (3.159) can be written

$$I(\mathbf{r}, v, \Omega, t) = \sum_{n=0}^\infty\left(\frac{2n+1}{4\pi}\right)[\psi_n]_{U=1}. \tag{3.164}$$

A spherical harmonic is defined as the product of U^n and $Y_n^m(\Omega)$, and thus Eq. (3.164) represents an expansion of the specific intensity of radiation in spherical harmonics. We note that the zeroth order spherical harmonic, ψ_0, is just

$$\psi_0 = \int_{4\pi} d\Omega\, I(\mathbf{r},\, \nu,\, \mathbf{\Omega},\, t), \tag{3.165}$$

and hence the inscattering term in Eq. (3.157) is proportional to ψ_0 only (no higher order spherical harmonics are involved, which is a direct consequence of the assumption that the scattering is isotropic).

With these definitions and notation out of the way, we now proceed to expand the equation of transfer in spherical harmonics. Use of Eq. (3.164) in Eq. (3.157) yields

$$\left[\frac{1}{c}\frac{\partial}{\partial t}\sum_{n=0}^{\infty}(2n+1)\,\psi_n(\nu)\right]_{U=1} + \left[\mathbf{U}\cdot\nabla_r\sum_{n=0}^{\infty}(2n+1)\,\psi_n(\nu)\right]_{U=1}$$

$$+\sigma(\nu)\left[\sum_{n=0}^{\infty}(2n+1)\,\psi_n(\nu)\right]_{U=1} = \sigma_a'(\nu)\,B(\nu) + \int_0^{\infty}d\nu'\,\frac{\nu}{\nu'}\,\sigma_{s0}(\nu'\rightarrow\nu)\,\psi_0(\nu'). \tag{3.166}$$

As in Eq. (3.166), it will be necessary to subscript the vector operators as we are dealing with two vectors, $\mathbf{r}$ and $\mathbf{U}$. In order to expand Eq. (3.166) in spherical harmonics, the only troublesome term is of the form $[\mathbf{U}\cdot\nabla_r\psi_n]_{U=1}$. We have the identity

$$[\mathbf{U}\cdot\nabla_r\psi_n]_{U=1} = \left[\mathbf{U}\cdot\nabla_r\psi_n - \frac{U^2}{2n+1}\,\nabla_U\cdot\nabla_r\psi_n\right]_{U=1} + \left[\frac{1}{2n+1}\,\nabla_U\cdot\nabla_r\psi_n\right]_{U=1}. \tag{3.167}$$

It can be shown (Davison, 1957d) that the term within the first bracket on the right hand side of Eq. (3.167) is a spherical harmonic of order $n+1$, and the term within the second bracket is a spherical harmonic of order $n-1$. Substituting Eq. (3.167) into Eq. (3.166), collecting spherical harmonics of order n, and multiplying by U^n, we find

$$\frac{1}{c}\frac{\partial\psi_0(\nu)}{\partial t} + \nabla_U\cdot\nabla_r\psi_1(\nu) + \sigma(\nu)\,\psi_0(\nu) = 4\pi\sigma_a'(\nu)\,B(\nu)$$

$$+\int_0^{\infty}d\nu'\,\frac{\nu}{\nu'}\,\sigma_{s0}(\nu'\rightarrow\nu)\,\psi_0(\nu'), \tag{3.168}$$

$$\frac{2n+1}{c}\frac{\partial\psi_n(\nu)}{\partial t} + \nabla_U\cdot\nabla_r\psi_{n+1}(\nu) + (2n+1)\,\sigma(\nu)\,\psi_n(\nu)$$

$$+(2n-1)\mathbf{U}\cdot\nabla_r\psi_{n-1}(\nu) - U^2\nabla_U\cdot\nabla_r\psi_{n-1}(\nu) = 0, \qquad n\geq 1. \tag{3.169}$$

Equations (3.168) and (3.169) represent the spherical harmonic expansion of the equation of transfer and are exact. Since ψ_n consists of $2n+1$ components, Eq. (3.169) represents $2n+1$ relationships between the components of ψ_{n+1}, ψ_n, and ψ_{n-1} (i.e., between the components I_{n+1}^m, I_n^m, and I_{n-1}^m). Symmetry considerations may reduce the number of nonzero components in Eq. (3.169), and the number of relationships is reduced accordingly.

The usual P–N approximation consists of considering the first $N+1$ equations of this infinite set, Eqs. (3.168) and (3.169), and setting

$$\nabla_U\cdot\nabla_r\psi_{N+1} = 0, \tag{3.170}$$

77

in the $(N+1)$th equation. This results in $N+1$ relationships for $N+1$ unknowns, ψ_0 through ψ_N. A more general truncation scheme is as follows. We again retain the first $N+1$ equations and use Eq. (3.170) to eliminate ψ_{N+1}. In addition, however, in the $(N+1)$th equation we replace ψ_{N-1} by $T_N\psi_{N-1}$, where $T_N \equiv T_N(\mathbf{r}, v, t)$, and is at this point an arbitrary function. The truncating equation of this generalized $P\text{–}N$ approximation is then

$$\frac{2N+1}{c}\frac{\partial\psi_N(v)}{\partial t} + (2N+1)\,\sigma(v)\,\psi_N(v) + (2N-1)\,\mathbf{U}\cdot\nabla_r T_N\psi_{N-1}(v)$$
$$- U^2\nabla_U\cdot\nabla_r T_N\psi_{N-1}(v) = 0. \tag{3.171}$$

If one sets $T_N = 1$, Eq. (3.171) reduces to the last equation in the usual $P\text{–}N$ approximation. On the other hand, if one sets

$$T_N = 1 + \frac{(N+1)}{N}\,R_N, \tag{3.172}$$

where $R_N \equiv R_N(\mathbf{r}, v, t)$ is given by Eq. (3.149), this modified $P\text{–}N$ approximation gives the correct asymptotic behavior in all orders N for all geometries. This was the primary reason for introducing this generalized truncation procedure. A more detailed discussion of this asymptotically correct $P\text{–}N$ approximation has been given by Pomraning (1965b).

The extension of the above analysis to a general (anisotropic) scattering law is straightforward. Equation (3.169), the nth spherical harmonic equation for isotropic scattering, is replaced with (Davison, 1957e)

$$\frac{2n+1}{c}\frac{\partial\psi_n(v)}{\partial t} + \nabla_U\cdot\nabla_r\psi_{n+1}(v) + (2n+1)\,\sigma(v)\,\psi_n(v) + (2n-1)\,\mathbf{U}\cdot\nabla_r\psi_{n-1}(v)$$

$$- U^2\nabla_U\cdot\nabla_r\psi_{n-1}(v) = (2n+1)\int_0^\infty dv'\,\frac{v}{v'}\,\sigma_{sn}(v' \to v)\,\psi_n(v'), \qquad n \geq 1. \tag{3.173}$$

Here $\sigma_{sn}(v' \to v)$ is the projection of the scattering kernel along the nth axis in P_n space [see Eqs. (2.36) and (2.37)]. The truncation procedure proceeds just as in the isotropic scattering case.

To complete the discussion of the $P\text{–}N$ method in general geometry, we need consider the initial and boundary conditions. Our discussion applies to both the isotropic and general anisotropic scattering equations. The initial conditions are straightforward. We expand the initial condition on the equation of transfer, Eq. (2.31), in surface harmonics as

$$I(\mathbf{r}, v, \mathbf{\Omega}, 0) = \Lambda(\mathbf{r}, v, \mathbf{\Omega}) = \frac{1}{4\pi}\sum_{n=0}^\infty\sum_{m=-n}^n A_n^m\Lambda_n^m(\mathbf{r}, v)\,Y_n^m(\mathbf{\Omega}). \tag{3.174}$$

The initial conditions on the $I_n^m(\mathbf{r}, v, t)$ are then given by

$$I_n^m(\mathbf{r}, v, 0) = \Lambda_n^m(\mathbf{r}, v); \qquad n = 0, 1, \ldots, N, \tag{3.175}$$

in the $P\text{–}N$ approximation. The boundary conditions are more involved. Denoting by p the number of conditions required, Davison (1957f) has shown that, for N odd,

$$p = p_0 + p_2 + \ldots + p_{N-1}, \tag{3.176}$$

where p_n is the largest number of linearly independent spherical harmonic components of order n which is compatible with the symmetry of the problem under consideration. He has suggested that these p conditions be of the Marshak form. Specifically, referring to the transport boundary condition, Eq. (2.29), Davison suggests

$$\int\limits_{\mathbf{n}\cdot\mathbf{\Omega}<0} d\Omega[I(\mathbf{r}_s, v, \mathbf{\Omega}, t) - \Gamma(\mathbf{r}_s, v, \mathbf{\Omega}, t)]\, Y_n^m(\mathbf{\Omega}) = 0, \tag{3.177}$$

as the P–N boundary conditions, where, as usual, $\mathbf{r}_s$ denotes a surface point and $\mathbf{n}$ is an outward normal vector at $\mathbf{r}_s$. In Eq. (3.177) we choose n and m corresponding to the $I_n^m(\mathbf{r}, v, t)$, which do not vanish from symmetry considerations, for n odd and all m corresponding to n for $n \leq N-2$, and for as many m as possible (so as to make a total of p conditions) for $n = N$ starting from the smallest $|m|$. In the classical P–N method ($T_N = 1$), one uses a truncated version of Eq. (3.159), with the upper limit on the summation over n taken as N, in Eq. (3.177) to obtain explicitly the p relationships among the I_n^m at the boundary of the system. In the asymptotic P–N method [T_N given by Eq. (3.172)], the form of $I(\mathbf{r}_s, v, \mathbf{\Omega}, t)$ to be used in Eq. (3.177) is more complex, and we refer the reader to the paper of Pomraning (1965b).

Since ψ_0 is just the angle integrated specific intensity [see Eq. (3.165)], the energy density in the P–N approximation is clearly given by

$$u(\mathbf{r}, t) = \frac{1}{c} \int\limits_0^\infty dv\psi_0(\mathbf{r}, v, t). \tag{3.178}$$

To obtain the radiative flux, we first note that the direction cosines of $\mathbf{\Omega}$ can be written in terms of the surface harmonics as

$$\Omega_x = \sin\theta \cos\varphi = -\tfrac{1}{2}[Y_1^1(\mathbf{\Omega}) + Y_1^{-1}(\mathbf{\Omega})], \tag{3.179}$$

$$\Omega_y = \sin\theta \sin\varphi = -\frac{1}{2i}[Y_1^1(\mathbf{\Omega}) - Y_1^{-1}(\mathbf{\Omega})], \tag{3.180}$$

$$\Omega_z = \cos\theta = Y_1^0(\mathbf{\Omega}). \tag{3.181}$$

Hence the x, y, and z components of the radiative flux are given by

$$F_x(\mathbf{r}, t) = -\tfrac{1}{2} \int\limits_0^\infty dv\,[I_1^1(\mathbf{r}, v, t) + I_1^{-1}(\mathbf{r}, v, t)], \tag{3.182}$$

$$F_y(\mathbf{r}, t) = -\frac{1}{2i} \int\limits_0^\infty dv\,[I_1^1(\mathbf{r}, v, t) - I_1^{-1}(\mathbf{r}, v, t)], \tag{3.183}$$

$$F_z(\mathbf{r}, t) = \int\limits_0^\infty dv I_1^0(\mathbf{r}, v, t). \tag{3.184}$$

In writing Eqs. (3.179) through (3.184) we have followed the usual convention of choosing the z axis as that axis which defines the polar angle θ, the x axis as that axis which defines the azimuthal angle φ, and the y axis such that the x, y, and z axes are orthogonal and form a right

The Equations of Radiation Hydrodynamics

handed coordinate system. Similarly, we have

$$\Omega_x \Omega_x = \tfrac{1}{3} Y_0^0(\mathbf{\Omega}) - \tfrac{1}{3} Y_2^0(\mathbf{\Omega}) + \tfrac{1}{12} Y_2^2(\mathbf{\Omega}) + \tfrac{1}{12} Y_2^{-2}(\mathbf{\Omega}), \tag{3.185}$$

$$\Omega_y \Omega_y = \tfrac{1}{3} Y_0^0(\mathbf{\Omega}) - \tfrac{1}{3} Y_2^0(\mathbf{\Omega}) - \tfrac{1}{12} Y_2^2(\mathbf{\Omega}) - \tfrac{1}{12} Y_2^{-2}(\mathbf{\Omega}), \tag{3.186}$$

$$\Omega_z \Omega_z = \tfrac{1}{3} Y_0^0(\mathbf{\Omega}) + \tfrac{2}{3} Y_2^0(\mathbf{\Omega}), \tag{3.187}$$

$$\Omega_x \Omega_y = \Omega_y \Omega_x = \frac{1}{12i} [Y_2^2(\mathbf{\Omega}) - Y_2^{-2}(\mathbf{\Omega})], \tag{3.188}$$

$$\Omega_x \Omega_z = \Omega_z \Omega_x = -\frac{1}{6} [Y_2^1(\mathbf{\Omega}) + Y_2^{-1}(\mathbf{\Omega})], \tag{3.189}$$

$$\Omega_y \Omega_z = \Omega_z \Omega_y = -\frac{1}{6i} [Y_2^1(\mathbf{\Omega}) - Y_2^{-1}(\mathbf{\Omega})]. \tag{3.190}$$

The pressure tensor components $p_{ij}(\mathbf{r}, t)$ then follow as

$$p_{xx}(\mathbf{r}, t) = \frac{1}{12c} \int_0^\infty d\nu (4I_0^0 - 4I_2^0 + I_2^2 + I_2^{-2}), \tag{3.191}$$

$$p_{yy}(\mathbf{r}, t) = \frac{1}{12c} \int_0^\infty d\nu (4I_0^0 - 4I_2^0 - I_2^2 - I_2^{-2}), \tag{3.192}$$

$$p_{zz}(\mathbf{r}, t) = \frac{1}{3c} \int_0^\infty d\nu (I_0^0 + 2I_2^0), \tag{3.193}$$

$$p_{xy}(\mathbf{r}, t) = p_{yx}(\mathbf{r}, t) = \frac{1}{12ic} \int_0^\infty d\nu (I_2^2 - I_2^{-2}), \tag{3.194}$$

$$p_{xz}(\mathbf{r}, t) = p_{zx}(\mathbf{r}, t) = -\frac{1}{6c} \int_0^\infty d\nu (I_2^1 + I_2^{-1}), \tag{3.195}$$

$$p_{yz}(\mathbf{r}, t) = p_{zy}(\mathbf{r}, t) = -\frac{1}{6ic} \int_0^\infty d\nu (I_2^1 - I_2^{-1}), \tag{3.196}$$

where $I_n^m \equiv I_n^m(\mathbf{r}, \nu, t)$ in these equations. It is again to be noted that

$$p_{xx} + p_{yy} + p_{zz} = u, \tag{3.197}$$

as is true in general [see Eq. (1.21)].

To show the utility of this rather abstract spherical harmonic analysis, it is useful to consider a specific geometry. As an example, we consider the form the general P–N equations take in the special case of spherical symmetry. The similar, but more complex, considerations in the case of cylindrical symmetry have been given by Pomraning (1965c).

We begin with the nth order spherical harmonic equation with anisotropic scattering in general geometric form, Eq. (3.173). In spherical geometry, the specific intensity of radiation depends only upon the radial coordinate r and the polar angle θ, defined as the angle between the position vector $\mathbf{r}$ and the direction vector $\boldsymbol{\Omega}$. Thus only the $m = 0$ term contributes in the expansion for ψ_n given by Eq. (3.163), and we have

$$\psi_n = U^n P_n(\mu) \, I_n(r, v, t), \tag{3.198}$$

where we have defined $\mu = \cos\theta$ and noted that $Y_n^0(\boldsymbol{\Omega}) = P_n(\mu)$, the nth order Legendre polynomial. We have also set $I_n^0(r, v, t)$ equal to $I_n(r, v, t)$ for notational simplicity. The orthogonality condition for the surface harmonics, when applied to Eq. (3.159), yields

$$I_n^m(\mathbf{r}, v, t) = \int\limits_{4\pi} d\Omega Y_n^{m*}(\boldsymbol{\Omega}) \, I(\mathbf{r}, v, \boldsymbol{\Omega}, t). \tag{3.199}$$

In the case of spherical symmetry, Eq. (3.199) reduces to

$$I_n(r, v, t) = 2\pi \int\limits_{-1}^{1} d\mu P_n(\mu) \, I(r, v, \mu, t), \tag{3.200}$$

a result similar to that obtained in plane geometry, Eq. (3.130).

It is clear from Eq. (3.173) that the only difficult terms in the reduction to spherical geometry are $\nabla_U \cdot \nabla_r \psi_n$ and $(2n+1)\mathbf{U} \cdot \nabla_r \psi_n - U^2 \nabla_U \cdot \nabla_r \psi_n$. To evaluate these terms, let us define a function $F_n \equiv F_n(r, U, \mu)$ as

$$F_n(r, U, \mu) = U^n P_n(\mu) \, f_n(r), \tag{3.201}$$

where $f_n(r)$ is an arbitrary function, and

$$U^2 \equiv U_x^2 + U_y^2 + U_z^2, \tag{3.202}$$

$$r^2 \equiv x^2 + y^2 + z^2, \tag{3.203}$$

$$\mu = \frac{x U_x + y U_y + z U_z}{rU}. \tag{3.204}$$

We will shortly, of course, identify $f_n(r)$ with $I_n(r, v, t)$ and $f_{N-1}(r)$ with $T_N(r, v, t) I_{N-1}(r, v, t)$. We first consider $\nabla_U \cdot \nabla_r F_n$, namely

$$\nabla_U \cdot \nabla_r F_n \equiv \sum_{i=1}^{3} \frac{\partial^2 F_n}{\partial x_i \, \partial U_i}, \tag{3.205}$$

where $x_1 = x$, $x_2 = y$, $x_3 = z$, and $U_1 = U_x$, $U_2 = U_y$, and $U_3 = U_z$. The chain rule of differentiation yields

$$\nabla_U \cdot \nabla_r F_n = n U^{n-1} P_n(\mu) \, f_n'(r) \sum_{i=1}^{3} \frac{\partial r}{\partial x_i} \frac{\partial U}{\partial U_i}$$

$$+ n U^{n-1} P_n'(\mu) \, f_n(r) \sum_{i=1}^{3} \frac{\partial \mu}{\partial x_i} \frac{\partial U}{\partial U_i} + U^n P_n'(\mu) \, f_n'(r) \sum_{i=1}^{3} \frac{\partial r}{\partial x_i} \frac{\partial \mu}{\partial U_i}$$

$$+ U^n P_n'(\mu) \, f_n(r) \sum_{i=1}^{3} \frac{\partial^2 \mu}{\partial x_i \, \partial U_i} + U^n P_n''(\mu) \sum_{i=1}^{3} \frac{\partial \mu}{\partial x_i} \frac{\partial \mu}{\partial U_i}, \tag{3.206}$$

where the prime indicates differentiation with respect to the argument. Computing the

The Equations of Radiation Hydrodynamics

derivatives needed in Eq. (3.206) from Eqs. (3.202) through (3.204), we find

$$\frac{\partial r}{\partial x_i} = \frac{x_i}{r} \; ; \qquad\qquad \frac{\partial U}{\partial U_i} = \frac{U_i}{U} , \tag{3.207}$$

$$\frac{\partial \mu}{\partial x_i} = \frac{U_i}{rU} - \frac{x_i\mu}{r^2} \; ; \quad \frac{\partial \mu}{\partial U_i} = \frac{x_i}{rU} - \frac{U_i\mu}{U^2} , \tag{3.208}$$

$$\frac{\partial^2 \mu}{\partial x_i\,\partial U_i} = \frac{1}{rU} - \frac{x_i^2}{Ur^3} - \frac{U_i^2}{rU^3} + \frac{x_iU_i\mu}{r^2U^2} . \tag{3.209}$$

Use of these results in Eq. (3.206) gives

$$\nabla_U\bullet\nabla_rF_n = U^{n-1}f_n'(r)[n\mu P_n(\mu)+(1+\mu^2)\,P_n'(\mu)]$$
$$+ U^{n-1}\frac{f_n(r)}{r}\,[n(1-\mu^2)\,P_n'(\mu)+(1+\mu^2)\,P_n'(\mu)-\mu(1-\mu^2)\,P_n''(\mu)]. \tag{3.210}$$

We now recall the differential equation satisfied by the Legendre polynomials, namely

$$(1-\mu^2)\frac{d^2P_n(\mu)}{d\mu^2} - 2\mu\frac{dP_n(\mu)}{d\mu} + n(n+1)P_n(\mu) = 0, \tag{3.211}$$

and the recurrence relationship

$$(1-\mu^2)\frac{dP_n(\mu)}{d\mu} = nP_{n-1}(\mu) - n\mu P_n(\mu), \tag{3.212}$$

as given, for example, in Abramowitz and Stegun (1964). We use Eq. (3.211) to eliminate $P_n''(\mu)$ in Eq. (3.210) and Eq. (3.212) to eliminate $P_n'(\mu)$ in the result. We find

$$\nabla_U\bullet\nabla_rF_n = U^{n-1}P_{n-1}(\mu)\left[n\frac{df_n(r)}{dr} + \frac{n(n+1)}{r}f_n(r)\right] \tag{3.213}$$

as the final result for $\nabla_U\bullet\nabla_rF_n$.

We next consider terms of the form $(2n+1)\mathbf{U}\bullet\nabla_rF_n - U^2\nabla_U\bullet\nabla_rF_n$. For this purpose we consider $\mathbf{U}\bullet\nabla_rF_n$, defined as

$$\mathbf{U}\bullet\nabla_rF_n \equiv \sum_{i=1}^{3} U_i\frac{\partial F_n}{\partial x_i} . \tag{3.214}$$

The chain rule of differentiation gives

$$\mathbf{U}\bullet\nabla_rF_n = U^nP_n(\mu)f_n'(r)\sum_{i=1}^{3} U_i\frac{\partial r}{\partial x_i} + U^nP_n'(\mu)f_n(r)\sum_{i=1}^{3} U_i\frac{\partial \mu}{\partial x_i} . \tag{3.215}$$

Use of Eqs. (3.207) and (3.208) in Eq. (3.215) yields

$$\mathbf{U}\bullet\nabla_rF_n = U^{n+1}\mu P_n(\mu)f_n'(r) + U^{n+1}(1-\mu^2)P_n'(\mu)f_n(r)/r. \tag{3.216}$$

Combining this result with Eq. (3.213) we get

$$(2n+1)\mathbf{U}\bullet\nabla_rF_n - U^2\nabla_U\bullet\nabla_rF_n$$
$$= U^{n+1}[(2n+1)\mu P_n(\mu)f_n'(r) + (2n+1)\,(1-\mu^2)P_n'(\mu)f_n(r)/r]$$
$$- U^{n+1}[nP_{n-1}(\mu)f_n'(r) + n(n+1)P_{n-1}(\mu)f_n(r)/r]. \tag{3.217}$$

Use of Eq. (3.212) to eliminate $P_n'(\mu)$ in Eq. (3.217) and use of the recurrence relationship (Abramowitz and Stegun, 1964)

$$(2n+1)\mu P_n(\mu) = (n+1)P_{n+1}(\mu)+nP_{n-1}(\mu) \tag{3.218}$$

to eliminate $\mu P_n(\mu)$ in the result yield

$$(2n+1)\mathbf{U}\cdot\nabla_r F_n - U^2\nabla_U\cdot\nabla_r F_n = U^{n+1}P_{n+1}(\mu)\left[(n+1)\frac{df_n(r)}{dr}-\frac{n(n+1)}{r}f_n(r)\right] \tag{3.219}$$

as our second result.

We use these results, namely Eqs. (3.213) and (3.219), in the nth order spherical harmonic equations with anisotropic scattering, Eqs. (3.168) and (3.173), identifying $f_n(r)$ with $I_n(r, v, t)$, and, in the truncating equation, identifying $f_{N-1}(r)$ with $T_N(r, v, t)I_{N-1}(r, v, t)$. This yields, as the spherical harmonic equations in spherical symmetry with anisotropic scattering,

$$\frac{1}{c}\frac{\partial I_0(v)}{\partial t}+\frac{1}{r^2}\frac{\partial}{\partial r}[r^2 I_1(v)]+\sigma(v)\,I_0(v)$$

$$= 4\pi\sigma_a'(v)\,B(v)+\int_0^\infty dv'\,\frac{v}{v'}\,\sigma_{s0}(v'\to v)\,I_0(v'), \tag{3.220}$$

$$\frac{2n+1}{c}\frac{\partial I_n(v)}{\partial t}+n\frac{\partial I_{n-1}(v)}{\partial r}-n(n-1)\frac{I_{n-1}(v)}{r}+(2n+1)\,\sigma(v)\,I_n(v)+(n+1)\frac{\partial I_{n+1}(v)}{\partial r}$$

$$+(n+1)(n+2)\frac{I_{n+1}(v)}{r} = (2n+1)\int_0^\infty dv'\,\frac{v}{v'}\,\sigma_{sn}(v'\to v)\,I_n(v'),\quad 1\le n\le N-1, \tag{3.221}$$

$$\frac{2N+1}{c}\frac{\partial I_N(v)}{\partial t}+N\frac{\partial}{\partial r}[T_N(v)I_{N-1}(v)]-N(N-1)\,T_N(v)\frac{I_{N-1}(v)}{r}$$

$$+(2N+1)\,\sigma(v)\,I_N(v) = (2N+1)\int_0^\infty dv'\,\frac{v}{v'}\,\sigma_{sN}(v'\to v)\,I_N(v'). \tag{3.222}$$

The boundary and initial conditions, as well as the expressions for the energy density, radiative flux, and pressure tensor, are the same in spherical geometry as in plane geometry, which have already been discussed (with the plane variable z interpreted as the radial variable r).

We have devoted this amount of detail to the P-N method in spherical geometry primarily because we felt it important to give explicitly an example of the use of the general spherical harmonic analysis. However, spherical geometry is an important geometry in practical applications as well. The similar reduction of the general spherical harmonic equations to the case of cylindrical symmetry, also important in practice, has been given by Pomraning (1965c).

The Equations of Radiation Hydrodynamics

6. The S–N (Discrete Ordinate) Approximation

One of the appealing features of the P–N method discussed in the last section is that it is in fact an entire set of approximations, which, by choosing N large enough, can be used to estimate the solution to the equation of transfer to within an arbitrarily small error. Another set of approximations with the same feature is the discrete ordinate or S–N method, with N again denoting the order of the approximation. For N infinite, the S–N solution is the exact solution to the equation of transfer, just as in the P–N method. In our discussion of the discrete ordinate method, we again do not treat the induced scattering terms, although the method can surely be applied to these terms at the expense of simplicity.

In plane geometry the relevant equation of transfer is Eq. (3.128). The basis of the S–N method as applied to this equation of transfer is extremely simple. One uses an integration quadrature scheme to approximate the integrals over the μ (angle) variable. If we consider an N point scheme, denoting the quadrature points by μ_j and the corresponding weights by w_j, we make the replacement

$$\int_{-1}^{1} d\mu' P_n(\mu') I(z, v, \mu', t) \rightarrow \sum_{-1}^{N} w_j P_n(\mu_j) I(z, v, \mu_j, t). \tag{3.223}$$

The equation of transfer then becomes

$$\frac{1}{c} \frac{\partial I(v, \mu)}{\partial t} + \mu \frac{\partial I(v, \mu)}{\partial z} + \sigma(v) I(v, \mu) = \sigma_a'(v) B(v)$$

$$+ \sum_{n=0}^{\infty} \frac{2n+1}{2} P_n(\mu) \int_0^{\infty} dv' \frac{v}{v'} \sigma_{sn}(v' \rightarrow v) \sum_{j=1}^{N} w_j P_n(\mu_j) I(v', \mu_j). \tag{3.224}$$

To obtain the S–N equations, one merely evaluates this equation at the quadrature points μ_i. That is, the Nth order discrete ordinate approximation in plane geometry consists of the N equations

$$\frac{1}{c} \frac{\partial I(v, \mu_i)}{\partial t} + \mu_i \frac{\partial I(v, \mu_i)}{\partial z} + \sigma(v) I(v, \mu_i) = \sigma_a'(v) B(v)$$

$$+ \sum_{n=0}^{\infty} \frac{2n+1}{2} P_n(\mu_i) \int_0^{\infty} dv' \frac{v}{v'} \sigma_{sn}(v' \rightarrow v) \sum_{-1}^{N} w_j P_n(\mu_j) I(v', \mu_j), \quad 1 \le i \le N, \tag{3.225}$$

for the N unknowns $I(z, v, \mu_i, t)$.

The N initial conditions required for the S–N approximation follow immediately from the initial condition on the equation of transfer, Eq. (2.31), by evaluating this condition at the quadrature points μ_i. We have

$$I(z, v, \mu_i, 0) = \Lambda(z, v, \mu_i), \quad 1 \le i \le N. \tag{3.226}$$

At the left hand face of the planar system, say $z = z_l$, we obtain the $N/2$ boundary conditions required (we assume N even with an equal number of positive and negative μ_i) by evaluating the transport boundary condition, Eq. (2.29), at the positive quadrature points. This gives

$$I(z_l, \nu, \mu_i, t) = \Gamma(z_l, \nu, \mu_i, t), \quad \mu_i > 0. \tag{3.227}$$

Similarly, at the right hand face, say $z = z_r$, we have

$$I(z_r, \nu, \mu_i, t) = \Gamma(z_r, \nu, \mu_i, t), \quad \mu_i < 0, \tag{3.228}$$

as the $N/2$ boundary conditions.

The quadrature scheme used in the S–N method is arbitrary, although it generally has quadrature points which occur in pairs, one being the negative of the other. This retains the symmetry of the exact equation of transfer and leads to an equal number of boundary conditions on each face of the system as just discussed. Since the quadrature scheme is arbitrary, it can be chosen to accurately integrate the expected angular dependence of the specific intensity. For example, if the angular dependence is highly peaked around $\mu = \pm 1$, one can use a quadrature which has its points concentrated near the endpoints of the μ range $-1 < \mu < 1$. Alternately, one could use a quadrature scheme which would integrate exactly the asymptotic distribution. The S–N method would then be an exact description in the asymptotic limit. If one has no *a priori* knowledge of the angular distribution, the usual choice is the Gauss–Legendre quadrature scheme, in which the μ_i are chosen as the zeros of the $(N+1)$th Legendre polynomial, i.e.,

$$P_{N+1}(\mu_i) = 0. \tag{3.229}$$

With this choice for the quadrature points and the associated weights it can be shown that the S–N method is closely related to the P–N method of one lower order (Lathrop, 1968) in systems with plane symmetry. Many aspects of the Gauss–Legendre S–N approximation have been examined in detail in plane geometry by Chandrasekhar (1960a).

The energy density, radiative flux, and pressure tensor according to the S–N method are obtained by again employing the same quadrature scheme used in deriving the S–N equations to perform the integrations over the μ variable. We find

$$u(z, t) = \frac{2\pi}{c} \int_0^\infty d\nu \sum_{i=1}^{N} w_i I(z, \nu, \mu_i, t), \tag{3.230}$$

$$F_z(z, t) = 2\pi \int_0^\infty d\nu \sum_{i=1}^{N} w_i \mu_i I(z, \nu, \mu_i, t), \tag{3.231}$$

$$p_{zz}(z, t) = \frac{2\pi}{c} \int_0^\infty d\nu \sum_{i=1}^{N} w_i \mu_i^2 I(z, \nu, \mu_i, t), \tag{3.232}$$

$$p_{xx}(z, t) = p_{yy}(z, t) = \frac{\pi}{c} \int_0^\infty d\nu \sum_{i=1}^{N} w_i(1 - \mu_i^2)\, I(z, \nu, \mu_i, t). \tag{3.233}$$

The other components of the radiative flux and pressure tensor are zero.

The modern S–N method in other geometries, in particular in curvilinear systems, is much more involved than it is in plane geometry. In particular, the angular quadrature is intimately associated with the finite differencing methods used to treat the spatial variable. Since such finite difference methods are outside the scope of this book, we shall make no attempt to discuss the S–N method in other geometries. For a discussion of this topic, we refer the reader to the excellent review article by Carlson and Lathrop (1968) and the references cited therein.

The P–N and S–N methods in some sense compete with each other in that both methods are rather general approximations to the equation of transfer which are capable of giving an arbitrarily small error. In practice, the S–N method is more widely used than the P–N method. This is primarily because the S–N method is more easily adaptable to large digital computers. However, in certain geometries the S–N method suffers from a defect not present in the P–N method, namely the so-called "ray effect", which distorts, in a qualitative as well as a quantitative sense, the S–N solution (Lathrop, 1968). In spite of this, the S–N method is the more popular of the two. Of course, the ray effect becomes less pronounced as N, the order of the approximation, is increased.

7. Equilibrium Diffusion Theory

Before passing on to a discussion of the multigroup method for treating the frequency variable in the equation of transfer, we consider what is generally referred to as equilibrium diffusion theory. This is a further simplification of the Eddington approximation discussed in Section 2 of this chapter, and really constitutes an approximate solution of the Eddington equations. That is, given the temperature distribution within a specified system, the equilibrium diffusion approximation provides an explicit expression giving the specific intensity as a function of all its variables, namely space, frequency, angle, and time. In particular, it gives the energy density, radiative flux, and pressure tensor as a function of space and time, as required in the equations of hydrodynamics when a radiation field is present. Although equilibrium diffusion theory corresponds to a very low order approximation in both frequency and angle, as well as in space and time, it is, because of its simplicity, a widely used approximation in radiation hydrodynamic problems. Surprisingly enough, in view of all the approximations made, it turns out to be a reasonably accurate description, giving gross features of the radiation flow correctly, in a qualitative and even a semi-quantitative sense.

We begin with the Eddington equation for the zeroth angular moment of the specific intensity, Eq. (3.20). The main assumption in this equation is that the specific intensity of radiation is almost isotropic, as expressed quantitatively by Eq. (3.1). We assume that the space and time dependences of $I_0(\mathbf{r}, \nu, t)$ are sufficiently weak that the left hand side of Eq. (3.20) can be taken as zero. This equation then becomes

$$\sigma_a'(\nu)[4\pi B(\nu) - I_0(\nu)] - \sigma_s(\nu) I_0(\nu) + \int_0^\infty d\nu' \, \frac{\nu}{\nu'} \sigma_{s0}(\nu' \to \nu) I_0(\nu')$$

$$+ \frac{c^2}{8\pi h} I_0(\nu) \int_0^\infty d\nu' \left[\frac{1}{\nu^2 \nu'} \sigma_{s0}(\nu' \to \nu) - \frac{1}{\nu'^3} \sigma_{s0}(\nu \to \nu') \right] I_0(\nu') = 0. \qquad (3.234)$$

If we further assume that the effective absorption coefficient $\sigma'_a(\mathbf{r}, \nu, t)$ and the scattering kernel $\sigma_{s0}(\mathbf{r}, \nu' \to \nu, t)$ are those appropriate to local thermodynamic equilibrium at some local temperature T, then Eq. (3.234) is identical to the equation describing the zeroth angular moment of the radiation field in complete thermodynamic equilibrium at temperature T. From thermodynamics we know the solution for the specific intensity must then be 4π times the Planck function, i.e.,

$$I_0(\mathbf{r}, \nu, t) = 4\pi B(\nu, T) = \frac{8\pi h \nu^3}{c^2} (e^{h\nu/kT} - 1)^{-1}, \tag{3.235}$$

where $T \equiv T(\mathbf{r}, t)$. Use of this result in the Eddington law of diffusion as given by Eq. (3.17) gives

$$\mathbf{I}_1(\mathbf{r}, \nu, t) = -\frac{4\pi}{3\sigma_{tr}(\mathbf{r}, \nu, t)} \nabla B(\nu, T). \tag{3.236}$$

The gradient operator acts on the Planck function through the temperature, and we can rewrite Eq. (3.236) as

$$\mathbf{I}_1(\mathbf{r}, \nu, t) = -\frac{4\pi}{3\sigma_{tr}(\mathbf{r}, \nu, t)} \frac{\partial B(\nu, T)}{\partial T} \nabla T(\mathbf{r}, t). \tag{3.237}$$

Use of Eqs. (3.235) and (3.237) in the Eddington representation of the specific intensity, Eq. (3.1), gives

$$I(\mathbf{r}, \nu, \mathbf{\Omega}, t) = B(\nu, T) - \frac{1}{\sigma_{tr}(\mathbf{r}, \nu, t)} \frac{\partial B(\nu, T)}{\partial T} \mathbf{\Omega} \cdot \nabla T(\mathbf{r}, t). \tag{3.238}$$

Equation (3.238) is the final result of the equilibrium diffusion approximation and allows an explicit calculation of the specific intensity once the temperature distribution is known.

The quantities of primary interest in radiation hydrodynamic problems are the radiative energy density, flux, and pressure tensor as defined by Eqs. (1.9), (1.16), and (1.19). In equilibrium diffusion theory these are given by, from Eq. (3.238),

$$u(\mathbf{r}, t) = \frac{4\pi}{c} \int_0^\infty d\nu B(\nu, T) = aT^4(\mathbf{r}, t), \tag{3.239}$$

$$\mathbf{F}(\mathbf{r}, t) = -\frac{4\pi}{3} \nabla T(\mathbf{r}, t) \int_0^\infty d\nu \, \frac{1}{\sigma_{tr}(\mathbf{r}, \nu, t)} \frac{\partial B(\nu, T)}{\partial T}, \tag{3.240}$$

$$p_{ii}(\mathbf{r}, t) = \frac{4\pi}{3c} \int_0^\infty d\nu B(\nu, T) = \frac{1}{3} aT^4(\mathbf{r}, t), \tag{3.241}$$

with $p_{ij}(\mathbf{r}, t) = 0$ for $i \neq j$ (i.e., the pressure tensor is diagonal). The constant a in Eqs. (3.239) and (3.241) is the radiation constant given by Eq. (1.13).

The Equations of Radiation Hydrodynamics

The expression for the radiative flux, Eq. (3.240), is frequently written in a somewhat different form. By grouping terms, this equation can be rewritten as

$$\mathbf{F}(\mathbf{r},\,t) = -\frac{4\pi}{3}\,\nabla T(\mathbf{r},\,t)\left[\int_0^\infty dv\,\frac{1}{\sigma_{\mathrm{tr}}(v)}\,\frac{\partial B(v,\,T)}{\partial T}\bigg/\int_0^\infty dv\,\frac{\partial B(v,\,T)}{\partial T}\right]\int_0^\infty dv\,\frac{\partial B(v,\,T)}{\partial T}\,. \tag{3.242}$$

If we recognize that

$$\int_0^\infty dv\,\frac{\partial B(v,\,T)}{\partial T} = \frac{ac}{\pi}\,T^3, \tag{3.243}$$

and define a mean or average (over frequency) transport coefficient $\sigma_R(\mathbf{r},\,t)$ as

$$\frac{1}{\sigma_R} = \frac{\displaystyle\int_0^\infty dv\,\frac{1}{\sigma_{\mathrm{tr}}(v)}\,\frac{\partial B(v,\,T)}{\partial T}}{\displaystyle\int_0^\infty dv\,\frac{\partial B}{\partial T}}, \tag{3.244}$$

Eq. (3.242) can be rewritten as

$$\mathbf{F}(\mathbf{r},\,t) = -\frac{ac}{3\sigma_R(\mathbf{r},\,t)}\,\nabla T^4(\mathbf{r},\,t). \tag{3.245}$$

The mean transport coefficient $\sigma_R(\mathbf{r},\,t)$ is generally referred to as the Rosseland mean, and is widely used in radiative transfer work as we shall discuss in the next section.

8. The Multigroup Method—The Planck and Rosseland Means

The generally accepted procedure for handling the frequency variable in the equation of transfer is the multigroup method, which really amounts to a discretization of the frequency variable. Rather than treating the frequency as a continuous variable, one assigns a given photon to one of G frequency groups, and all photons within a given group are treated the same, assigning average properties, such as the absorption coefficient, to these photons.

To introduce the multigroup method, let us consider the equation of transfer with no scattering, and with an absorption coefficient σ_a' which is independent of frequency. We then have

$$\frac{1}{c}\,\frac{\partial I(v,\,\boldsymbol{\Omega})}{\partial t} + \boldsymbol{\Omega}\cdot\nabla I(v,\,\boldsymbol{\Omega}) = \sigma_a'[B(v)-I(v,\,\boldsymbol{\Omega})]. \tag{3.246}$$

According to Eqs. (1.9), (1.16), and (1.19), the radiative energy density, flux, and pressure tensor, the quantities of particular interest in radiation hydrodynamic problems, all involve integrals of the specific intensity over the frequency variable. Thus it makes some sense to integrate the equation of transfer, Eq. (3.246), over frequency. We find, since σ_a' is independent of frequency,

$$\frac{1}{c}\frac{\partial I(\mathbf{\Omega})}{\partial t} + \mathbf{\Omega}\cdot\nabla I(\mathbf{\Omega}) = \sigma_a'\left[\frac{ac}{4\pi}T^4 - I(\mathbf{\Omega})\right], \tag{3.247}$$

where we have defined

$$I(\mathbf{r},\mathbf{\Omega},t) \equiv \int_0^\infty dv I(\mathbf{r},v,\mathbf{\Omega},t). \tag{3.248}$$

Equation (3.247) is exact and is referred to as the gray, or one group, equation of transfer since all photons are treated together in a single frequency group extending from $v = 0$ to $v = \infty$.

Let us again consider the equation of transfer with no scattering but with an absorption coefficient which depends upon the photon frequency, i.e.,

$$\frac{1}{c}\frac{\partial I(v,\mathbf{\Omega})}{\partial t} + \mathbf{\Omega}\cdot\nabla I(v,\mathbf{\Omega}) = \sigma_a'(v)[B(v) - I(v,\mathbf{\Omega})]. \tag{3.249}$$

The gray equation of transfer associated with Eq. (3.249) is generally taken as, in analogy to Eq. (3.247),

$$\frac{1}{c}\frac{\partial I(\mathbf{\Omega})}{\partial t} + \mathbf{\Omega}\cdot\nabla I(\mathbf{\Omega}) = \bar{\sigma}_a'\left[\frac{ac}{4\pi}T^4 - I(\mathbf{\Omega})\right], \tag{3.250}$$

where $I(\mathbf{\Omega}) \equiv I(\mathbf{r},\mathbf{\Omega},t)$ is again defined by Eq. (3.248). Here $\bar{\sigma}_a'$ is some kind of mean absorption coefficient averaged over frequency. If $\bar{\sigma}_a'$ is allowed to be a function of space and time only, as is generally the case in practice, Eq. (3.250) is an approximate equation for $I(\mathbf{r},\mathbf{\Omega},t)$ no matter what choice is made for $\bar{\sigma}_a'$. This is easily seen by integrating Eq. (3.249) over all frequency. One indeed finds a result like Eq. (3.250), but with the important difference that $\bar{\sigma}_a'$ is a function of $\mathbf{\Omega}$ as well as $\mathbf{r}$ and t. In fact, $\bar{\sigma}_a'$ must be defined as

$$\bar{\sigma}_a'(\mathbf{r},\mathbf{\Omega},t) \equiv \frac{\displaystyle\int_0^\infty dv\,\sigma_a'(\mathbf{r},v,t)[B(v,T) - I(\mathbf{r},v,\mathbf{\Omega},t)]}{\displaystyle\int_0^\infty dv[B(v,T) - I(\mathbf{r},v,\mathbf{\Omega},t)]} \tag{3.251}$$

for Eq. (3.250) to be an exact one group equation. Of course, for Eq. (3.250) to be useful, one must know $\bar{\sigma}_a'$ which, according to Eq. (3.251), involves the unknown specific intensity. (If the specific intensity were known, the radiative transfer problem would be solved.) We shall return to this point shortly. We re-emphasize at this time, however, that in practice $\bar{\sigma}_a'$ is universally chosen as independent of $\mathbf{\Omega}$, which in general implies an approximation.

With this introduction, we now consider the equation of transfer with scattering and construct the more general multigroup, rather than the one group, equations. We shall discuss the utility of the multigroup equations, even though one may only be interested in one group results (the specific intensity integrated over all frequency), following our derivation of the multigroup equations. The relevant equation of transfer is Eq. (2.170), which can be written

$$\frac{1}{c}\frac{\partial I(v,\Omega)}{\partial t}+\Omega\cdot\nabla I(v,\Omega)=\sigma_a'(v)\,[B(v)-I(v,\Omega)]$$

$$-\sigma_s(v)\,I(v,\Omega)+\int_{4\pi}d\Omega'\int_0^\infty dv'\,\frac{v}{v'}\,\sigma_s(v'\to v,\Omega'\cdot\Omega)\,I(v',\Omega')$$

$$+\frac{c^2}{2h}I(v,\Omega)\int_{4\pi}d\Omega'\int_0^\infty dv'\left[\frac{1}{v^2v'}\,\sigma_s(v'\to v,\Omega'\cdot\Omega)\right.$$

$$\left.-\frac{1}{v'^3}\,\sigma_s(v\to v',\Omega\cdot\Omega')\right]I(v',\Omega'). \tag{3.252}$$

We divide the frequency range into G groups with boundaries $v=v_0=0,\ v_1,\ v_2,\ \ldots,$ $v_{G-2},\ v_{G-1},\ v_G=\infty$, and define the gth group specific intensity as

$$I_g(\mathbf{r},\Omega,t)=\int_{v_{g-1}}^{v_g}dv I(\mathbf{r},v,\Omega,t),\qquad 1\le g\le G. \tag{3.253}$$

Integration of Eq. (3.252) over the gth group yields

$$\frac{1}{c}\frac{\partial I_g(\Omega)}{\partial t}+\Omega\cdot\nabla I_g(\Omega)=\sigma_{ag}'(\Omega)\left[b_g\frac{ac}{4\pi}T^4-I_g(\Omega)\right]$$

$$-\sigma_{sg}(\Omega)\,I_g(\Omega)+\int_{4\pi}d\Omega'\,\sigma_{s,\,g'\to g}(\Omega,\Omega')\,I_{g'}(\Omega')$$

$$+I_g(\Omega)\int_{4\pi}d\Omega'\sum_{g'=1}^{G}A_{g'\to g}(\Omega,\Omega')\,I_{g'}(\Omega'),\qquad 1\le g\le G, \tag{3.254}$$

where we have defined b_g as that fraction of $acT^4/4\pi$ lying within the gth group, i.e.,

$$b_g=\frac{\int_{v_{g-1}}^{v_g}dvB(v)}{\int_0^\infty dvB(v)}=\frac{\int_{v_{g-1}}^{v_g}dvB(v)}{\dfrac{ac}{4\pi}T^4}. \tag{3.255}$$

Equation (3.254) is exact providing we define the mean interaction coefficients as

$$\sigma_{ag}'(\Omega)=\frac{\int_{v_{g-1}}^{v_g}dv\sigma_a'(v)[B(v)-I(v,\Omega)]}{\int_{v_{g-1}}^{v_g}dv[B(v)-I(v,\Omega)]}, \tag{3.256}$$

$$\sigma_{sg}(\Omega)=\frac{\int_{v_{g-1}}^{v_g}dv\,\sigma_s(v)\,I(v,\Omega)}{\int_{v_{g-1}}^{v_g}dv I(v,\Omega)}, \tag{3.257}$$

$$\sigma_{s,\,g'\to g}(\Omega,\,\Omega') = \frac{\displaystyle\int_{\nu_{g'-1}}^{\nu_{g'}} d\nu' \int_{\nu_{g-1}}^{\nu_g} d\nu \,\frac{\nu}{\nu'}\,\sigma_s(\nu' \to \nu,\,\Omega\cdot\Omega')\,I(\nu',\,\Omega')}{\displaystyle\int_{\nu_{g'-1}}^{\nu_{g'}} d\nu'\,I(\nu',\,\Omega')}\,,\qquad(3.258)$$

$A_{g'\to g}(\Omega,\,\Omega')$

$$= \frac{\displaystyle\int_{\nu_{g'-1}}^{\nu_{g'}} d\nu' \int_{\nu_{g-1}}^{\nu_g} d\nu \,\frac{c^2}{2h\nu'}\left[\frac{1}{\nu^2}\,\sigma_s(\nu' \to \nu,\,\Omega'\cdot\Omega)-\frac{1}{\nu'^2}\,\sigma_s(\nu \to \nu',\,\Omega\cdot\Omega')\right]I(\nu',\,\Omega')\,I(\nu,\,\Omega)}{\displaystyle\int_{\nu_{g'-1}}^{\nu_{g'}} d\nu' \int_{\nu_{g-1}}^{\nu_g} d\nu\,I(\nu',\,\Omega')\,I(\nu,\,\Omega)}\,.$$

$$(3.259)$$

The G equations given by Eq. (3.254) represent the general form of the multigroup equations. These equations are coupled through the scattering interaction and must be solved simultaneously.

For the multigroup equations to be useful, one must be able to compute or estimate the group constants defined by Eqs. (3.256) through (3.259). An exact calculation of these constants involves a complete knowledge of the specific intensity which, of course, is unknown. The underlying assumption in the multigroup method is that these group constants, since they are homogeneous functionals of the specific intensity, are relatively insensitive to the weighting function $I(\mathbf{r},\,\nu,\,\Omega,\,t)$ used in computing these averages over frequency. Hence one hopes that a relatively crude estimate for the specific intensity will lead to reasonably accurate group constants. As the group width approaches zero, of course, the group constants become increasingly less dependent upon the estimate made for $I(\mathbf{r},\,\nu,\,\Omega,\,t)$. This is the reason that a multigroup formulation of the frequency variable is preferable to a one group, or gray, treatment, even though the ultimate goal may be one group results (energy density, flux, and pressure tensor).

To evaluate the group constants involving the scattering kernel, Eqs. (3.257) through (3.259), the specific intensity is generally taken as the Planck function at the local material temperature. This insures correctness as one approaches equilibrium; i.e., at thermodynamic equilibrium the specific intensity is in fact given by the Planck function. Away from equilibrium the only justification for the use of the Planck function is that the scattering coefficient is a relatively smooth function of frequency, and hence the choice of the weighting function in these group constants is not crucial as long as a reasonable function is used. In the case of σ'_{ag}, the average absorption coefficient, more care should be taken. Absorption coefficients encountered in practice are generally complex and widely varying functions of frequency, and the use of different weighting functions can lead to quite different results. In practice, σ'_{ag} is generally taken as either a group Rosseland or group Planck mean.

The Rosseland mean, similar to that introduced in the last section, follows from the assumption that the specific intensity is given by the equilibrium diffusion approximation, i.e.,

$$B(v, T) - I(\mathbf{r}, v, \mathbf{\Omega}, t) = -\frac{1}{\sigma_{\mathrm{tr}}(\mathbf{r}, v, t)} \frac{\partial B(v, T)}{\partial T} \mathbf{\Omega} \cdot \nabla T(\mathbf{r}, t).$$

(3.260)

Use of this result in Eq. (3.256) yields the Rosseland result

$$\sigma'_{ag} = \frac{\displaystyle\int_{v_{g-1}}^{v_g} dv \, \frac{\sigma'_a(v)}{\sigma_{\mathrm{tr}}(v)} \frac{\partial B(v, T)}{\partial T}}{\displaystyle\int_{v_{g-1}}^{v_g} dv \, \frac{1}{\sigma_{\mathrm{tr}}(v)} \frac{\partial B(v, T)}{\partial T}}.$$

(3.261)

In the absence of scattering we have $\sigma'_a(v) = \sigma_{\mathrm{tr}}(v)$, and Eq. (3.261) reduces to a group Rosseland mean of the form given by Eq. (3.244) of the last section.

The Planck mean is appropriate in the case of time independent radiative transfer in an optically thin, emission dominated system. To show this, we consider the time independent integral equation of transfer given by Eq. (2.78). In the special case of no radiation impinging upon the surface of the system ($\Gamma = 0$), which is what is meant by emission dominated, we have

$$I(\mathbf{r}, \mathbf{\Omega}) = \int_0^{|\mathbf{r} - \mathbf{r}_s|} ds' Q(\mathbf{r} - s'\mathbf{\Omega}, \mathbf{\Omega}) \exp\left[-\int_0^{s'} ds'' \sigma(\mathbf{r} - s''\mathbf{\Omega}) \right].$$

(3.262)

The frequency variable in Eqs. (2.78) and (3.262) has been suppressed for ease of notation. From Eq. (3.262) it follows that

$$I(\mathbf{r}, \mathbf{\Omega}) \le \int_0^{|\mathbf{r} - \mathbf{r}_s|} ds' Q(\mathbf{r} - s'\mathbf{\Omega}, \mathbf{\Omega}) = \int_0^{|\mathbf{r} - \mathbf{r}_s|} ds' \left[\frac{Q(\mathbf{r} - s'\mathbf{\Omega}, \mathbf{\Omega})}{\sigma(\mathbf{r} - s'\mathbf{\Omega})} \right] \sigma(\mathbf{r} - s'\mathbf{\Omega}),$$

(3.263)

and hence

$$I(\mathbf{r}, \mathbf{\Omega}) \le \left[\frac{Q}{\sigma} \right]_{\max} \tau(\mathbf{r}, \mathbf{r}_s).$$

(3.264)

Here $[Q/\sigma]_{\max}$ is the maximum value of Q/σ along the straight line path $\mathbf{r} - s'\mathbf{\Omega}$, and $\tau(\mathbf{r}, \mathbf{r}_s)$ is the optical depth between $\mathbf{r}$, the point in question, and $\mathbf{r}_s$, the surface point, as defined by Eq. (2.79). From the definition of Q, Eq. (2.69), it is clear that Q/σ is a bounded function, and hence Eq. (3.264) shows that $I(\mathbf{r}, \mathbf{\Omega})$ approaches zero as $\tau(\mathbf{r}, \mathbf{r}_s)$ goes to zero, i.e., as the system becomes optically thin. In the optically thin case we then have

$$B(v, T) - I(\mathbf{r}, v, \mathbf{\Omega}) \approx B(v, T),$$

(3.265)

and Eq. (3.256) gives

$$\sigma'_{ag} = \frac{\displaystyle\int_{v_{g-1}}^{v_g} dv \sigma'_a(v) \, B(v, T)}{\displaystyle\int_{v_{g-1}}^{v_g} dv B(v, T)}.$$

(3.266)

This type of average is generally referred to as the Planck mean, or Planck averaged, absorption coefficient.

As the above discussion indicates, the use of the Rosseland and Planck mean absorption coefficients is only strictly appropriate in limiting circumstances. Nevertheless, one or the other mean is generally used in the multigroup method. For realistic absorption coefficients, these two means can differ by an order of magnitude or more, and thus the results of the multigroup method can vary widely depending upon which mean is used. In truth, neither mean is correct in general. In the next section we present a new multigroup method which involves generalizations of the Planck and Rosseland means which occur together in the multigroup equations. As we shall see, this assures the proper behavior in both the equilibrium diffusion and optically thin limits.

For completeness, we give the expressions for the radiative energy density, radiative flux, and radiative pressure tensor in the multigroup approximation. From Eqs. (1.9), (1.16), and (1.19) we have

$$u(\mathbf{r}, t) = \frac{1}{c} \sum_{g=1}^{G} \int_{4\pi} d\Omega I_g(\mathbf{r}, \Omega, t), \tag{3.267}$$

$$\mathbf{F}(\mathbf{r}, t) = \sum_{g=1}^{G} \int_{4\pi} d\Omega \Omega I_g(\mathbf{r}, \Omega, t), \tag{3.268}$$

$$p(\mathbf{r}, t) = \frac{1}{c} \sum_{g=1}^{G} \int_{4\pi} d\Omega \Omega \Omega I_g(\mathbf{r}, \Omega, t). \tag{3.269}$$

9. A New Multigroup Method

As was pointed out in the last section, the use of either the Planck or Rosseland mean in the multigroup equations is incorrect except in limiting cases. In this section we derive a multigroup approximation which attempts to correct this defect of the classical multigroup method. In particular, we will show that in a certain sense the Planck and Rosseland means are only appropriate if the temperature is a weak function of space, and we shall derive generalizations of these means to strong temperature gradients. Secondly, and probably more important, the multigroup equations we shall develop here involve both the generalized Planck and Rosseland means simultaneously. This allows a multigroup method which behaves properly in both the equilibrium diffusion and optically thin limits.

For simplicity, we consider the time independent equation of transfer in plane geometry with no scattering. The generalization of this method to include scattering in general geometry should be straightforward, following the so-called B–N method used in neutron transport theory (Hurwitz and Zweifel, 1955; Wilkins *et al.*, 1955; Pomraning, 1965d). We also restrict the discussion to the development of a gray, or one group, equation of transfer, but the extension of the method to the multigroup equations will be self-evident.

The equation of transfer we shall consider is, then,

$$\mu \frac{\partial I(\nu, \mu)}{\partial z} = \sigma(\nu) \left[B(\nu, T) - I(\nu, \mu) \right], \tag{3.270}$$

93

where $\sigma(z, \nu) = \sigma'_a(z, \nu)$ is the absorption coefficient (corrected as always for the induced emission). We integrate this equation over all frequency, defining separate average coefficients for emission and absorption. We find

$$\mu \frac{\partial I(\mu)}{\partial z} = \sigma_P \frac{ac}{4\pi} T^4 - \bar{\sigma}(\mu)\, I(\mu), \qquad (3.271)$$

where $I(z, \mu)$ is the one group, or frequency integrated, specific intensity

$$I(z, \mu) = \int_0^\infty d\nu I(z, \nu, \mu). \qquad (3.272)$$

Here $\sigma_P(z)$ is the Planck mean given by

$$\sigma_P(z) = \frac{\displaystyle\int_0^\infty d\nu \sigma(z, \nu)\, B(\nu, T)}{\displaystyle\int_0^\infty d\nu B(\nu, T)}, \qquad (3.273)$$

and $\bar{\sigma}(z, \mu)$ is an average absorption coefficient defined as

$$\bar{\sigma}(z, \mu) = \frac{\displaystyle\int_0^\infty d\nu \sigma(\nu)\, I(z, \nu, \mu)}{\displaystyle\int_0^\infty d\nu I(z, \nu, \mu)}. \qquad (3.274)$$

The frequency independent equation of transfer, Eq. (3.271), is exact. However, to compute $\bar{\sigma}(z, \mu)$ one needs know $I(z, \nu, \mu)$, which, of course, is presumed unknown. Hence in practice one must replace $I(z, \nu, \mu)$ in Eq. (3.274) by a weight function $I(z, \nu, \mu)$ which should be an approximation in some sense to $I(z, \nu, \mu)$. The question of what is an appropriate weight function constitutes the entire gray opacity, or mean absorption coefficient, problem. It is important to note that in Eq. (3.271) we have introduced two mean absorption coefficients, σ_P and $\bar{\sigma}$, whereas in the multigroup method of the last section a single mean coefficient was used. If $I(z, \nu, \mu)$ were known exactly, then both treatments are exact, and hence equivalent. In practice, of course, one must use an approximation to $I(z, \nu, \mu)$ for the purpose of computing the mean absorption coefficients, and then Eq. (3.271), containing two means, has an advantage; namely, it is an exact equation in the limit of an optically thin system. In this case, $I(z, \mu)$ is much less than $acT^4/4\pi$, as was shown in the last section, and hence the choice made for $\bar{\sigma}$ is irrelevant. The only important thing is that the emission coefficient be given by the Planck mean, and this is the case in Eq. (3.271) no matter what choice is made for $\bar{\sigma}$.

However, we note that in general $\bar{\sigma}$ depends upon angle, which detracts from the appeal of this result as a gray equation of transfer. Let us therefore consider an alternate to Eq. (3.271). As an intermediate step in the development, we expand the specific intensity in Eq. (3.270) in Legendre polynomials according to

$$I(z, \nu, \mu) = \sum_{n=0}^\infty \frac{2n+1}{4\pi} I_n(z, \nu)\, P_n(\mu). \qquad (3.275)$$

This yields the infinite set of equations

$$\frac{\partial I_1(\nu)}{\partial z} = \sigma(\nu)[4\pi B(\nu) - I_0(\nu)], \tag{3.276}$$

$$n\frac{\partial I_{n-1}(\nu)}{\partial z} + (2n+1)\,\sigma(\nu)\,I_n(\nu) + (n+1)\frac{\partial I_{n+1}(\nu)}{\partial z} = 0, \qquad n \geq 1. \tag{3.277}$$

We define the gray moments as

$$I_n(z) = \int_0^\infty d\nu I_n(z,\,\nu). \tag{3.278}$$

Integration of Eqs. (3.276) and (3.277) over all frequency yields the gray moment equations

$$\frac{dI_1}{\partial z} = \sigma_P ac T^4 - \sigma_0 I_0, \tag{3.279}$$

$$n\frac{dI_{n-1}}{dz} + (2n+1)\,\sigma_n I_n + (n+1)\frac{dI_{n+1}}{dz} = 0, \qquad n \geq 1. \tag{3.280}$$

Here $\sigma_P(z)$ is again the Planck mean defined by Eq. (3.273), and the $\sigma_n(z)$ are given by

$$\sigma_n(z) = \frac{\displaystyle\int_0^\infty d\nu\sigma(z,\,\nu)\,I_n(z,\,\nu)}{\displaystyle\int_0^\infty d\nu I_n(z,\,\nu)}. \tag{3.281}$$

As before, since the $I_n(z,\,\nu)$ are not known one must use an approximation, say $\tilde{I}_n(z,\,\nu)$, in Eq. (3.281) to obtain the moment absorption coefficients.

To proceed further, we note that the $I_n(z)$ defined by Eqs. (3.279) and (3.280) represent a converging sequence as a function of n. That is, for n large enough $I_n(z)$ is arbitrarily small. If one sets to zero $I_n(z)$ for $n > N$, Eqs. (3.279) and (3.280) represent the P–N or spherical harmonic approximation discussed in Section 5 of this chapter. In the present context we do not envision making the P–N approximation. We shall, however, use the known convergence of the P–N method to argue in the following way. Since for large enough n the $I_n(z)$ are arbitrarily small, their coefficients in Eq. (3.280), namely $\sigma_n(z)$, are unimportant. That is, as long as $\sigma_n(z)$ is a number of reasonable magnitude, its precise value is irrelevant for large enough n. In view of this, we replace $\sigma_n(z)$ in Eq. (3.280) by $\sigma^*(z)$ for $n > N$. That is, from $n = N+1$ to $n = \infty$ we use a common mean absorption coefficient, independent of n. At present we shall not specify $\sigma^*(z)$. The important point here is that $\sigma^*(z)$ is completely arbitrary for N large enough. One might also note that in practice, even without the supporting argument just made, one must stop calculating the $\sigma_n(z)$ at some point, say $n = N$, or else Eq. (3.281) implies an infinite number of calculations.

With this introduction of $\sigma^*(z)$, Eqs. (3.279) and (3.280) become

$$\frac{dI_1}{dz} = \sigma_P ac T^4 - \sigma_0 I_0, \tag{3.282}$$

$$n\frac{dI_{n-1}}{dz} + (2n+1)\,\sigma_n I_n + (n+1)\frac{dI_{n+1}}{dz} = 0, \qquad 1 \leq n \leq N, \tag{3.283}$$

$$n\frac{dI_{n-1}}{dz} + (2n+1)\,\sigma^* I_n + (n+1)\frac{dI_{n+1}}{dz} = 0, \qquad n > N. \tag{3.284}$$

The Equations of Radiation Hydrodynamics

Now, it is easily verified that Eqs. (3.282) through (3.284) are the Legendre moment equations corresponding to the gray equation of transfer

$$\mu \frac{\partial I(\mu)}{\partial z} = \sigma_P \frac{ac}{4\pi} T^4 - \sigma^* I(\mu) + \int_{4\pi} d\Omega' \sigma_e(\mathbf{\Omega} \cdot \mathbf{\Omega}') \, I(\mu'). \tag{3.285}$$

Here $\sigma_e(z, \mathbf{\Omega} \cdot \mathbf{\Omega}')$ is an effective scattering kernel given by

$$\sigma_e(z, \mathbf{\Omega} \cdot \mathbf{\Omega}') = \sum_{n=0}^{N} \frac{2n+1}{4\pi} \, [\sigma^*(z) - \sigma_n(z)] \, P_n(\mathbf{\Omega} \cdot \mathbf{\Omega}'). \tag{3.286}$$

Once $\sigma^*(z)$ and the $\sigma_n(z)$ have been specified, points we shall consider shortly, Eq. (3.285) is the final result in the reduction from the multifrequency to a gray equation of transfer. Of course, one must also specify N, which should be based on a compromise between simplicity (N small) and accuracy (N large).

Concerning the choice of $\sigma^*(z)$, we recall that it is completely arbitrary if N is large enough. In practice, however, N is likely to be chosen as a quite small integer, so the choice of $\sigma^*(z)$ is of some importance. Two reasonable choices for $\sigma^*(z)$ are

$$\sigma^*(z) = \sigma_0(z), \tag{3.287}$$

and

$$\sigma^*(z) = \sigma_N(z). \tag{3.288}$$

The rational behind Eq. (3.287) is that this choice makes the $n = 0$ (isotropic) component of the effective scattering kernel disappear, and hence the reduction to the gray equation introduces a scattering-like term with only higher angular components. This seems reasonable since one expects the nongray effects associated with frequency dependent transport to be most important in highly nondiffusion problems, i.e., in problems with a strong angular dependence. The reasoning leading to Eq. (3.288) is that $\sigma^*(z)$ is meant as an approximation to $\sigma_n(z)$ for $n > N$, and one might expect that $\sigma_N(z)$ would be a closer approximation than $\sigma_0(z)$, simply because the index is larger. We shall show, in fact, that this is indeed so in a certain limiting case. These two choices for $\sigma^*(z)$ have also been suggested by Bell *et al.* (1967) in a related development of the S–N (discrete ordinate) equations in neutron transport theory.

As the simplest example of our considerations thus far, let us set $N = 1$. Equation (3.285) becomes

$$\mu \frac{\partial I(\mu)}{\partial z} = \sigma_P \frac{ac}{4\pi} T^4 - \sigma^* I(\mu) + \frac{1}{2} (\sigma^* - \sigma_0) \int_{-1}^{1} d\mu' I(\mu') + \frac{3}{2} \mu (\sigma^* - \sigma_1) \int_{-1}^{1} d\mu' \, \mu' I(\mu'). \tag{3.289}$$

In particular, with $\sigma^* = \sigma_0$ Eq. (3.289) becomes

$$\mu \frac{\partial I(\mu)}{\partial z} = \sigma_P \frac{ac}{4\pi} T^4 - \sigma_0 I(\mu) + \frac{3}{2} \mu (\sigma_0 - \sigma_1) \int_{-1}^{1} d\mu' \, \mu' I(\mu'), \tag{3.290}$$

96

whereas $\sigma^* = \sigma_1$ gives

$$\mu\frac{\partial I(\mu)}{\partial z} = \sigma_P\frac{ac}{4\pi}T^4 - \sigma_1 I(\mu) - \frac{1}{2}(\sigma_0 - \sigma_1)\int_{-1}^{1} d\mu' I(\mu'). \tag{3.291}$$

With σ_0 and σ_1 chosen as we shall discuss shortly, either Eq. (3.290) or (3.291) appears to be more accurate than the conventional gray equation, discussed in the last section, using the Planck or Rosseland mean (Pomraning, 1971). Further, Eq. (3.291) seems to be the more accurate of the two.

We now turn to the final problem of estimating the moment absorption coefficients, $\sigma_n(z)$. If one assumes an equilibrium diffusion representation for the specific intensity, i.e.,

$$I(z, v, \mu) = B(v, T) - \frac{\mu}{\sigma(z, v)}\mu\frac{\partial B(v, T)}{\partial T}\frac{dT(z)}{dz}, \tag{3.292}$$

one has

$$I_0(z, v) = 4\pi B(v, T), \tag{3.293}$$

$$I_1(z, v) = -\frac{4\pi}{3\sigma(z, v)}\frac{\partial B(v, T)}{\partial T}\frac{dT(z)}{dz}. \tag{3.294}$$

Equation (3.281) then gives

$$\sigma_0(z) = \frac{\displaystyle\int_0^\infty dv\sigma(v)\,B(v, T)}{\displaystyle\int_0^\infty dv B(v, T)}, \tag{3.295}$$

$$\sigma_1(z) = \frac{\displaystyle\int_0^\infty dv\,\frac{\partial B(v, T)}{\partial T}}{\displaystyle\int_0^\infty dv\,\frac{1}{\sigma(v)}\frac{\partial B(v, T)}{\partial T}}. \tag{3.296}$$

That is, in this limit σ_0 is just the Planck mean and σ_1 is the Rosseland mean. Since equilibrium diffusion theory is correct in the limit of a zero temperature gradient, Eqs. (3.295) and (3.296) imply that the use of the Planck and Rosseland means are only strictly appropriate if the temperature is a weak function of space. We now undertake the task of generalizing these means, as well as estimating σ_n for $n > 1$, to the case of strong temperature gradients.

For this purpose, we construct an asymptotic solution to the multifrequency equation of transfer, Eq. (3.270). We consider a system occupying $0 < z < a$ and assume no incident radiation impinging on either surface (this assumption is not necessary, but simplifies the analysis somewhat). Then

$$I(0, v, \mu) = 0, \qquad \mu > 0, \tag{3.297}$$

$$I(a, v, \mu) = 0, \qquad \mu < 0. \tag{3.298}$$

In Section 3 of this chapter, an asymptotic solution to the equation of transfer was obtained

The Equations of Radiation Hydrodynamics

by dealing with the differential form of the equation, i.e., Eq. (3.270) in this case. Here, however, we choose to first cast Eq. (3.270), together with the boundary conditions Eqs. (3.297) and (3.298), into integral form. Working with the integral equation has the advantage that it is clear what is neglected in constructing an asymptotic solution, namely edge effects.

The integral equation corresponding to Eqs. (3.270), (3.297), and (3.298) can be obtained as a special case of Eq. (2.78) or by integrating Eq. (3.270) directly. In either event, the result is

$$I(\tau, \nu, \mu) = \int_0^\tau d\tau' \frac{B(\tau')}{\mu} e^{(\tau'-\tau)/\mu}, \qquad \mu > 0, \tag{3.299}$$

$$I(\tau, \nu, \mu) = -\int_\tau^b d\tau' \frac{B(\tau')}{\mu} e^{(\tau'-\tau)/\mu}, \qquad \mu < 0. \tag{3.300}$$

Here we have defined the optical depth τ as

$$\tau \equiv \int_0^z dz' \sigma(z', \nu), \tag{3.301}$$

and

$$b \equiv \int_0^a dz' \sigma(z', \nu). \tag{3.302}$$

Obviously both τ and b are functions of frequency. If we concentrate on a depth τ far from the edges of the system (we seek an asymptotic solution), i.e.,

$$1 \ll \tau; \quad 1 \ll (b-\tau), \tag{3.303}$$

and expand $B(\tau')$ in Eqs. (3.299) and (3.300) in a Taylor series about $\tau' = \tau$, then Eqs. (3.299) and (3.300) lead to, upon integrating term by term,

$$I(\tau, \nu, \mu) = \left[\frac{1}{1+\mu D}\right] B(\tau), \quad \text{all} \quad \mu, \tag{3.304}$$

where the operator D is defined by

$$D = \partial/\partial\tau. \tag{3.305}$$

The operator $(1+\mu D)^{-1}$ in Eq. (3.304) is just a shorthand notation for its Taylor series expansion.

By algebraic manipulation, Eq. (3.304) can be rewritten as

$$I(\tau, \nu, \mu) = (1-\mu D)\left[1+\frac{\mu^2 D^2}{1-\mu^2 D^2}\right] B(\tau). \tag{3.306}$$

Equilibrium diffusion theory, discussed in Section 7 of this chapter, corresponds to neglecting the $\mu^2 D^2/(1-\mu^2 D^2)$ operator in Eq. (3.306). Physically this means that the temperature is a sufficiently weak function of the spatial variable that one can neglect its second and higher derivatives. In the present treatment we relax this restriction and assume, for the

98

purposes of treating this operator, that $B(\tau)$ in the vicinity of a particular τ can be represented by an arbitrary linear combination of a growing and a decaying exponential in τ (or as a sum of a sine and cosine in τ). That is, we assume that $B(\tau)$ locally satisfies a Helmholtz equation

$$D^2 B(\tau) = \alpha^2 B(\tau), \tag{3.307}$$

where $\alpha^2 > 0$ corresponds to exponentials and $\alpha^2 < 0$ corresponds to trigonometric functions. If we further assume that α^2 is a slowly varying function of space, then, by successive differentiations,

$$D^{2n} B(\tau) = \alpha^{2n} B(\tau), \qquad n = 1, 2, \ldots, \tag{3.308}$$

and Eq. (3.306) becomes

$$I(\tau, \nu, \mu) = \left[\frac{1}{1 - \mu^2 \alpha^2} \right] (1 - \mu D)\, B(\tau). \tag{3.309}$$

The special case $\alpha = 0$ corresponds to the equilibrium diffusion approximation. It should by noted that Eq. (3.309) only makes sense for $\alpha^2 < 1$. That is, an asymptotic solution onle exists for these restricted values of α^2. This is connected with the neglect of surface terms in passing from Eqs. (3.299) and (3.300) to Eq. (3.304). If the source $B(\tau)$ grows faster than e^τ for any frequency, one cannot neglect surface terms as we have done, and if surface effects cannot be neglected an asymptotic solution is nonextant. Reintroducing the variable z rather than τ, we have from Eq. (3.309)

$$I(z, \nu, \mu) = \left(\frac{1}{1 - \mu^2 \alpha^2} \right) \left(1 - \frac{\mu}{\sigma} \frac{\partial T}{\partial z} \frac{\partial}{\partial T} \right) B(\nu, T). \tag{3.310}$$

Equation (3.310) is our final result for the asymptotic distribution, except that we need specify α, which in general is a function of space and frequency, although the spatial dependence is assumed weak.

We now turn to this task; namely, given an arbitrary temperature distribution, we seek α. From Eq. (3.307) we find

$$\alpha^2 = \left(\frac{1}{B} \frac{\partial B}{\partial T} \right) \frac{\partial^2 T}{\partial \tau^2} + \left(\frac{1}{B} \frac{\partial^2 B}{\partial T^2} \right) \left(\frac{\partial T}{\partial \tau} \right)^2. \tag{3.311}$$

If we assume σ is a slowly varying function of space, which is consistent with our prior assumption that α is a slowly varying function of space, Eq. (3.311) becomes

$$\sigma^2 \alpha^2 = \left(\frac{1}{B} \frac{\partial B}{\partial T} \right) \frac{\partial^2 T}{\partial z^2} + \left(\frac{1}{B} \frac{\partial^2 B}{\partial T^2} \right) \left(\frac{dT}{dz} \right)^2. \tag{3.312}$$

The reason we desire the asymptotic distribution, and hence α, is for use in computing frequency averaged absorption coefficients. This presupposes that σ as a function of frequency is rapidly varying. Otherwise, any reasonable average such as the Planck or Rosseland mean would be acceptable, and there would be no need for any of the present analysis. Accordingly, we argue that the main frequency variation of α in Eq. (3.312) comes from σ, not B. This is particularly true if one is forming multigroup absorption coefficients. Hence we write Eq. (3.312) as

$$\sigma^2\alpha^2 = \text{constant in frequency} = \varepsilon^2, \tag{3.313}$$

and our asymptotic solution, Eq. (3.310), becomes

$$I(z,\ \nu,\ \mu) = \left(\frac{1}{1-\mu^2\varepsilon^2/\sigma^2}\right)\left(1-\frac{\mu}{\sigma}\frac{dT}{dz}\frac{\partial}{\partial T}\right)B(\nu,\ T). \tag{3.314}$$

We determine ε by returning to Eq. (3.307) which, with Eq. (3.313) and the assumption that σ is a slowly varying function of space, yields

$$\frac{\partial^2 B(T,\ \nu)}{\partial z^2} = \varepsilon^2 B(T,\ \nu). \tag{3.315}$$

Integrating Eq. (3.315) over all frequency, we find

$$\varepsilon^2 = \frac{1}{T^4}\frac{\partial^2 T^4}{\partial z^2}. \tag{3.316}$$

Equation (3.314), with ε^2 given by Eq. (3.316), constitutes our final result for the asymptotic solution to the equation of transfer.

The motivation in calculating this asymptotic result was that it would constitute $I(z,\ \nu,\ \mu)$, the approximate solution for the specific intensity needed to compute the moment absorption coefficients defined by Eq. (3.281). Forming the Legendre moments of Eq. (3.314), we find that the $I_n(z,\ \nu)$ required in Eq. (3.281) are given by

$$I_{2n}(z,\ \nu) = 2\pi B(\nu,\ T)\int_{-1}^{1} d\mu P_{2n}(\mu)\left(\frac{1}{1-\mu^2\varepsilon^2/\sigma^2}\right), \qquad n = 0,\ 1,\ \dots, \tag{3.317}$$

$$I_{2n+1}(z,\ \nu) = -2\pi\frac{dT}{dz}\frac{\partial B(\nu,\ T)}{\partial T}\frac{1}{\sigma}\int_{-1}^{1} d\mu P_{2n+1}(\mu)\left(\frac{\mu}{1-\mu^2\varepsilon^2/\sigma^2}\right), \qquad n = 0,\ 1,\ \dots. \tag{3.318}$$

If we define

$$A_n(\xi) = \frac{1}{2}\int_{-1}^{1} d\mu P_n(\mu)\left(\frac{1}{1-\mu\xi}\right), \tag{3.319}$$

it is easily shown that the use of Eqs. (3.317) and (3.318) in Eq. (3.281) gives

$$\sigma_n(z) = \frac{\displaystyle\int_{0}^{\infty} d\nu\sigma(z,\ \nu)\,w_n(\nu,\ T)A_n(\varepsilon/\sigma)}{\displaystyle\int_{0}^{\infty} d\nu w_n(\nu,\ T)A_n(\varepsilon/\sigma)}, \tag{3.320}$$

where

$$w_n(\nu,\ T) = \begin{cases} B(\nu,\ T), & n\ \text{even}, \\ \partial B(\nu,\ T)/\partial T, & n\ \text{odd}. \end{cases} \tag{3.321}$$

An explicit calculation easily shows that $A_0(\xi)$ is given by

$$A_0(\xi) = \frac{1}{2\xi} \ln \left(\frac{1+\xi}{1-\xi}\right). \tag{3.322}$$

The $A_n(\xi)$ for $n \geq 1$ follow from a recurrence relation which is easily derived from the recurrence relation for Legendre polynomials, Eq. (3.218). The result is

$$A_{n+1}(\xi) = \left(\frac{2n+1}{n+1}\right) \frac{1}{\xi} A_n(\xi) - \left(\frac{n}{n+1}\right) A_{n-1}(\xi) - \frac{1}{\xi} \delta_{n0}, \quad n \geq 0, \tag{3.323}$$

where δ_{nm} is the Kronecker delta function.

If ε^2 is negative, then ε is pure imaginary and we set $\varepsilon = i\delta$. Let us define

$$\bar{A}_n(\eta) = i^n A_n(i\eta). \tag{3.324}$$

We have

$$\bar{A}_0(\eta) = \frac{1}{\eta} \tan^{-1}(\eta), \tag{3.325}$$

and

$$\bar{A}_{n+1}(\eta) = \left(\frac{2n+1}{n+1}\right) \frac{1}{\eta} \bar{A}_n(\eta) + \left(\frac{n}{n+1}\right) \bar{A}_{n-1}(\eta) - \frac{1}{\eta} \delta_{n0}, \quad n \geq 0. \tag{3.326}$$

It is clear from Eqs. (3.325) and (3.326) that $A_n(\eta)$ is real if η is real. Thus for ε^2 negative, Eq. (3.320) becomes

$$\sigma_n(z) = \frac{\displaystyle\int_0^\infty dv\sigma(z, v)\, w_n(v, T)\, \bar{A}_n(\delta/\sigma)}{\displaystyle\int_0^\infty dv w_n(v, T)\, \bar{A}_n(\delta/\sigma)}, \tag{3.327}$$

with $w_n(v, T)$ still given by Eq. (3.321).

Let us consider the limit of small ε (or small δ). It is easily shown that

$$A_n(\xi) \xrightarrow[\xi \to 0]{} \frac{(n!)^2 2^n}{(2n+1)!} \xi^n, \tag{3.328}$$

and Eq. (3.320) then gives

$$\sigma_n = \frac{\displaystyle\int_0^\infty dv w_n(v, T)\sigma^{1-n}}{\displaystyle\int_0^\infty dv w_n(v, T)\sigma^{-n}}, \quad n = 0, 1, \ldots. \tag{3.329}$$

In particular, in this limit σ_0 is the Planck mean σ_P, and σ_1 is the Rosseland mean σ_R, i.e.,

$$\sigma_0(\varepsilon = 0) = \sigma_P = \frac{\displaystyle\int_0^\infty dv\sigma(v)\, B(v, T)}{\displaystyle\int_0^\infty dv B(v, T)}, \tag{3.330}$$

$$\sigma_1(\varepsilon = 0) = \sigma_R = -\frac{\displaystyle\int_0^\infty dv\,\frac{\partial B(v, T)}{\partial T}}{\displaystyle\int_0^\infty dv\,\frac{1}{\sigma(v)}\frac{\partial B(v, T)}{\partial T}}. \tag{3.331}$$

Because of the limiting behavior of σ_1 to the Rosseland mean, one can easily show that the gray equation of transfer, Eq. (3.285), upon making the equilibrium diffusion approximation, will give the correct diffusion relationship, i.e., the mean involved will be the Rosseland mean. As was argued earlier, Eq. (3.285) is also correct in the optically thin, emission dominated, case since the emission coefficient is the Planck mean. Since σ_0 and σ_1 limit to the Planck and Rosseland means in the limit of a zero temperature gradient, these quantities can be considered as generalizations, to nonzero temperature gradients, of these classical frequency averaged absorption coefficients.

One can show, in this limit of small ε, that the σ_n defined by Eq. (3.329) obey a certain inequality. We use the Schwartz inequality

$$(\varphi, \varphi)\,(\psi, \psi) \geq (\varphi, \psi)^2, \tag{3.332}$$

where φ and ψ are arbitrary functions of frequency and the parentheses indicate the inner product over the interval $0 < v < \infty$. We set

$$\varphi = \sqrt{\sigma^{2-n}w}\,; \quad \psi = \sqrt{\sigma^{-n}w'}, \tag{3.333}$$

and Eq. (3.332) gives

$$\frac{(\sigma^{2-n}, w)}{(\sigma^{1-n}, \sqrt{ww'})} \geq \frac{(\sigma^{1-n}, \sqrt{ww'})}{(\sigma^{-n}, w')}. \tag{3.334}$$

If we assume that the frequency dependences of w and w' (i.e., of B and $\partial B/\partial T$) are weak compared to that of σ, then the choice of w and w' is irrelevant and we can take $w = w'$ in Eq. (3.334). This gives

$$\sigma_{n-1} \geq \sigma_n. \tag{3.335}$$

In particular, for $n = 1$ this gives the well known result that $\sigma_P \geq \sigma_R$.

As an example of these generalized Planck and Rosseland means, let us take $\sigma(v)$ as the Elsasser band, which corresponds to an infinite array of equally spaced, equal strength Lorentz lines. We have (Elsasser, 1938)

$$\sigma(v) = \frac{S}{d}\frac{\sinh (2\pi h/d)}{[\cosh (2\pi h/d) - \cos (2\pi v/d)]}. \tag{3.336}$$

Here S and h are the total area and halfwidth at half maximum, respectively, of a single line, and d is the spacing between line centers. We choose a distance scale such that the minimum value of $\sigma(v)$ over v is unity. Then

$$\sigma(v) = \frac{\cosh \beta + 1}{\cosh \beta - \cos (2\pi v/d)}, \tag{3.337}$$

where $\beta = 2\pi h/d$. We further take the spacing d sufficiently small so that B and $\partial B/\partial T$ are in effect constant in frequency over a distance d. Then, in the limit of small ε, in which case Eq. (3.329) applies, we find

$$\sigma_n(\varepsilon = 0) = (\cosh \beta + 1) \frac{\int_0^{2\pi} dx(\cosh \beta - \cos x)^{n-1}}{\int_0^{2\pi} dx(\cosh \beta - \cos x)^n}. \tag{3.338}$$

These integrals are easily evaluated and we find

$$\sigma_0(\varepsilon = 0) = \sigma_P = \frac{\cosh \beta + 1}{\sinh \beta}, \tag{3.339}$$

$$\sigma_n(\varepsilon = 0) = \frac{\cosh \beta + 1}{\sinh \beta} \frac{P_{n-1}(\coth \beta)}{P_n(\coth \beta)}, \quad n \geq 1, \tag{3.340}$$

where $P_n(z)$ is the usual Legendre polynomial. In particular, Eqs. (3.339) and (3.340) give

$$\frac{\sigma_0(\varepsilon = 0)}{\sigma_1(\varepsilon = 0)} = \frac{\sigma_P}{\sigma_R} = \coth \beta, \tag{3.341}$$

a well known result for the Elsasser band.

Figures 3.2 through 3.5 show some results for σ_0 and σ_1 for a general temperature gradient (ε^2 not necessarily small) corresponding to the Elsasser band, as calculated numerically from Eqs. (3.320) and (3.327) for $n = 0$ and 1. The quantities actually plotted in these figures are the ratio of σ_0 to the Planck mean, and the ratio of σ_1 to the Rosseland mean, i.e.,

$$\frac{\sigma_0}{\sigma_P} = \frac{\sigma_0}{\sigma_0(\varepsilon = 0)} = \frac{\sinh \beta}{\cosh \beta + 1} \sigma_0, \tag{3.342}$$

$$\frac{\sigma_1}{\sigma_R} = \frac{\sigma_1}{\sigma_1(\varepsilon = 0)} = \frac{\cosh \beta}{\cosh \beta + 1} \sigma_1. \tag{3.343}$$

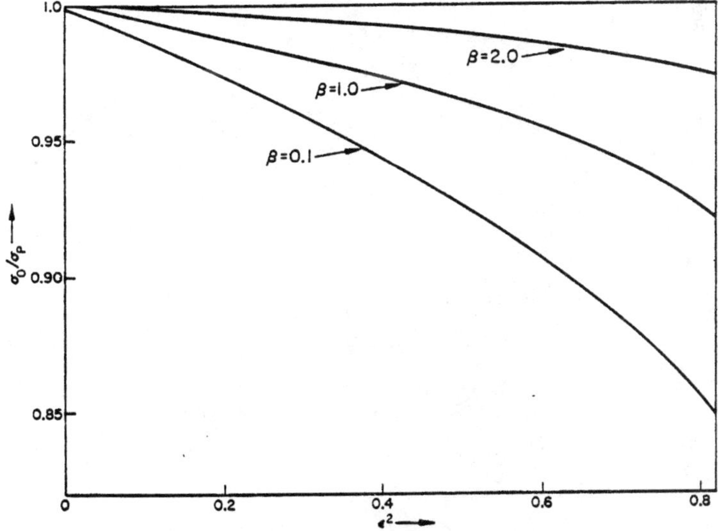

FIG. 3.2. Elsasser band, σ_0 for positive ε^2.

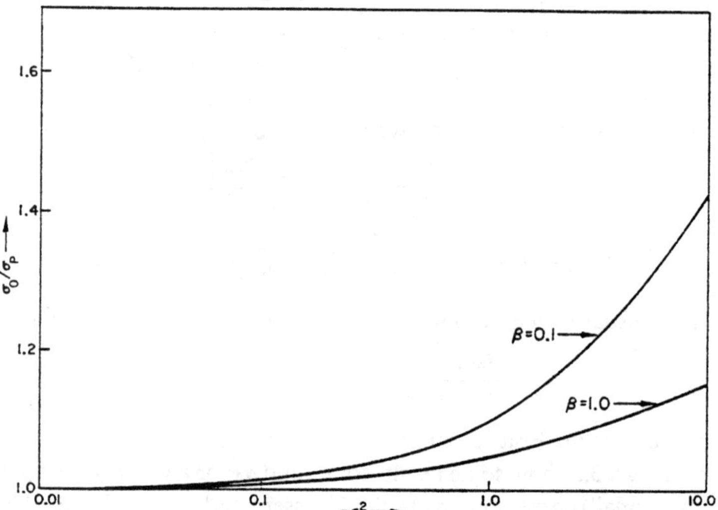

Fig. 3.3. Elsasser band, σ_0 for negative ε^2.

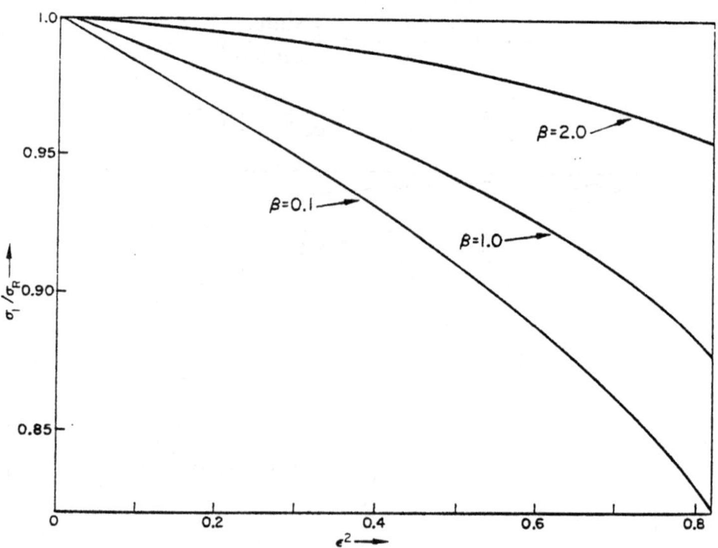

Fig. 3.4. Elsasser band, σ_1 for positive ε^2.

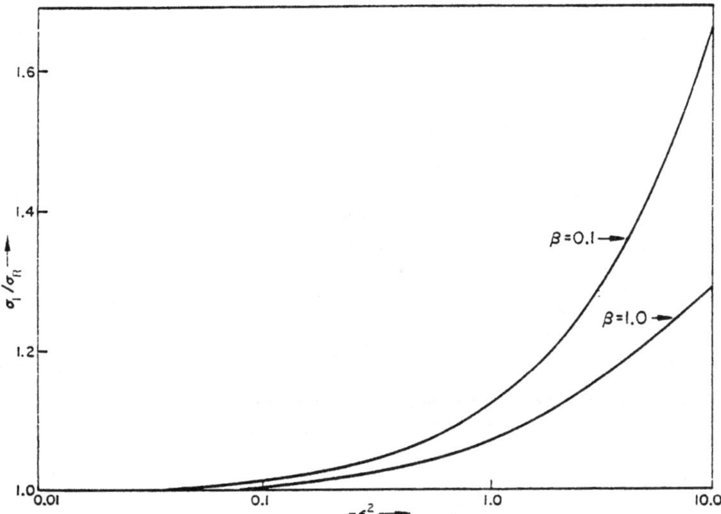

FIG. 3.5. Elsasser band, σ_1 for negative ε^2.

It can be seen that σ_1, the generalized Rosseland mean, depends somewhat more strongly on the temperature gradient than does σ_0, the generalized Planck mean. We also note that for $\varepsilon^2 > 0$, both σ_0 and σ_1 are less than the Planck and Rosseland means, respectively, whereas for $\varepsilon^2 < 0$, $\sigma_0 > \sigma_P$ and $\sigma_1 > \sigma_R$. This behavior is not restricted to the Elsasser band, but can be shown to be true for any $\sigma(\nu)$.

IV

The Representation of Polarized Light in the Equation of Transfer

---

1. Introduction

Thus far our discussions of the equation of transfer have not included the state of polarization in the description of the radiation field. As we shall show in this chapter, four parameters are required to specify the state of polarization of a beam of light. Further, the state of polarization, and hence these four parameters, in general change when a photon undergoes a scattering event. Accordingly, a proper description of photon transport including polarization effects involves four coupled equations of transfer. The single equation of transfer discussed in Chapter II can be considered as the result of averaging this set of four equations over polarization states, assuming the light to be natural (unpolarized). This averaging process, however, introduces an error. It is the purpose of this chapter to eliminate this error by formulating the radiative transfer equations with a proper accounting of the state of polarization.

The need for four parameters to describe the state of polarization of a general beam of light is easily demonstrated. The intensity of the beam clearly constitutes one parameter and is, in fact, the only parameter used in our discussions thus far. In addition, at each space point and for a given frequency and direction of propagation, we shall show that the most general mixture of light can be regarded as a mixture of an elliptically polarized stream and an independent stream of natural (unpolarized) light. The proportion of each of these two components (the degree of polarization) constitutes the second parameter. The remaining two parameters are needed to describe the ellipse associated with the elliptically polarized component. One of these two parameters suffices to specify the orientation of the ellipse, i.e., the angle between the major (or minor) axis and a fixed axis in space. This is generally referred to as a specification of the plane of polarization. The final parameter is the ellipticity of the ellipse, i.e., the ratio of the two axes of the ellipse. Because of the diverse nature of these four quantities, it is convenient to formulate the equation of transfer in terms of an equivalent set of quantities called the Stokes parameters. All four Stokes parameters have the dimensions of the specific intensity.

Chandrasekhar (1960b) was the first to formulate radiative transfer with a proper accounting of the state of polarization. For the most part, we shall follow his treatment very closely.

However, we shall include more detail and in some cases approach the required algebra somewhat differently in an attempt to make the discussion more clear. We also introduce an obvious generalization of the treatment of Chandrasekhar by including time and frequency dependences, absorption, and the emission term in the equation of transfer. We shall not, however, treat induced processes (stimulated emission and scattering) as discussed in Section 9 of Chapter II.

The next two sections of this chapter are devoted to a discussion of the Stokes representation of polarized light. Section 2 discusses elliptically polarized light, and Section 3 deals with a general mixture of light. The final two sections incorporate the Stokes formalism into the equation of radiative transfer. Section 4 gives the analysis for a general scattering process, and Section 5 considers, as an important and concrete example, the case of Thomson (Rayleigh) scattering. In this last section we compare some of the diffusion characteristics of the equation of transfer with Thomson scattering with those of the (approximate) equation which results from the neglect of polarization effects (as in Chapter II). This gives some idea of the importance of a proper accounting of the state of polarization insofar as energy transfer is concerned.

2. Elliptically Polarized Light

At a point **r** in space, we consider the plane perpendicular to the propagation direction of a photon (beam of light). At any instant the electric (and the magnetic) vector associated with this light beam will lie in this plane. We resolve this vector into perpendicular directions l and r. If we assume that the electric vector vibrates in a regular (sinusoidal) manner, the most general form we can write is

$$\xi_l(t) = \xi_l^0 \sin(\omega t - \varepsilon_l), \tag{4.1}$$

$$\xi_r(t) = \xi_r^0 \sin(\omega t - \varepsilon_r), \tag{4.2}$$

where ω is the circular frequency ($\omega = 2\pi\nu$), and ξ_l^0, ξ_r^0, ε_l, ε_r are constants. Here $\underline{\xi}(t)$, with components $\xi_l(t)$ and $\xi_r(t)$, is not exactly the electric vector, but is proportional to it. We choose the proportionality constant as follows. We know the intensity of a light beam is proportional to the square of the electric vector. Hence at a given instant the intensity is proportional to $\xi_l^2(t) + \xi_r^2(t)$. Averaged over one period of vibration, the mean intensity is then proportional to $(\xi_l^0)^2 + (\xi_r^0)^2$. We choose the proportionality constant between ξ and the electric vector such that the mean (over a period) intensity of the light beam is given by

$$I = (\xi_l^0)^2 + (\xi_r^0)^2 = I_l + I_r. \tag{4.3}$$

The sole reason for this choice is that it leads to results, such as Eq. (4.3), which do not involve unnecessary constants.

Let us obtain a parametric equivalent to Eqs. (4.1) and (4.2) by eliminating the independent variable (time) between these two equations. We find

$$\left[\frac{\xi_l(t)}{\xi_l^0}\right]^2 - 2\cos(\varepsilon_l - \varepsilon_r)\frac{\xi_l(t)\xi_r(t)}{\xi_l^0\xi_r^0} + \left[\frac{\xi_r(t)}{\xi_r^0}\right]^2 = \sin^2(\varepsilon_l - \varepsilon_r). \tag{4.4}$$

The Equations of Radiation Hydrodynamics

Equation (4.4) is the equation of an ellipse in the variables ξ_l and ξ_r (the polarization ellipse), and shows that the tip of the vector $\xi(t)$ defined by Eqs. (4.1) and (4.2) traces out an ellipse with time. However, because of the cross product term $\xi_l(t)\,\xi_r(t)$ in Eq. (4.4), the axes of this ellipse do not coincide with the l and r axes (Fig. 4.1). Let us define χ as the angle between the l axis and any one of the axes of the ellipse. Of course, χ is not unique. There exists an infinite number of values of χ, all differing by an integral multiple of $\pi/2$.

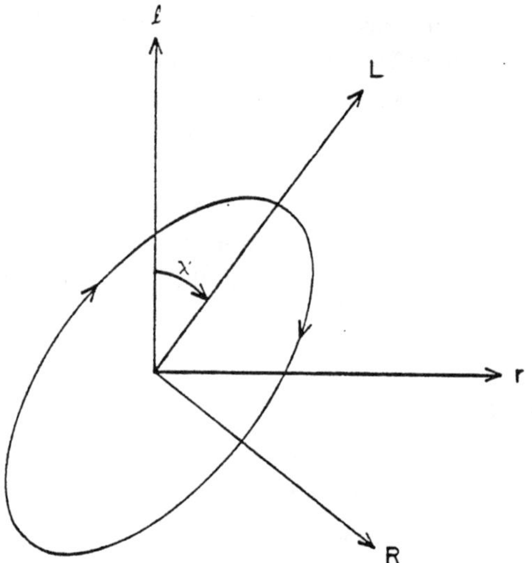

Fig. 4.1. The polarization ellipse.

We compute χ by considering a rotation of the axes l and r to new axes L and R such that the L and R axes coincide with the axes of the ellipse. The required rotation is then the angle χ (Fig. 4.1). A point ξ with coordinates ξ_l and ξ_r can also be described by the coordinates ξ_L and ξ_R. The relationship between these two descriptions corresponding to an angle of rotation χ is given by

$$\xi_l = \xi_L \cos \chi - \xi_R \sin \chi, \tag{4.5}$$
$$\xi_r = \xi_L \sin \chi + \xi_R \cos \chi. \tag{4.6}$$

Use of Eqs. (4.5) and (4.6) in Eq. (4.4) gives

$$\xi_L^2(t) \left[\frac{\cos^2 \chi}{(\xi_l^0)^2} - \frac{\sin 2\chi}{\xi_l^0 \xi_r^0} \cos(\varepsilon_l - \varepsilon_r) + \frac{\sin^2 \chi}{(\xi_r^0)^2} \right]$$
$$+ \xi_L(t)\,\xi_R(t) \left\{ \left[\frac{1}{(\xi_r^0)^2} - \frac{1}{(\xi_l^0)^2} \right] \sin 2\chi - 2 \left[\frac{\cos(\varepsilon_l - \varepsilon_r)}{\xi_l^0 \xi_r^0} \right] \cos 2\chi \right\}$$
$$+ \xi_R^2(t) \left[\frac{\sin^2 \chi}{(\xi_l^0)^2} + \frac{\sin 2\chi}{\xi_l^0 \xi_r^0} \cos(\varepsilon_l - \varepsilon_r) + \frac{\cos^2 \chi}{(\xi_r^0)^2} \right] = \sin^2(\varepsilon_l - \varepsilon_r). \tag{4.7}$$

The axes of this ellipse will coincide with the L and R axes if the cross product term $\xi_L(t)\xi_R(t)$

108

vanishes, i.e., if

$$\left[\frac{1}{(\xi_r^0)^2} - \frac{1}{(\xi_l^0)^2}\right]\sin 2\chi - 2\left[\frac{\cos(\varepsilon_l-\varepsilon_r)}{\xi_l^0\xi_r^0}\right]\cos 2\chi = 0. \tag{4.8}$$

This gives

$$\tan 2\chi = \frac{2\xi_l^0\xi_r^0 \cos(\varepsilon_l-\varepsilon_r)}{(\xi_l^0)^2 - (\xi_r^0)^2} \tag{4.9}$$

as the equation determining χ, the angle between the l axis and one of the axes of the ellipse. If χ is a solution to Eq. (4.9), then so is $\chi+n\pi/2$, where n is an arbitrary integer, positive or negative. As was remarked earlier, this infinity of solutions is consistent with the definition of χ; there are an infinite number of rotations of the l and r axes which will make the rotated axes coincide with the axes of the ellipse.

The other quantity of interest for the polarization ellipse is the ellipticity, i.e., the ratio of the axes of the ellipse. We define an angle β such that the tangent of β is this ratio of axes. We suppose that the absolute value of β lies between 0 and $\pi/2$ (this covers all ratios from zero to infinity), and that the sign of β depends upon whether the polarization is right or left handed. That is, the sign of β depends upon whether the tip of the electric vector traces, with time, a clockwise or a counterclockwise arc.

Considering the coefficients of $\xi_L^2(t)$ and $\xi_R^2(t)$ in Eq. (4.7), we can immediately write

$$\tan^2\beta = \frac{(\xi_r^0)^2\cos^2\chi - \xi_l^0\xi_r^0\cos(\varepsilon_l-\varepsilon_r)\sin 2\chi + (\xi_l^0)^2\sin^2\chi}{(\xi_r^0)^2\sin^2\chi + \xi_l^0\xi_r^0\cos(\varepsilon_l-\varepsilon_r)\sin 2\chi + (\xi_l^0)^2\cos^2\chi}, \tag{4.10}$$

where χ in this equation is a solution of Eq. (4.9). We note that the right hand side of Eq. (4.10) could just as well have been taken as its inverse, depending upon which ratio of axes (major/minor or minor/major) we choose to work with. Equation (4.10) as written actually covers both cases, for if χ is replaced with $\chi+\pi/2$ (an equally valid angle) the right hand side of Eq. (4.10) goes to its inverse. Now, from Eq. (4.10) we have

$$\beta = \tan^{-1}\eta, \tag{4.11}$$

where η is given by

$$\eta = \pm\left[\frac{(\xi_r^0)^2\cos^2\chi - \xi_l^0\xi_r^0\cos(\varepsilon_l-\varepsilon_r)\sin 2\chi + (\xi_l^0)^2\sin^2\chi}{(\xi_r^0)^2\sin^2\chi + \xi_l^0\xi_r^0\cos(\varepsilon_l-\varepsilon_r)\sin 2\chi + (\xi_l^0)^2\cos^2\chi}\right]^{1/2}. \tag{4.12}$$

The $\pm$ sign occurs because of the need to take the square root in passing from Eq. (4.10) to Eqs. (4.11) and (4.12). Using the trigonometric identity

$$\sin 2\beta = \frac{2\eta}{1+\eta^2}, \tag{4.13}$$

we find

$$\sin 2\beta = \pm\frac{\{[(\xi_l^0)^2+(\xi_r^0)^2]^2 - [\{(\xi_l^0)^2-(\xi_r^0)^2\}\cos 2\chi + 2\xi_l^0\xi_r^0\cos(\varepsilon_l-\varepsilon_r)\sin 2\chi]^2\}^{1/2}}{(\xi_l^0)^2+(\xi_r^0)^2}. \tag{4.14}$$

Use of Eq. (4.9) for $\tan 2\chi$ in Eq. (4.14) gives, after a good deal of algebra, the simple result

$$\sin 2\beta = \pm\frac{2\xi_l^0\xi_r^0 \sin(\varepsilon_l-\varepsilon_r)}{(\xi_l^0)^2+(\xi_r^0)^2}. \tag{4.15}$$

Finally, we note that the sign in Eq. (4.15) is arbitrary since the sign of β only serves as a measure of the sense of polarization (right or left handed), and one has a free choice of convention concerning the sense of polarization. For concreteness, we choose the positive sign and write

$$\sin 2\beta = \frac{2\xi_l^0 \xi_r^0 \sin(\varepsilon_l - \varepsilon_r)}{(\xi_l^0)^2 + (\xi_r^0)^2} \tag{4.16}$$

as the equation giving the ellipticity of the polarization ellipse.

Equation (4.16) has two solutions in the range $-\pi/2 \le \beta \le \pi/2$. Either both values of β are positive with a sum of $\pi/2$, or both values are negative with a sum of $-\pi/2$. In either event, if we denote the two solutions of Eq. (4.16) by β_1 and β_2, we have

$$\tan \beta_1 = \frac{1}{\tan \beta_2}. \tag{4.17}$$

Since $\tan \beta$ is the ratio of the two axes of the ellipse, the two solutions merely reflect the fact that two such axes ratios (major/minor or minor/major) are possible, with one ratio being the inverse of the other. These two solutions correspond to two different values of the angle χ differing by 90°.

Given ξ_l^0, ξ_r^0, ε_l, and ε_r, the constants defining the regular vibrations of the electric vector as in Eqs. (4.1) and (4.2), the quantities I, χ, and β can be computed from Eqs. (4.3), (4.9), and (4.16), and represent a complete description of the polarization ellipse. We now show how these three quantities are related to the Stokes parameters. The four Stokes parameters for the case we have been discussing are defined as

$$I = (\xi_l^0)^2 + (\xi_r^0)^2 = I_l + I_r, \tag{4.18}$$

$$Q = (\xi_l^0)^2 - (\xi_r^0)^2 = I_l - I_r, \tag{4.19}$$

$$U = 2\xi_l^0 \xi_r^0 \cos(\varepsilon_l - \varepsilon_r), \tag{4.20}$$

$$V = 2\xi_l^0 \xi_r^0 \sin(\varepsilon_l - \varepsilon_r). \tag{4.21}$$

Let us express Q, U, and V in terms of the three ellipse parameters I, χ, and β. From Eq. (4.16) we find

$$\cos 2\beta = \pm \frac{\{[(\xi_l^0)^2 - (\xi_r^0)^2]^2 + 4(\xi_l^0)^2 (\xi_r^0)^2 \cos^2(\varepsilon_l - \varepsilon_r)\}^{1/2}}{(\xi_l^0)^2 + (\xi_r^0)^2}, \tag{4.22}$$

and from Eq. (4.9) we obtain

$$\sin 2\chi = \pm \frac{2\xi_l^0 \xi_r^0 \cos(\varepsilon_l - \varepsilon_r)}{\{[(\xi_l^0)^2 - (\xi_r^0)^2]^2 + 4(\xi_l^0)^2(\xi_r^0)^2 \cos^2(\varepsilon_l - \varepsilon_r)\}^{1/2}}, \tag{4.23}$$

$$\cos 2\chi = \pm \frac{(\xi_l^0)^2 - (\xi_r^0)^2}{\{[(\xi_l^0)^2 - (\xi_r^0)^2] + 4(\xi_l^0)^2(\xi_r^0)^2 \cos^2(\varepsilon_l - \varepsilon_r)\}^{1/2}}. \tag{4.24}$$

The $\pm$ signs in Eqs. (4.22) through (4.24) arise from the square root operations. From Eqs. (4.16) and (4.22) through (4.24) it is easily shown that Q, U, and V can be written

$$Q = I \cos 2\beta \cos 2\chi, \tag{4.25}$$

$$U = I \cos 2\beta \sin 2\chi, \tag{4.26}$$

$$V = I \sin 2\beta, \tag{4.27}$$

which are the desired representations.

In using Eqs. (4.22) through (4.24) to obtain Eqs. (4.25) and (4.26), it may appear that the $\pm$ signs have been inadvertently dropped. Actually, the $\pm$ signs are implicitly contained in the right hand sides of Eqs. (4.25) and (4.26). As we have discussed earlier, neither χ nor β are unique, although all allowable values of χ and β describe the same polarization ellipse. By choosing different allowable values of χ and β, one can alter the signs of the right hand sides of Eqs. (4.25) and (4.26). Of course, in the end one must choose values to make Eqs. (4.25) and (4.26) true identities. For example, if one chooses from Eq. (4.16) [which is identical to Eq. (4.27)] that value of β which is, in absolute value, the smaller of the two allowable solutions, then $\cos 2\beta > 0$. Accordingly, one must then select a value of χ from Eq. (4.9) such that $\cos 2\chi$ and Q have the same sign.

One can also write U and V in terms of Q, χ, and β. From Eqs. (4.16) and (4.22) through (4.24) we obtain

$$U = Q \tan 2\chi, \tag{4.28}$$

$$V = Q \tan 2\beta \sec 2\chi. \tag{4.29}$$

From Eqs. (4.27) and (4.28) it is clear that the ellipticity of the ellipse and the plane of polarization can be expressed in terms of the Stokes parameters as

$$\sin 2\beta = V/I, \tag{4.30}$$

$$\tan 2\chi = U/Q. \tag{4.31}$$

Further, it can readily be verified that in the case being discussed here, namely elliptically polarized light, we have the identity

$$I^2 = Q^2 + U^2 + V^2. \tag{4.32}$$

Hence for elliptically polarized light only three of the four Stokes parameters are independent. This is consistent with our discussion in Section 1 of this chapter. For partially polarized light, four parameters are needed to completely specify the state of polarization. In the present discussion, however, one of these parameters was specified from the outset, namely the ratio of the intensity of natural light to the intensity of elliptically polarized light. (This ratio is zero since we are dealing with pure elliptically polarized light.)

In writing the time dependence of the electric vector, Eqs. (4.1) and (4.2), we have taken the amplitudes ξ_l^0 and ξ_r^0 as well as the phases ε_l and ε_r to be constants, independent of time. This is an idealization which does not occur in practice. (In practice, one does not observe monochromatic light with an infinitely long wave train.) These amplitudes and phases are, in fact, time dependent due to continual fluctuations and variations. For approximately monochromatic light, however, the time scale for these variations is very much longer than a period corresponding to one vibration. Because of the very high frequency of the electromagnetic oscillations which represent light, the amplitudes and phases may change irregularly millions of times per second, and still remain essentially constant for millions of vibrations. However, in the case of elliptically polarized light, the variations of the amplitudes and phases are not independent. Although these variations may be irregular, the difference of the phases $\delta \equiv \varepsilon_l - \varepsilon_r$ and the ratio of the amplitudes $R \equiv \xi_l^0/\xi_r^0$ must be time independent, subject to no variations, for the light to be elliptically polarized. This is clear from the expressions for χ and β, Eqs. (4.9) and (4.16). If δ and R are time independent, then these equations show that χ and β are constant in time. That is, although the intensity will vary irregularly with time, the orientation and ellipticity of the polarization ellipse will be independent of time.

Since the amplitudes and phases of the light can vary irregularly millions of times per second, the apparent properties of the light are time averaged, or mean, properties. Thus the apparent intensities I_l and I_r in the directions l and r will be given by mean values

$$I_l = \overline{(\xi_l^0)^2}; \quad I_r = \overline{(\xi_r^0)^2}. \tag{4.33}$$

In analogy to our earlier results, we define the Stokes parameters in terms of time averages or means as

$$I = \overline{(\xi_l^0)^2} + \overline{(\xi_r^0)^2}, \tag{4.34}$$

$$Q = \overline{(\xi_l^0)^2} - \overline{(\xi_r^0)^2}, \tag{4.35}$$

$$U = 2\overline{(\xi_l^0 \xi_r^0)} \cos \delta, \tag{4.36}$$

$$V = 2\overline{(\xi_l^0 \xi_r^0)} \sin \delta. \tag{4.37}$$

By averaging our previous results of this section over time, recognizing that $\delta \equiv \varepsilon_l - \varepsilon_r$ and $R \equiv \xi_l^0 / \xi_r^0$, and hence χ and β, are time independent for elliptically polarized light, we find that the previous representation of the Stokes parameters in terms of χ and β still hold. We summarize these results here for convenience. We have, with I_l and I_r defined in a mean sense according to Eq. (4.33), for elliptically polarized light,

$$I = I_l + I_r, \tag{4.38}$$

$$Q = I \cos 2\beta \cos 2\chi = I_l - I_r, \tag{4.39}$$

$$U = I \cos 2\beta \sin 2\chi = Q \tan 2\chi, \tag{4.40}$$

$$V = I \sin 2\beta = Q \tan 2\beta \sec 2\chi. \tag{4.41}$$

Finally, it is clear that in this time averaged sense one still has the identity

$$I^2 = Q^2 + U^2 + V^2, \tag{4.42}$$

as is evident from Eqs. (4.34) through (4.37) or Eqs. (4.38) through (4.41).

At this point, the utility of the Stokes parameters is not entirely clear. For elliptically polarized light these parameters are merely a restatement of the parameters I, χ, and β which describe the polarization ellipse. In the next section, we analyze an arbitrary (partially polarized) beam of light. It is here that the Stokes parameters are of prime importance.

Before doing this, however, let us show the relationship between the Stokes parameters and the special cases of linear and circular polarization. For linear polarization, the ellipticity of the polarization ellipse is zero (or infinite), and since $\tan \beta$ is defined as the ellipticity [see the paragraph above Eq. (4.10)], we have $\beta = 0$ (or $\pm \pi/2$). From Eqs. (4.25) through (4.27) we then have $Q = I \cos 2\chi$, $U = I \sin 2\chi$, and $V = 0$. If we normalize the specific intensity I to unity, the general Stokes vector $s \equiv [I; Q; U; V]$ takes the form $s = [1; \cos 2\chi; \sin 2\chi; 0]$ for linearly polarized light. For vibrations purely along the l axis (vertical polarization), $\chi = 0$ and $s = [1; 1; 0; 0]$. For vibrations purely along the r axis (horizontal polarization), $\chi = \pi/2$ and $s = [1; -1; 0; 0]$. For vibrations along the 45 degree line, $\chi = \pi/4$ and $s = [1; 0; 1; 0]$. These are the most common examples of linear polarization. For circular polarization, the ellipticity is unity and hence $\beta = \pi/4$. From Eqs. (4.25) through (4.27), we have $Q = 0$, $U = 0$, and $V = I$. Thus the normalized Stokes vector for circular polarization is $s = [1; 0; 0; 1]$. For circular polarization with the opposite sense of polarization, corresponding to $\beta = -\pi/4$, we have $s = [1; 0; 0; -1]$.

3. Arbitrarily (Partially) Polarized Light

To analyze an arbitrarily polarized beam of light, we again represent the instantaneous vibrations in the beam along the l and r axes by

$$\xi_l(t) = \xi_l^0 \sin(\omega t - \varepsilon_l), \tag{4.43}$$

$$\xi_r(t) = \xi_r^0 \sin(\omega t - \varepsilon_r). \tag{4.44}$$

As discussed in the last section, the amplitudes ξ_l^0 and ξ_r^0, and the phases ε_l and ε_r, are not constant in time, but are subject to continual and irregular variations. However, the time scale of these variations is long compared to the period corresponding to a single vibration. This is necessarily so for us to speak of an (approximately) time independent circular frequency for the light beam. These irregular variations of ξ_l^0, ξ_r^0, ε_l, and ε_r are not all independent. Certain correlations persist among them, and it is the character of these correlations which determine the state of polarization of a beam of light. As an example, if the variations are such that ξ_l^0/ξ_r^0 and $\varepsilon_l - \varepsilon_r$ are constant in time, subject to no variations, then the light is elliptically (completely) polarized, as we have discussed in the last section. In general, however, such is not the case and in this section we consider the more general case of partially polarized light.

Experimentally, one can measure the intensity of the light beam in the transverse plane (the plane perpendicular to the direction of propagation of the beam) at any angle ψ, measured, say, with respect to the l axis. The other degree of freedom available to the experimentalist is to retard the measurement of one component of the vibrations, say ξ_l, as compared to the other component ξ_r. The analysis of a general beam of light is equivalent to asking what are the consequences of these two experimental degrees of freedom.

If we retard the ξ_l component of the vibrations by a constant time ε/ω, the experimentally measured light beam will be described by, from Eqs. (4.43) and (4.44),

$$\xi_l(t) = \xi_l^0 \sin(\omega t - \varepsilon_l - \varepsilon), \tag{4.45}$$

$$\xi_r(t) = \xi_r^0 \sin(\omega t - \varepsilon_r). \tag{4.46}$$

Resolving the vibrations represented by Eqs. (4.45) and (4.46) along a direction making an angle ψ with the l axis, we find

$$\xi_\psi(t) = \xi_l(t) \cos \psi + \xi_r(t) \sin \psi, \tag{4.47}$$

or, using Eqs. (4.45) and (4.46) for $\xi_l(t)$ and $\xi_r(t)$,

$$\xi_\psi(t) = [\xi_l^0 \cos(\varepsilon_l + \varepsilon) \cos \psi + \xi_r^0 \cos \varepsilon_r \sin \psi] \sin \omega t$$
$$- [\xi_l^0 \sin(\varepsilon_l + \varepsilon) \cos \psi + \xi_r^0 \sin \varepsilon_r \sin \psi] \cos \omega t. \tag{4.48}$$

The momentary intensity (by which we mean the intensity averaged over a single vibrational period) is proportional to $\xi_\psi^2(t)$, time averaged over one period. We find

$$I_{\mathrm{mom}}(\psi; \varepsilon) = (\xi_l^0)^2 \cos^2 \psi + (\xi_r^0)^2 \sin^2 \psi$$
$$+ 2\xi_l^0 \xi_r^0 (\cos \delta \cos \varepsilon - \sin \delta \sin \varepsilon) \sin \psi \cos \psi, \tag{4.49}$$

where we have defined δ as the phase difference

$$\delta \equiv \varepsilon_l - \varepsilon_r. \tag{4.50}$$

113

Recalling that ξ_l^0, ξ_r^0, and δ in general depend upon time, we obtain the apparent intensity in the direction ψ corresponding to a retardation ε by averaging Eq. (4.49) over time. We find, keeping ψ and ε constant in the averaging process,

$$I(\psi;\varepsilon) = \overline{(\xi_l^0)^2}\cos^2\psi + \overline{(\xi_r^0)^2}\sin^2\psi$$
$$+\left[\overline{(\xi_l^0\xi_r^0\cos\delta)}\cos\varepsilon - \overline{(\xi_l^0\xi_r^0\sin\delta)}\sin\varepsilon\right]\sin 2\psi. \tag{4.51}$$

From Eq. (4.51) it is clear that the intensities in the directions $l(\psi = 0)$ and $r(\psi = \pi/2)$ are independent of the retardation ε and given by

$$I(0;\varepsilon) = \overline{(\xi_l^0)^2}, \tag{4.52}$$
$$I(\pi/2;\varepsilon) = \overline{(\xi_r^0)^2}. \tag{4.53}$$

If we define the four parameters

$$I_l = \overline{(\xi_l^0)^2}, \tag{4.54}$$
$$I_r = \overline{(\xi_r^0)^2}, \tag{4.55}$$
$$U = 2\overline{(\xi_l^0\xi_r^0\cos\delta)}, \tag{4.56}$$
$$V = 2\overline{(\xi_l^0\xi_r^0\sin\delta)}, \tag{4.57}$$

we can write Eq. (4.51) as

$$I(\psi;\varepsilon) = I_l\cos^2\psi + I_r\sin^2\psi + \tfrac{1}{2}(U\cos\varepsilon - V\sin\varepsilon)\sin 2\psi. \tag{4.58}$$

Alternately, we can introduce the set of four Stokes parameters

$$I = \overline{(\xi_l^0)^2} + \overline{(\xi_r^0)^2}, \tag{4.59}$$
$$Q = \overline{(\xi_l^0)^2} - \overline{(\xi_r^0)^2}, \tag{4.60}$$
$$U = 2\overline{(\xi_l^0\xi_r^0\cos\delta)}, \tag{4.61}$$
$$V = 2\overline{(\xi_l^0\xi_r^0\sin\delta)}, \tag{4.62}$$

and write Eq. (4.51) as

$$I(\psi;\varepsilon) = \tfrac{1}{2}[I + Q\cos 2\psi + (U\cos\varepsilon - V\sin\varepsilon)\sin 2\psi]. \tag{4.63}$$

It should be noted that Eqs. (4.59) through (4.62), the Stokes parameters defined here for a general beam of light, are in agreement with the Stokes parameters introduced in the last section, Eqs. (4.34) through (4.37), since for elliptically polarized light the phase difference δ is a constant.

From Eq. (4.63) it is clear that the total intensity I, and the additional parameters Q, U, and V, completely determine, from an experimental point of view, the character of an arbitrary beam of light. That is, two beams of light with the same Stokes parameters are optically equivalent since experimentally they cannot be distinguished.

We now show that when several beams of light are mixed, each Stokes parameter of the mixture is the sum of the respective Stokes parameters for the separate beams. This is true in general only if the separate beams are independent, by which we mean the separate beams have no permanent phase relations among themselves. We shall see precisely what this means in the course of the analysis. It suffices to consider only two beams of light. We have already pointed out in Section 10 of Chapter II that two beams of light with different frequencies cannot interfere. Hence in the present analysis we consider both beams of light to have the same frequency. We represent the first beam of light, with the l component retarded by an amount ε, by

$$\xi_{l1}(t) = \xi_{l1}^0 \sin (\omega t - \varepsilon_{l1} - \varepsilon), \tag{4.64}$$

$$\xi_{r1}(t) = \xi_{r1}^0 \sin (\omega t - \varepsilon_{r1}). \tag{4.65}$$

Similarly, the vibrations of the second beam, also retarded by ε, are described by

$$\xi_{l2}(t) = \xi_{l2}^0 \sin (\omega t - \varepsilon_{l2} - \varepsilon), \tag{4.66}$$

$$\xi_{r2}(t) = \xi_{r2}^0 \sin (\omega t - \varepsilon_{r2}). \tag{4.67}$$

Since the amplitudes at a given instant add, the vibrations of the (retarded) composite beam, whose components we denote by $\eta_l(t)$ and $\eta_r(t)$, are represented by

$$\eta_l(t) = \xi_{l1}^0 \sin (\omega t - \varepsilon_{l1} - \varepsilon) + \xi_{l2}^0 \sin (\omega t - \varepsilon_{l2} - \varepsilon), \tag{4.68}$$

$$\eta_r(t) = \xi_{r1}^0 \sin (\omega t - \varepsilon_{r1}) + \xi_{r2}^0 \sin (\omega t - \varepsilon_{r2}). \tag{4.69}$$

Considering the amplitude in a direction which makes an angle ψ with the l axis, we have

$$\eta_\psi(t) = [\xi_{l1}^0 \sin (\omega t - \varepsilon_{l1} - \varepsilon) + \xi_{l2}^0 \sin (\omega t - \varepsilon_{l2} - \varepsilon)] \cos \psi$$
$$+ [\xi_{r1}^0 \sin (\omega t - \varepsilon_{r1}) + \xi_{r2}^0 \sin (\omega t - \varepsilon_{r2})] \sin \psi. \tag{4.70}$$

Squaring $\eta_\psi(t)$ and averaging over a single period, we find that the momentary intensity along the ψ direction, corresponding to a retardation of the lth component by an amount ε, is given by

$$I_{\text{mom}}(\psi; \varepsilon) = [(\xi_{l1}^0)^2 + (\xi_{l2}^0)^2] \cos^2 \psi + [(\xi_{r1}^0)^2 + (\xi_{r2}^0)^2] \sin^2 \psi$$
$$+ 2\xi_{l1}^0 \xi_{r1}^0 (\cos \delta_1 \cos \varepsilon - \sin \delta_1 \sin \varepsilon) \sin \psi \cos \psi$$
$$+ 2\xi_{l2}^0 \xi_{r2}^0 (\cos \delta_2 \cos \varepsilon - \sin \delta_2 \sin \varepsilon) \sin \psi \cos \psi$$
$$+ 2\xi_{l1}^0 \xi_{l2}^0 \cos (\varepsilon_{l1} - \varepsilon_{l2}) \cos^2 \psi$$
$$+ 2\xi_{r1}^0 \xi_{r2}^0 \cos (\varepsilon_{r1} - \varepsilon_{r2}) \sin^2 \psi$$
$$+ 2[\xi_{l1}^0 \xi_{r2}^0 \cos (\varepsilon + \varepsilon_{l1} - \varepsilon_{r2}) + \xi_{l2}^0 \xi_{r1}^0 \cos (\varepsilon + \varepsilon_{l2} - \varepsilon_{r1})] \sin \psi \cos \psi. \tag{4.71}$$

Here we have defined the phase differences

$$\delta_1 \equiv \varepsilon_{l1} - \varepsilon_{r1}, \tag{4.72}$$

$$\delta_2 \equiv \varepsilon_{l2} - \varepsilon_{r2}. \tag{4.73}$$

To obtain the final expression for $I(\psi; \varepsilon)$, the apparent intensity, we need to compute the time average of Eq. (4.71). The last three terms in this equation, since they involve mixed indices (namely, both indices one and two), are assumed to average to zero. This is what is meant by the two beams of light being independent. In particular, it is assumed that no permanent phase relations exist between the two beams, and hence the phase differences $\varepsilon_{l1} - \varepsilon_{l2}$, $\varepsilon_{r1} - \varepsilon_{r2}$, $\varepsilon_{l1} - \varepsilon_{r2}$, and $\varepsilon_{l2} - \varepsilon_{r1}$ fluctuate irregularly in time. Further, the cosine functions involving these phase differences will then fluctuate irregularly in time, being positive and negative for approximately equal times. This is the justification required to set the time average of the terms with mixed indices to zero. [See Eq. (4.97) for the precise definition of the time average. The point here is that the time interval T is essentially infinite, but the integral is finite because of the irregular behavior of the cosine functions. Thus the integral divided by T, i.e., the time average, is zero.] Hence, for independent beams of light, the time average of Eq. (4.71) can be written

$$I(\psi; \varepsilon) = \tfrac{1}{2}\{(I_1+I_2)+(Q_1+Q_2)\cos 2\psi$$
$$+[(U_1+U_2)\cos \varepsilon-(V_1+V_2)\sin \varepsilon]\sin 2\psi\}, \tag{4.74}$$

where I, Q, U, and V, appropriately subscripted, are the Stokes parameters for each individual beam of light.

Equation (4.74) shows that for a beam of light resulting from a mixture of several independent beams, the Stokes parameters are given by

$$I = \sum_i I_i, \tag{4.75}$$

$$Q = \sum_i Q_i, \tag{4.76}$$

$$U = \sum_i U_i, \tag{4.77}$$

$$V = \sum_i V_i, \tag{4.78}$$

where the subscript i refers to the ith beam. In particular, for a mixture of independent elliptically polarized beams, we have, from Eqs. (4.38) through (4.41),

$$I = \sum_i I_i, \tag{4.79}$$

$$Q = \sum_i I_i \cos 2\beta_i \cos 2\chi_i, \tag{4.80}$$

$$U = \sum_i I_i \cos 2\beta_i \sin 2\chi_i, \tag{4.81}$$

$$V = \sum_i I_i \sin 2\beta_i, \tag{4.82}$$

where I_i, χ_i, and β_i define the intensity, plane of polarization, and ellipticity of the ith beam.

Let us now analyze natural (unpolarized) light. The experimental definition of natural light is that when resolved in any direction ψ and subject to any retardation ε, the result is the same, namely one-half the total intensity of the beam. That is, for natural light we have

$$I(\psi; \varepsilon) = \tfrac{1}{2} I \quad \text{(for any } \psi \text{ and } \varepsilon\text{).} \tag{4.83}$$

From Eq. (4.63) we see that Eq. (4.83) holds if and only if

$$Q = U = V = 0. \tag{4.84}$$

That is, for natural light all of the Stokes parameters except the total intensity I are zero.

Let us consider the possibility of representing natural light as the mixture of two independent elliptically polarized beams of light. We let χ_1 and β_1 define the plane of polarization and ellipticity of the first beam, and similarly χ_2 and β_2 describe the ellipse associated with the second beam. Further, the ratio of the intensities of the two beams is taken as $1 : q$.

Since the two beams are assumed independent, the Stokes parameters are additive, and from Eqs. (4.80) through (4.82) we find that $Q = U = V = 0$ (i.e., the combined beam is natural light) if

$$Q = 0 = \cos 2\beta_1 \cos 2\chi_1 + q \cos 2\beta_2 \cos 2\chi_2, \tag{4.85}$$

$$U = 0 = \cos 2\beta_1 \sin 2\chi_1 + q \cos 2\beta_2 \sin 2\chi_2, \tag{4.86}$$

$$V = 0 = \sin 2\beta_1 + q \sin 2\beta_2. \tag{4.87}$$

Transposing the last term of each of the above equations, squaring, and adding together the results, we find $q^2 = 1$. Since q is by definition positive, we have

$$q = 1. \tag{4.88}$$

One solution of Eq. (4.87) is then

$$-\beta_2 = \beta_1. \tag{4.89}$$

Of course, Eq. (4.87) has an infinity of other solutions because of the multibranched nature of the inverse sine function. However, since β_1 and β_2 are restricted to lie between $-\pi/2$ and $+\pi/2$, only one other solution is possible. This is

$$-\beta_2 = \pm \frac{\pi}{2} - \beta_1, \tag{4.90}$$

where the $+$ sign is used if β_1 is positive and the $-$ sign is used if β_1 is negative. As argued in Section 2 of this chapter [the paragraph below Eq. (4.16)], β_1 and $\pm\pi/2 - \beta_1$ actually correspond to two representations of the same ellipse, depending upon which ratio of axes (major/minor or minor/major) one chooses to work with (recall that this ratio is given by $\tan \beta$). Hence we can disregard Eq. (4.90) and consider Eq. (4.89) as the relationship between the ellipticities of the two beams of polarized light. With Eqs. (4.88) and (4.89), Eqs. (4.85) and (4.86) give

$$\cos 2\chi_1 = -\cos 2\chi_2, \tag{4.91}$$

$$\sin 2\chi_1 = -\sin 2\chi_2. \tag{4.92}$$

Equations (4.91) and (4.92) show that the two values of χ differ by 90 degrees, and we write

$$\chi_1 - \chi_2 = \pm \frac{\pi}{2}. \tag{4.93}$$

It might also be noted that Eqs. (4.85) through (4.87) are also satisfied by $q = 1, \beta_1 = -\beta_2 = \pm\pi/4$ (circularly polarized light). However, this is a special case of the general solution we have already found.

Thus we have shown that natural light can be considered as a mixture of two independent elliptically polarized beams of light of equal intensity ($q = 1$). The polarization ellipses associated with these two beams are similar, their major axes perpendicular to each other ($\chi_1 - \chi_2 = \pm\pi/2$), and the sense of polarization (clockwise or counterclockwise) of one beam is contrary to that of the other ($\beta_1 = -\beta_2$). This is shown schematically in Fig. 4.2. It should be emphasized that β_1 and χ_1 are arbitrary. Once they have been chosen, however, β_2 and χ_2 follow from Eqs. (4.89) and (4.93).

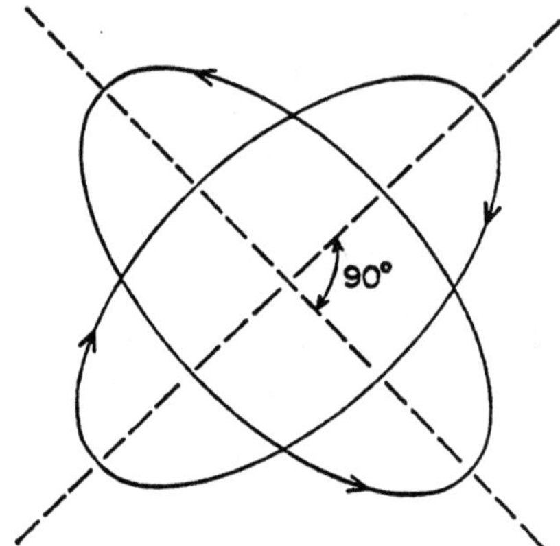

FIG. 4.2. Natural light.

Two beams of light with polarization parameters (β, χ) and $(-\beta, \chi \pm \pi/2)$ are said to be oppositely polarized. Hence an alternate way of stating our result is that natural light is equivalent to a mixture of any two independent, oppositely polarized beams of light of equal intensity.

Having considered both elliptically (completely) polarized light (Section 2 of this chapter) and natural (unpolarized) light, we now analyze arbitrarily (partially) polarized light. We first prove the result that

$$I^2 \geq Q^2 + U^2 + V^2, \tag{4.94}$$

for a general beam of light, with the equality holding only for an elliptically (completely) polarized beam. From the definitions of the Stokes parameters, Eqs. (4.59) through (4.62), we find that

$$I^2 - Q^2 = 4\overline{(\xi_l^0)^2} \times \overline{(\xi_r^0)^2}, \tag{4.95}$$

$$U^2 + V^2 = 4\left[\overline{(\xi_l^0 \xi_r^0 \cos \delta)}^2 + \overline{(\xi_l^0 \xi_r^0 \sin \delta)}^2\right]. \tag{4.96}$$

Now, given a function $f(t)$, the time average over an interval T is defined as

$$\bar{f} = \frac{1}{T} \int_0^T dt\, f(t). \tag{4.97}$$

Equations (4.95) and (4.96) then yield, with explicit indication of the time dependences of ξ_l^0, ξ_r^0, and δ,

$$I^2 - Q^2 - U^2 - V^2 = \frac{4}{T^2} \int_0^T dt \int_0^T dt'\, [\xi_l^0(t)]^2\, [\xi_r^0(t')]^2$$

$$- \frac{4}{T^2} \int_0^T dt \int_0^T dt'\, \xi_l^0(t)\xi_r^0(t)\xi_l^0(t')\xi_r^0(t')[\cos \delta(t) \cos \delta(t') + \sin \delta(t) \sin \delta(t')]. \tag{4.98}$$

Interchanging the integration variables t and t' in half of the first integral, and using a simple identity to combine the trigonometric terms, we find that Eq. (4.98) can be rewritten

$$I^2 - Q^2 - U^2 - V^2 = \frac{2}{T^2} \int_0^T dt \int_0^T dt' \{ [\xi_l^0(t) \, \xi_r^0(t')]^2 + [\xi_l^0(t') \, \xi_r^0(t)]^2$$
$$- 2\xi_l^0(t) \, \xi_r^0(t') \, \xi_l^0(t') \, \xi_r^0(t) \cos [\delta(t) - \delta(t')] \}. \tag{4.99}$$

One more algebraic reordering in Eq. (4.99) gives

$$I^2 - Q^2 - U^2 - V^2 = \frac{2}{T^2} \int_0^T dt \int_0^T dt' \{ \xi_l^0(t) \, \xi_r^0(t') - \xi_l^0(t') \, \xi_r^0(t) \cos [\delta(t) - \delta(t')] \}^2$$
$$+ \frac{2}{T^2} \int_0^T dt \int_0^T dt' \{ \xi_l^0(t') \, \xi_r^0(t) \sin [\delta(t) - \delta(t')] \}^2. \tag{4.100}$$

The right hand side of Eq. (4.100), being the sum of two integrals, each of which has a non-negative integrand, must be either positive or zero. The only case which gives zero is if each integrand vanishes, i.e., if

$$\delta(t) = \delta(t'), \tag{4.101}$$

and

$$\xi_l^0(t) \, \xi_r^0(t') = \xi_l^0(t') \, \xi_r^0(t), \tag{4.102}$$

for all t and t'. This implies that both the phase difference δ and the ratio of amplitudes ξ_l^0/ξ_r^0 are constant in time. As discussed in Section 2 of this chapter, these are just the conditions for the light to be elliptically polarized.

Thus we have shown, for a general beam of light, that

$$I^2 \geq Q^2 + U^2 + V^2, \tag{4.103}$$

with the equality holding only in the case of elliptically polarized light. We have also shown in Section 2 of this chapter that the equation $I^2 = Q^2 + U^2 + V^2$ is always true for elliptically polarized light. Thus it follows that the existence of this equality is both a necessary and sufficient condition for a beam of light to be elliptically polarized.

With Eq. (4.103) established, we now show how a general beam of light, with Stokes parameters $[I; Q; U; V]$, can be resolved into two independent beams, one of which is natural (unpolarized) light and the other elliptically polarized light. This decomposition is accomplished by inspection. Since the Stokes parameters for independent beams are additive it is clear that a mixture of two beams with Stokes parameters

$$\text{Beam 1: } [I - (Q^2 + U^2 + V^2)^{1/2}; \, 0; \, 0; \, 0], \tag{4.104}$$
$$\text{Beam 2: } [(Q^2 + U^2 + V^2)^{1/2}; \, Q; \, U; \, V] \tag{4.105}$$

is equivalent to the original beam with Stokes parameters $[I; Q; U; V]$. Further, in view of our previous discussion, the first beam represents natural light and the second beam elliptically polarized light. The ellipticity and plane of polarization of this second beam are given by, from Eqs. (4.30) and (4.31),

$$\sin 2\beta = \frac{V}{(Q^2+U^2+V^2)^{1/2}}, \qquad (4.106)$$

$$\tan 2\chi = \frac{U}{Q}. \qquad (4.107)$$

These two formulae always have a solution and hence it is always possible to represent a general beam of light as a superposition of a natural beam and an independent elliptically polarized beam. Moreover, this representation is unique since the apparent multiple solutions of Eqs. (4.106) and (4.107) really represent the same situation [see the discussions in the paragraphs below Eqs. (4.16) and (4.27)].

Equations (4.104) and (4.105) give the decomposition of a general beam of light into a beam of natural light and an independent beam of elliptically polarized light. An alternate way of decomposing a general beam of light is to resolve it into two independent beams of elliptically polarized light. This decomposition follows directly from Eqs. (4.104) and (4.105). We have previously shown that natural light is equivalent to a mixture of any two independent beams of oppositely polarized light of equal intensity. Hence the natural light given by Eq. (4.104) is equivalent to two independent polarized beams, each of intensity

$$\tfrac{1}{2}[I-(Q^2+U^2+V^2)^{1/2}], \qquad (4.108)$$

in the states of polarization (β, χ) and $(-\beta, \chi\pm\pi/2)$. Here β and χ are arbitrary. However, as we shall see, it is useful in the present context to choose β and χ according to Eqs. (4.106) and (4.107). Thus our original beam of light with Stokes parameters $[I; Q; U; V]$ is equivalent to the sum of three independent elliptically polarized beams. From Eqs. (4.104), (4.105), and (4.108) we see that the intensities and states of polarization of these beams are given by

$$\text{Beam 1a: } \tfrac{1}{2}[I-(Q^2+U^2+V^2)^{1/2}]; \qquad (-\beta, \chi\pm\pi/2), \qquad (4.109)$$

$$\text{Beam 1b: } \tfrac{1}{2}[I-(Q^2+U^2+V^2)^{1/2}]; \qquad (\beta, \chi), \qquad (4.110)$$

$$\text{Beam 2: } (Q^2+U^2+V^2)^{1/2}; \qquad (\beta, \chi). \qquad (4.111)$$

Since the two beams given by Eqs. (4.110) and (4.111) are in the same state of polarization, we can combine them, by adding their intensities, into a single beam.

Hence our final result is that a general (partially polarized) beam of light with Stokes parameters $[I; Q; U; V]$ is equivalent to two independent elliptically polarized beams, with intensities and states of polarization given by

$$\text{Beam I: } \tfrac{1}{2}[I-(Q^2+U^2+V^2)^{1/2}]; \qquad (-\beta, \chi\pm\pi/2), \qquad (4.112)$$

$$\text{Beam II: } \tfrac{1}{2}[I+(Q^2+U^2+V^2)^{1/2}]; \qquad (\beta, \chi). \qquad (4.113)$$

Here β and χ are given by, repeating Eqs. (4.106) and (4.107),

$$\sin 2\beta = \frac{V}{(Q^2+U^2+V^2)^{1/2}}, \qquad (4.114)$$

$$\tan 2\chi = \frac{U}{Q}. \qquad (4.115)$$

It would be useful to end this section with a discussion of the physical meaning of the Stokes parameters I, Q, U, and V. However, aside from the specific intensity I (or its components I_l and I_r), there seems to be no such simple meaning which can be attached to the Stokes parameters. One should be careful, however, not to interpret U and V as radiation intensities because of the phase relationships associated with their definitions [see Eqs. (4.61) and (4.62)].

4. The Equation of Transfer for (Partially) Polarized Light

Having discussed the necessary background material on the analytic representation of partially polarized light by means of the Stokes parameters, we are now in a position to formulate the equation of transfer including the effects of polarization. The crucial point in this formulation is the treatment of the scattering process, since a scattering event in general changes all four Stokes parameters. Thus in the general case one would expect four equations of transfer, one for each Stokes parameter, with the added complexity that these four equations would be fully coupled through the scattering term.

As a reference point, we repeat here the equation of transfer derived in Chapter II. We have, repeating Eq. (2.17),

$$\frac{1}{c}\frac{\partial I(v, \Omega)}{\partial t}+\Omega\cdot\nabla I(v, \Omega) = S(v)-\sigma(v)\,I(v, \Omega)$$

$$+\int_0^\infty dv'\int_{4\pi} d\Omega'\,\frac{v}{v'}\,\sigma_s(v'\to v, \Omega'\cdot\Omega)\,I(v', \Omega'). \tag{4.116}$$

If we decompose $\sigma_s(v'\to v, \Omega'\cdot\Omega)$ according to Eq. (1.33), i.e.,

$$\sigma_s(v'\to v, \Omega'\cdot\Omega) = \sigma_s(v')\,K(v'\to v, \Omega'\cdot\Omega), \tag{4.117}$$

we then have

$$\frac{1}{c}\frac{\partial I(v, \Omega)}{\partial t}+\Omega\cdot\nabla I(v, \Omega) = S(v)-\sigma(v)\,I(v, \Omega)$$

$$+\int_0^\infty dv'\int_{4\pi} d\Omega'\,\sigma_s(v')\,\frac{v}{v'}\,K(v'\to v, \Omega'\cdot\Omega)\,I(v', \Omega'). \tag{4.118}$$

The kernel $K(v'\to v, \Omega'\cdot\Omega)$ is normalized such that [see Eq. (1.34)]

$$\int_0^\infty dv\int_{4\pi} d\Omega K(v'\to v, \Omega'\cdot\Omega) = 2\pi\int_0^\infty dv\int_{-1}^1 d\mu_0 K(v'\to v, \mu_0) = 1. \tag{4.119}$$

In Eq. (4.118) $\sigma(v)$ is the sum of the absorption and scattering coefficients, i.e.,

$$\sigma(v) = \sigma_a(v)+\sigma_s(v), \tag{4.120}$$

and $S(v)$ is the source term due to spontaneous emission of photons. Equation (4.118) is an equation of transfer for the total intensity of radiation which does not properly account for the state of polarization of the radiation. It is the intent of this section to generalize Eq.

(4.118) to take polarization into account, using the Stokes formalism discussed in the last two sections of this chapter.

In our formulation we shall use the intensities at right angles to each other, I_l and I_r, in addition to the parameters U and V, rather than the equivalent Stokes parameters I, Q, U, and V. Thus we characterize the radiation field with frequency ν at a point in space $\mathbf{r}$ at time t by the four quantities

$$I_l \equiv I_l(\mathbf{r}, \nu, \mathbf{\Omega}, t), \tag{4.121}$$

$$I_r \equiv I_r(\mathbf{r}, \nu, \mathbf{\Omega}, t), \tag{4.122}$$

$$U \equiv U(\mathbf{r}, \nu, \mathbf{\Omega}, t), \tag{4.123}$$

$$V \equiv V(\mathbf{r}, \nu, \mathbf{\Omega}, t). \tag{4.124}$$

As usual, the argument $\mathbf{\Omega}$ implies a polar angle θ and a corresponding azimuthal angle φ referred to some coordinate system at the point under consideration. The directions l and r, both perpendicular to $\mathbf{\Omega}$, refer to directions in the meridian plane (the plane containing the z axis and the vector $\mathbf{\Omega}$) and at right angles to it, respectively (Fig. 4.3). If we take the matter

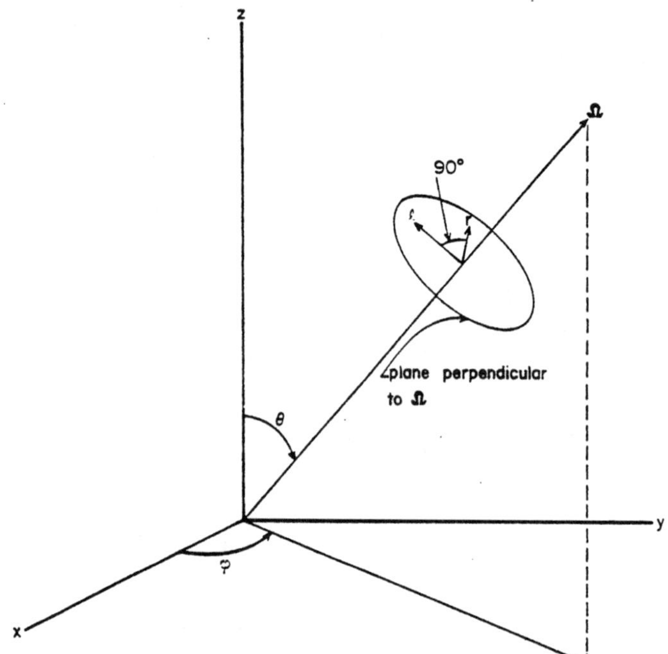

FIG. 4.3. The definition of the specific intensity of radiation.

to be isotropic (no preferred direction), then the absorption and scattering coefficients will be independent of the state of polarization of the photon. By arguments analogous to those used in Chapter II, we are then led to the equation of transfer

$$\frac{1}{c}\frac{\partial \mathbf{I}(\nu, \mathbf{\Omega})}{\partial t} + \mathbf{\Omega} \cdot \nabla \mathbf{I}(\nu, \mathbf{\Omega}) = \mathbf{S}(\nu) - \sigma(\nu)\,\mathbf{I}(\nu, \mathbf{\Omega}) + \text{inscattering}. \tag{4.125}$$

Here $\mathbf{I} \equiv \mathbf{I}(\mathbf{r}, \nu, \mathbf{\Omega}, t)$ is a four component vector $[I_l; I_r; U; V]$, and $\mathbf{S} \equiv \mathbf{S}(\mathbf{r}, \nu, t)$ is a source vector with components $[S_l; S_r; S_U; S_V]$ describing the polarization state of the photons arising from spontaneous emission. Since the matter is assumed isotropic, the source will be unpolarized and we have

$$S_l = S_r = \tfrac{1}{2}S; \quad S_U = S_V = 0, \tag{4.126}$$

where S denotes the intensity of the source as in Eq. (4.118).

The difficulty in completing Eq. (4.125) is the inscattering term. What we require is a generalization, to include polarization effects, of the inscattering term in Eq. (4.118). One might expect this generalization to be of the form

$$\text{inscattering} = \int_0^\infty dv' \int_{4\pi} d\Omega' \sigma_s(v') \frac{v}{v'} \mathbf{K}(v' \to v, \Omega' \to \Omega) \mathbf{I}(v', \Omega'). \tag{4.127}$$

Here $\mathbf{K} \equiv \mathbf{K}(\mathbf{r}, v' \to v, \Omega' \to \Omega, t)$ is a four by four matrix kernel which, if nondiagonal, couples the individual equations of transfer for the four parameters I_l, I_r, U, and V. It might be noted that we have allowed this matrix kernel to depend upon Ω and Ω' separately rather than upon $\Omega \cdot \Omega'$, the cosine of the scattering angle, alone. Equation (4.127) is, as we shall see, a proper description of the inscattering term, and the matrix $\mathbf{K}$, which we need derive, will in fact depend upon Ω and Ω' separately, even though the matter is isotropic. The reason for this will become clear as we go through the analysis.

The basic physics of the scattering interaction is most conveniently described if one refers the parameters I_l, I_r, U, and V to directions l and r lying in the plane of scattering and perpendicular to it, respectively. It must be remembered that this is *not* the plane of reference used to define the vector $\mathbf{I}$ in Eq. (4.125). The consequence of introducing these two planes of reference [the meridian plane as in Eq. (4.125) and the plane of scattering] is what leads to the algebraic difficulty in defining the inscattering term in Eq. (4.125). Now, using the plane of scattering as the reference plane, we define the four by four matrix $\mathbf{R}$, with components $r_{ij} \equiv r_{ij}(\mathbf{r}, v' \to v, \Omega' \cdot \Omega, t)$, such that the term

$$\sigma_s(v') \frac{v}{v'} \mathbf{R}(v' \to v, \Omega' \cdot \Omega) \mathbf{I}^s(v', \Omega') \, dv' \, d\Omega', \tag{4.128}$$

when integrated over all v' and Ω', is the desired inscattering term, aside from the difficulty associated with the two planes of reference. The superscript s in Eq. (4.128) indicates that this equation has the parameters I_l, I_r, U, and V referred to directions l and r defined in the plane of scattering. The factor v/v' in this equation, as in the inscattering term in Eq. (4.118), merely reflects the fact that all equations of radiative transfer are conventionally written in terms of the specific intensity (which contains an energy factor hv) rather than the distribution function for the number of photons; whereas the scattering coefficient $\sigma_s(v')$ is defined with reference to the number of photons scattered. We note that the components of the matrix $\mathbf{R}$ depend only upon $\Omega \cdot \Omega'$ rather than upon Ω and Ω' separately. This follows from the assumption that the matter is isotropic (i.e., has no preferred direction associated with it). Since the plane of reference in Eq. (4.128) also has no preferred direction in space (it is the plane of scattering), the scattering description in this plane of reference must depend only upon the scattering angle, whose cosine is $\Omega \cdot \Omega'$. This is the reason that the scattering plane is the natural plane of reference for the directions l and r for describing the basic physics, reflected in the matrix kernel $\mathbf{R}$, of the scattering interaction.

The Equations of Radiation Hydrodynamics

Prior to integrating Eq. (4.128) over all ν' and Ω' and using the result as the inscattering term in Eq. (4.125), we must transform it from the scattering plane of reference to the meridian plane of reference. To do this, it is necessary that we digress momentarily and consider the law of transformation for the parameters I_l, I_r, U, and V for a rotation of the l and r axes.

Before doing this, however, we note that the integral of Eq. (4.128) over all ν' and Ω' is the correct inscattering term for Eq. (4.125) in one special case, namely natural light. In this case, the plane of reference is irrelevant since $I_l = I_r = \frac{1}{2}I$ for natural light in all planes of reference. The equation of transfer discussed in Chapter II, namely that which neglects polarization effects, follows from the present considerations by using the integral of Eq. (4.128) over ν' and Ω' as the inscattering term in Eq. (4.125). Then, of the four equations contained in Eq. (4.125), the first two, namely those for I_l and I_r, are added together, setting $I_l = I_l^s = I_r = I_r^s = \frac{1}{2}I$ and $U^s = V^s = 0$. In general $U = U^s = V = V^s = 0$ will not satisfy the remaining two equations, nor will $Q \equiv I_l - I_r = I_l^s - I_r^s = 0$ satisfy the difference of the first two equations. This constitutes the approximation in neglecting the effects of polarization in the equation of transfer. The result of this approximation is the scalar equation of transfer of the usual form

$$\frac{1}{c}\frac{\partial I(\nu, \Omega)}{\partial t} + \Omega \cdot \nabla I(\nu, \Omega) = S(\nu) - \sigma(\nu)\, I(\nu, \Omega)$$

$$+ \int_0^\infty d\nu' \int_{4\pi} d\Omega' \sigma_s(\nu')\, \frac{\nu}{\nu'}\left[\frac{1}{2}\sum_{i,\,j=1}^{2} r_{ij}(\nu' \to \nu, \Omega' \cdot \Omega)\right] I(\nu', \Omega'). \quad (4.129)$$

Comparison of this result with Eq. (4.118) shows that the scalar kernel K is related to the components of the polarization matrix kernel $\mathbf{R}$ according to

$$K(\nu' \to \nu, \Omega' \cdot \Omega) = \frac{1}{2}\sum_{i,\,j=1}^{2} r_{ij}(\nu' \to \nu, \Omega' \cdot \Omega). \quad (4.130)$$

From Eq. (4.119) we then conclude

$$\pi \sum_{i,\,j=1}^{2}\int_0^\infty d\nu \int_{-1}^{1} d\mu_0 r_{ij}(\nu' \to \nu, \mu_0) = 1, \quad (4.131)$$

which serves to normalize the matrix $\mathbf{R}$.

We now consider the law of transformation of the Stokes parameters for a rotation of the l and r axes. We have previously shown that a general beam of light can be considered as the sum of two independent elliptically polarized beams. Thus the desired law of transformation can be obtained by considering the case of an elliptically polarized beam, for which the Stokes parameters are given by Eqs. (4.38) through (4.41). We consider a rotation of axes through an angle φ to axes l' and r' (Fig. 4.4). This has the effect of reducing the angle χ, defining the plane of polarization, by an amount φ. The total intensity I and the ellipticity β are, of course, unaffected by this rotation. Denoting with a prime the Stokes parameters referred to the l' and r' axes, we find from Eqs. (4.38) through (4.41) that

$$I' = I, \quad (4.132)$$

$$\begin{aligned}
Q' &= I\cos 2\beta \cos(2\chi - 2\varphi)\\
&= I\cos 2\beta(\cos 2\chi \cos 2\varphi + \sin 2\chi \sin 2\varphi)\\
&= Q\cos 2\varphi + U\sin 2\varphi, \quad (4.133)
\end{aligned}$$

124

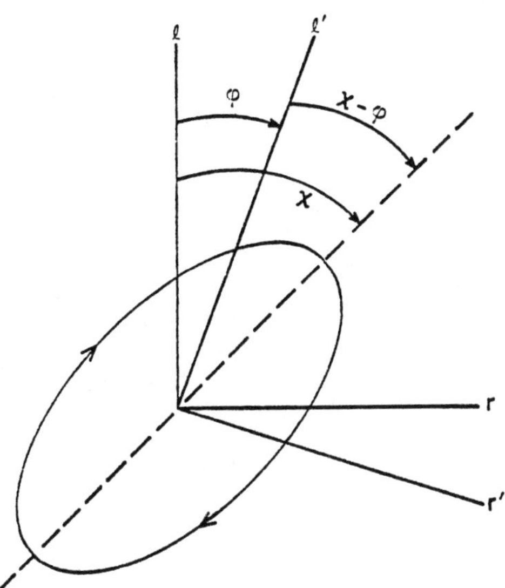

FIG. 4.4. Rotation of the l and r axes.

$$U' = I \cos 2\beta \sin (2\chi - 2\varphi)$$
$$= I \cos 2\beta(-\cos 2\chi \sin 2\varphi + \sin 2\chi \cos 2\varphi)$$
$$= -Q \sin 2\varphi + U \cos 2\varphi, \tag{4.134}$$
$$V' = V. \tag{4.135}$$

Thus we conclude that if $\mathbf{I} = [I; Q; U; V]$ denotes a vector whose components are the four Stokes parameters I, Q, U, and V which characterize an arbitrarily (partially) polarized beam of light, then the effect of a rotation of the l and r axes through an angle φ as in Fig. 4.4 is to subject $\mathbf{I}$ to the linear transformation

$$\mathbf{I}' = \mathbf{M}(\varphi)\,\mathbf{I}, \tag{4.136}$$

where the matrix $\mathbf{M}(\varphi)$ is given by

$$\mathbf{M}(\varphi) = \begin{bmatrix} 1 & 0 & 0 & 0 \\ 0 & \cos 2\varphi & \sin 2\varphi & 0 \\ 0 & -\sin 2\varphi & \cos 2\varphi & 0 \\ 0 & 0 & 0 & 1 \end{bmatrix}. \tag{4.137}$$

It is clear from physical considerations that $\mathbf{M}(\varphi)$ must satisfy the group relations

$$\mathbf{M}(\varphi_1)\,\mathbf{M}(\varphi_2) = \mathbf{M}(\varphi_1 + \varphi_2), \tag{4.138}$$
$$\mathbf{M}^{-1}(\varphi) = \mathbf{M}(-\varphi), \tag{4.139}$$

as can easily be verified. We also note that the matrix $\mathbf{M}$ has the property

$$\mathbf{M}^T(\varphi) = \mathbf{M}(-\varphi), \tag{4.140}$$

where $\mathbf{M}^T(\varphi)$ denotes the transpose of $\mathbf{M}(\varphi)$ (i.e., the matrix resulting from the interchange of rows and columns).

Rather than working with the Stokes parameters I, Q, U, and V, one can ask for the law of transformation of the parameters I_l, I_r, U, and V for a rotation of the axes through an angle φ. Recalling that

$$I = I_l + I_r, \tag{4.141}$$

$$Q = I_l - I_r, \tag{4.142}$$

we find that Eqs. (4.132) through (4.135) can be written

$$I'_l = I_l \cos^2 \varphi + I_r \sin^2 \varphi + \tfrac{1}{2} U \sin 2\varphi, \tag{4.143}$$

$$I'_r = I_l \sin^2 \varphi + I_r \cos^2 \varphi - \tfrac{1}{2} U \sin 2\varphi, \tag{4.144}$$

$$U' = -I_l \sin 2\varphi + I_r \sin 2\varphi + U \cos 2\varphi, \tag{4.145}$$

$$V' = V. \tag{4.146}$$

Thus if $\mathsf{I} = [I_l; I_r; U; V]$ is the four component vector describing a general beam of light, the effect of a rotation of the l and r axes through an angle φ as in Fig. 4.4 is to subject I to the linear transformation

$$\mathsf{I}' = \mathsf{L}(\varphi)\,\mathsf{I}, \tag{4.147}$$

where the rotation matrix $\mathsf{L}(\varphi)$ is given by

$$\mathsf{L}(\varphi) = \begin{bmatrix} \cos^2 \varphi & \sin^2 \varphi & \tfrac{1}{2}\sin 2\varphi & 0 \\ \sin^2 \varphi & \cos^2 \varphi & -\tfrac{1}{2}\sin 2\varphi & 0 \\ -\sin 2\varphi & \sin 2\varphi & \cos 2\varphi & 0 \\ 0 & 0 & 0 & 1 \end{bmatrix}. \tag{4.148}$$

As with the $\mathsf{M}(\varphi)$ matrix, the $\mathsf{L}(\varphi)$ matrix must satisfy the group relations

$$\mathsf{L}(\varphi_1)\mathsf{L}(\varphi_2) = \mathsf{L}(\varphi_1+\varphi_2), \tag{4.149}$$

$$\mathsf{L}^{-1}(\varphi) = \mathsf{L}(-\varphi). \tag{4.150}$$

The analogue of Eq. (4.140) for the $\mathsf{L}(\varphi)$ matrix is

$$\mathsf{L}^T(\varphi) = \mathsf{Q}^{-1}\mathsf{L}(-\varphi)\,\mathsf{Q}, \tag{4.151}$$

where the Q matrix is given by

$$\mathsf{Q} = \begin{bmatrix} 1 & 0 & 0 & 0 \\ 0 & 1 & 0 & 0 \\ 0 & 0 & 2 & 0 \\ 0 & 0 & 0 & 2 \end{bmatrix}. \tag{4.152}$$

The matrix Q, which will occur again in the analysis to come, clearly arises here because we are describing the vector I by $[I_l; I_r; U; V]$ rather than by the Stokes vector $[I; Q; U; V]$ [i.e., no matrix Q is present in Eq. (4.140)].

We are now in a position to return to the equation of transfer, Eq. (4.125), and define the inscattering term. We need transform Eq. (4.128) from the scattering plane of reference to the meridian plane of reference [the plane used in Eq. (4.125)], and integrate the result over all ν' and Ω'. The three pertinent planes of reference are shown in Fig. 4.5. The plane $OP'P$ is the scattering plane, the plane $OP'Z$ is the meridian plane for the beam of light prior to the scattering event, and the plane OPZ is the meridian plane for the light following the scattering. To transform Eq. (4.128), defined with respect to the scattering plane of reference, to the meridian plane of reference following the scattering, we need apply the linear operator $\mathsf{L}(\pi-\psi)$ to Eq. (4.128), where ψ is the angle in the spherical triangle $P'PZ$ between the planes

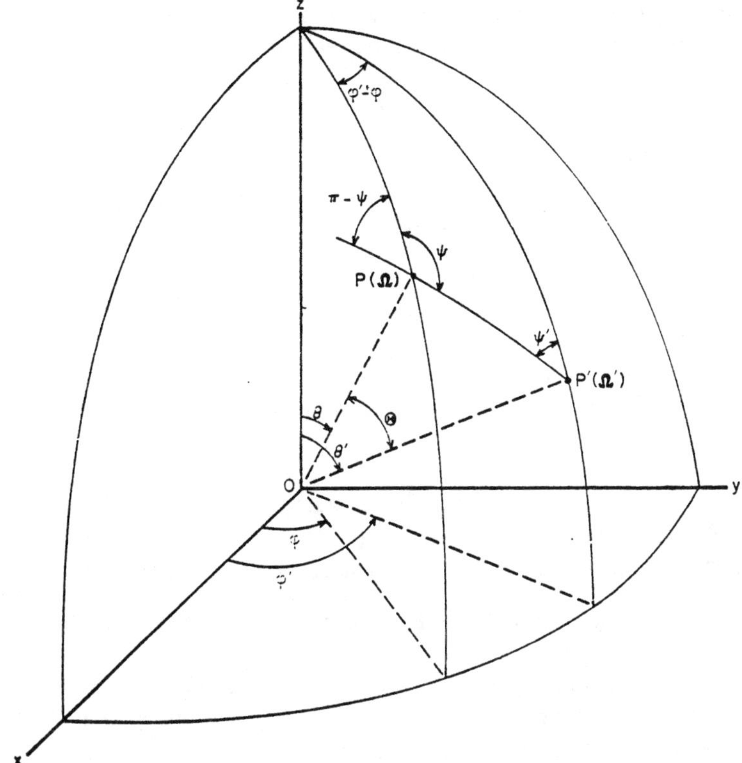

FIG. 4.5. Transformation between the plane of scattering and the meridian planes.

$OP'P$ (scattering plane) and OPZ (meridian plane after scattering). We use the matrix **L** rather than **M** since we are formulating the equation of transfer in terms of the intensity vector $\mathbf{I} = [I_l; I_r; U; V]$ rather than in terms of the Stokes parameters. Hence the inscattering term is written

$$\text{inscattering} = \int_0^\infty d\nu' \int_{4\pi} d\Omega' \sigma_s(\nu') \frac{\nu}{\nu'} \mathbf{L}(\pi-\psi)\, \mathbf{R}(\nu' \to \nu, \Omega'\cdot\Omega)\, \mathbf{I}^s(\nu', \Omega'), \quad (4.153)$$

and the equation of transfer, Eq. (4.125), becomes

$$\frac{1}{c}\frac{\partial \mathbf{I}(\nu, \Omega)}{\partial t} + \Omega \cdot \nabla \mathbf{I}(\nu, \Omega) = \mathbf{S}(\nu) - \sigma(\nu)\, \mathbf{I}(\nu, \Omega)$$

$$+ \int_0^\infty d\nu' \int_{4\pi} d\Omega' \sigma_s(\nu') \frac{\nu}{\nu'} \mathbf{L}(\pi-\psi)\, \mathbf{R}(\nu' \to \nu, \Omega'\cdot\Omega)\, \mathbf{I}^s(\nu', \Omega'). \quad (4.154)$$

Equation (4.154), although perfectly correct, contains two intensity vectors, namely **I**, referred to the meridian plane, and **I**^{s}, referred to the plane of scattering. However, just as we applied the transformation $\mathbf{L}(\pi-\psi)$ to Eq. (4.128), we can write

$$\mathbf{I} = \mathbf{L}(\psi')\, \mathbf{I}^s, \quad (4.155)$$

127

where ψ' is the angle in the spherical triangle $P'PZ$ between the plane $OP'P$ (scattering plane) and the plane $OP'Z$ (meridian plane before scattering). Inverting Eq. (4.155) we find

$$\mathbf{I}^s = \mathbf{L}^{-1}(\psi')\,\mathbf{I}, \tag{4.156}$$

and making use of Eq. (4.150) we obtain

$$\mathbf{I}^s = \mathbf{L}(-\psi')\,\mathbf{I}. \tag{4.157}$$

Use of Eq. (4.157) in Eq. (4.154) gives

$$\frac{1}{c}\frac{\partial \mathbf{I}(\nu, \mathbf{\Omega})}{\partial t} + \mathbf{\Omega}\cdot\nabla\,\mathbf{I}(\nu, \mathbf{\Omega}) = \mathbf{S}(\nu) - \sigma(\nu)\,\mathbf{I}(\nu, \mathbf{\Omega})$$

$$+ \int_0^\infty d\nu' \int_{4\pi} d\mathbf{\Omega}'\sigma_s(\nu')\frac{\nu}{\nu'}\,\mathbf{L}(\pi-\psi)\,\mathbf{R}(\nu' \to \nu, \mathbf{\Omega}'\cdot\mathbf{\Omega})\,\mathbf{L}(-\psi')\,\mathbf{I}(\nu',\mathbf{\Omega}') \tag{4.158}$$

as the final form of the equation of transfer including polarization effects.

Comparison of the inscattering term in Eq. (4.158) with Eq. (4.127) shows that the matrix kernel $\mathbf{K}$ describing scattering of (partially) polarized light is of the form

$$\mathbf{K}(\nu' \to \nu, \mathbf{\Omega}' \to \mathbf{\Omega}) = \mathbf{L}(\pi-\psi)\,\mathbf{R}(\nu' \to \nu, \mathbf{\Omega}'\cdot\mathbf{\Omega})\,\mathbf{L}(-\psi'). \tag{4.159}$$

Because of the angles ψ and ψ' (which depend upon $\mathbf{\Omega}$ and $\mathbf{\Omega}'$) appearing in Eq. (4.159), the kernel $\mathbf{K}$ is not a function of $\mathbf{\Omega}\cdot\mathbf{\Omega}'$ alone, but depends upon $\mathbf{\Omega}$ and $\mathbf{\Omega}'$ separately. However, the matrix $\mathbf{K}$ does have a certain symmetry property if the matrix $\mathbf{R}$ is such that

$$\mathbf{R}^T = \mathbf{Q}^{-1}\mathbf{R}\mathbf{Q}, \tag{4.160}$$

where the superscript T again denotes the transpose, and $\mathbf{Q}$ is the diagonal matrix given by Eq. (4.152). To show the symmetry of $\mathbf{K}$ (actually $\mathbf{Q}^{-1}\mathbf{K}$), let us consider the transpose of the matrix $\mathbf{Q}^{-1}\mathbf{K}(\mathbf{\Omega}' \to \mathbf{\Omega})$. (We drop all frequency arguments for notational convenience.) We have

$$[\mathbf{Q}^{-1}\mathbf{K}(\mathbf{\Omega}' \to \mathbf{\Omega})]^T = [\mathbf{Q}^{-1}\mathbf{L}(\pi-\psi)\,\mathbf{R}(\mathbf{\Omega}'\cdot\mathbf{\Omega})\,\mathbf{L}(-\psi')]^T$$

$$= \mathbf{L}^T(-\psi')\,\mathbf{R}^T(\mathbf{\Omega}'\cdot\mathbf{\Omega})\,\mathbf{L}^T(\pi-\psi)(\mathbf{Q}^{-1})^T. \tag{4.161}$$

Using Eq. (4.151) for $\mathbf{L}^T(\varphi)$, using Eq. (4.160) for $\mathbf{R}^T$, and recognizing that $(\mathbf{Q}^{-1})^T = \mathbf{Q}^{-1}$ since $\mathbf{Q}^{-1}$ is diagonal, we find

$$[\mathbf{Q}^{-1}\mathbf{K}(\mathbf{\Omega}' \to \mathbf{\Omega})]^T = \mathbf{Q}^{-1}\mathbf{L}(\psi')\,\mathbf{R}(\mathbf{\Omega}'\cdot\mathbf{\Omega})\,\mathbf{L}(\psi-\pi). \tag{4.162}$$

We now use the fact that an interchange of $\mathbf{\Omega}$ and $\mathbf{\Omega}'$ changes the angle ψ' to $\pi-\psi$ and ψ to $\pi-\psi'$. Hence Eq. (4.162) gives

$$[\mathbf{Q}^{-1}\mathbf{K}(\mathbf{\Omega} \to \mathbf{\Omega}')]^T = \mathbf{Q}^{-1}\mathbf{L}(\pi-\psi)\,\mathbf{R}(\mathbf{\Omega}'\cdot\mathbf{\Omega})\,\mathbf{L}(-\psi'). \tag{4.163}$$

Finally, we recognize that the right hand side of Eq. (4.163) is just $\mathbf{Q}^{-1}\mathbf{K}(\mathbf{\Omega}' \to \mathbf{\Omega})$ [see Eq. (4.159)], and hence we have shown that

$$[\mathbf{Q}^{-1}\mathbf{K}(\mathbf{\Omega} \to \mathbf{\Omega}')]^T = \mathbf{Q}^{-1}\mathbf{K}(\mathbf{\Omega}' \to \mathbf{\Omega}). \tag{4.164}$$

We reiterate that Eq. (4.164) holds only if the basic scattering matrix $\mathbf{R}$ has the symmetry displayed in Eq. (4.160). Equation (4.160) is just an expression of Helmholtz's principle of reciprocity for single scattering with proper allowance made for the polarization of the light (Chandrasekhar, 1960c).

The matrix $\mathbf{Q}$ which enters both Eqs. (4.160) and (4.164) is present because we have defined the intensity vector as $\mathbf{I} = [I_l; I_r; U; V]$. Had we instead worked throughout with the Stokes vector $\mathbf{I} = [I; Q; U; V]$, this factor $\mathbf{Q}$ would not have appeared. We have seen the occurrence of the matrix $\mathbf{Q}$ for this reason earlier [compare Eqs. (4.140) and (4.151)].

The equation of transfer we have derived, Eq. (4.158), is really four equations of transfer for the four parameters I_l, I_r, U, and V describing the (partially) polarized light. In general these four equations are fully coupled through the inscattering term. However, in the case of systems with plane (or spherical) symmetry these four equations reduce to two because the symmetry of the situation demands that $U = V = 0$. This is easily argued by returning to Eqs. (4.112) through (4.115) which give the decomposition of a general (partially polarized) beam of light into two independent elliptically polarized beams. In planar (or spherical) systems the plane of polarization, described by the angle χ, must either be along the meridian plane $(\chi = 0)$ or perpendicular to it $(\chi = \pm\pi/2)$ in order to maintain the symmetry. Equation (4.115) then gives $U = 0$. Further, a positive or negative value of β, the ellipticity, implies a certain sense of polarization (clockwise or counterclockwise), either of which constitutes an asymmetry and is therefore inconsistent with plane (or spherical) symmetry. Thus we must have $\beta = 0$ and Eq. (4.114) then gives $V = 0$. Since $\beta = 0$ corresponds to a value of zero (or infinity) for the ratio of axes of the polarization ellipse, we see that in situations with plane (or spherical) symmetry, the light is the sum of two independent plane polarized beams—one polarized along the meridian plane and the other at right angles to it. Thus for systems with this symmetry, one need deal with only two coupled equations of transfer for I_l and I_r, or, equivalently, for I and Q. In making the above argument, we have assumed that the initial and boundary conditions on the equation of transfer do not destroy the symmetry. Specifically, we have assumed that $U = V = 0$ and that I_l and I_r are azimuthally independent at the initial time and at the boundaries of the system.

In the next section of this chapter we consider Thomson (Rayleigh) scattering as a concrete example of the analysis of this section. There we will see explicitly the symmetry of the matrix $\mathbf{K}$ expressed by Eq. (4.164), and the reduction of the general equation of transfer, Eq. (4.158), to only two equations for I_l and I_r in the case of problems with plane symmetry.

Before leaving this section, however, we consider explicit expressions, in terms of Ω and Ω', for the angles ψ and ψ' appearing in the kernel $\mathbf{K}$ [see Eq. (4.159)]. To this end, let us compute a unit vector normal to the meridian plane OPZ (Fig. 4.5). This is accomplished by computing the vector cross product of the vector Ω and a vector in the z direction, say the unit vector $\mathbf{k}$, and then normalizing the result to unit length. The vector Ω is given in terms of θ and φ as

$$\Omega = (\sin\theta\cos\varphi)\,\mathbf{i} + (\sin\theta\sin\varphi)\,\mathbf{j} + (\cos\theta)\,\mathbf{k}, \qquad (4.165)$$

where $\mathbf{i}$, $\mathbf{j}$, and $\mathbf{k}$ represent unit vectors along the x, y, and z axes, respectively. Thus $\mathbf{n}(OPZ)$, a unit vector normal to the plane OPZ, is given by

$$\mathbf{n}(OPZ) = \frac{\mathbf{k}\times\Omega}{|\mathbf{k}\times\Omega|} = (-\sin\varphi)\,\mathbf{i} + (\cos\varphi)\,\mathbf{j}. \qquad (4.166)$$

The Equations of Radiation Hydrodynamics

Similarly, a unit vector normal to the scattering plane OPP' (Fig. 4.5), which we denote by $\mathbf{n}(OPP')$, is found by computing the vector cross product of $\boldsymbol{\Omega}$ and $\boldsymbol{\Omega}'$, and normalizing the result to unit length. The vector $\boldsymbol{\Omega}'$ is given by the right hand side of Eq. (4.165) with primes on θ and φ. We find

$$\mathbf{n}(OPP') = \frac{\boldsymbol{\Omega}' \times \boldsymbol{\Omega}}{|\boldsymbol{\Omega}' \times \boldsymbol{\Omega}|} = \frac{1}{N}\{(\sin\theta'\sin\varphi'\cos\theta - \sin\theta\sin\varphi\cos\theta')\,\mathbf{i}$$

$$+ (\sin\theta\cos\varphi\cos\theta' - \sin\theta'\cos\varphi'\cos\theta)\,\mathbf{j}$$

$$+ [\sin\theta\sin\theta'\sin(\varphi-\varphi')]\,\mathbf{k}\}, \tag{4.167}$$

where N is the positive root of N^2, given by

$$N^2 = \sin^2\theta'\cos^2\theta + \sin^2\theta\cos^2\theta' + \sin^2\theta\sin^2\theta'\sin^2(\varphi-\varphi')$$

$$- 2\sin\theta\sin\theta'\cos\theta\cos\theta'\cos(\varphi-\varphi'). \tag{4.168}$$

Since both $\mathbf{n}(OPZ)$ and $\mathbf{n}(OPP')$ are unit vectors, their dot product gives directly the cosine of the angle between the meridian plane OPZ and the scattering plane OPP'. This is the cosine of the angle ψ (Fig. 4.5). Thus

$$\cos\psi = \mathbf{n}(OPZ)\cdot\mathbf{n}(OPP') = \frac{1}{N}[\sin\theta\cos\theta' - \sin\theta'\cos\theta\cos(\varphi-\varphi')]. \tag{4.169}$$

One can obtain an alternate expression for the factor N by using the explicit expression for the cosine of the scattering angle Θ, i.e.,

$$\cos\Theta = \boldsymbol{\Omega}\cdot\boldsymbol{\Omega}' = \cos\theta\cos\theta' + \sin\theta\sin\theta'\cos(\varphi-\varphi'). \tag{4.170}$$

Computing $\sin^2\Theta = 1 - \cos^2\Theta$ from Eq. (4.170), we find

$$\sin\Theta = N, \tag{4.171}$$

where N is the positive root of Eq. (4.168). Thus Eq. (4.169) can be written

$$\cos\psi = \frac{\sin\theta\cos\theta' - \sin\theta'\cos\theta\cos(\varphi-\varphi')}{\sin\Theta}. \tag{4.172}$$

An expression for $\cos\psi'$, where ψ' is the angle between the meridian plane $OP'Z$ and the scattering plane $OP'P$ (Fig. 4.5), is obtained in a similar manner as

$$\cos\psi' = \frac{\sin\theta'\cos\theta - \sin\theta\cos\theta'\cos(\varphi-\varphi')}{\sin\Theta}. \tag{4.173}$$

Since both ψ and ψ' lie between 0 and π, Eqs. (4.172) and (4.173) completely determine these two angles.

It might also be noted that Eqs. (4.172) and (4.173) follow directly from the application of the sine–cosine law (relating two angles and three sides) of spherical trigonometry to the spherical triangle $P'PZ$ of Fig. 4.5 (see, for example, Bauer and Brooke, 1932). Other relations of this type also follow from spherical trigonometry. Figure 4.6 shows the spherical triangle of interest, with angles $\varphi'-\varphi$, ψ, and ψ', and with sides Θ, θ', and θ. For example,

130

from the relation between one side and three angles we find

$$\sin \psi \sin \psi' \cos \Theta - \cos \psi \cos \psi' = \cos(\varphi' - \varphi). \tag{4.174}$$

From the relation between two sides and three angles we obtain

$$\sin \psi \cos \psi' \cos \Theta + \cos \psi \sin \psi' = \cos \theta' \sin(\varphi' - \varphi), \tag{4.175}$$

and

$$\sin \psi' \cos \psi \cos \Theta + \cos \psi' \sin \psi = \cos \theta \sin(\varphi' - \varphi). \tag{4.176}$$

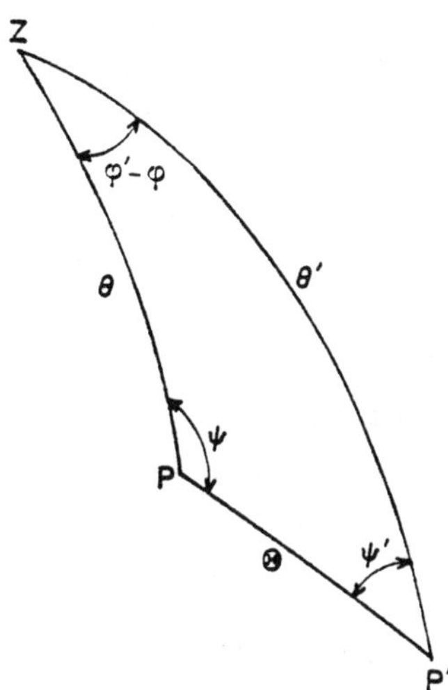

FIG. 4.6. The spherical triangle $P'PZ$.

Finally, from the relation between one side and three angles, coupled with the law of sines and the law of cosines, one can derive a relation between all three sides and all three angles of the form

$$\sin \psi \sin \psi' - \cos \psi \cos \psi' \cos \Theta = \sin \theta \sin \theta' + \cos \theta \cos \theta' \cos(\varphi' - \varphi). \tag{4.177}$$

Many other relations of this type can be found, but Eqs. (4.174) through (4.177) are ones we shall find particularly useful in the next section.

5. The Special Case of Thomson (Rayleigh) Scattering

A scattering interaction of some importance in radiative transfer problems is Thomson (Rayleigh) scattering. We shall indicate the physics behind this type of scattering in Chapter VII. For the moment we consider only the mathematical details of Thomson scattering as it enters Eq. (4.158), the equation of transfer including polarization effects. This serves as a concrete example of the general considerations of the last section of this chapter.

The Equations of Radiation Hydrodynamics

According to Chandrasekhar (1960b), the basic scattering kernel, described by the matrix $\mathbf{R}(\nu' \to \nu, \boldsymbol{\Omega}' \cdot \boldsymbol{\Omega})$, for Thomson scattering is given by

$$\mathbf{R}(\nu' \to \nu, \boldsymbol{\Omega}' \cdot \boldsymbol{\Omega}) = \frac{3}{8\pi}\,\delta(\nu - \nu')\begin{bmatrix} \cos^2\Theta & 0 & 0 & 0 \\ 0 & 1 & 0 & 0 \\ 0 & 0 & \cos\Theta & 0 \\ 0 & 0 & 0 & \cos\Theta \end{bmatrix}, \qquad (4.178)$$

where $\Theta \equiv \cos^{-1}(\boldsymbol{\Omega}' \cdot \boldsymbol{\Omega})$ is the scattering angle and $\delta(\nu - \nu')$ is the Dirac delta function expressing the fact that the photon frequency is not changed by the scattering event. Hence the matrix kernel $\mathbf{K}(\nu' \to \nu, \boldsymbol{\Omega}' \to \boldsymbol{\Omega})$, describing Thomson scattering of (partially) polarized light, follows from Eq. (4.159) as

$$\mathbf{K}(\nu' \to \nu, \boldsymbol{\Omega}' \to \boldsymbol{\Omega}) = \frac{3}{8\pi}\,\delta(\nu - \nu')\begin{bmatrix} \cos^2\psi & \sin^2\psi & -\frac{1}{2}\sin 2\psi & 0 \\ \sin^2\psi & \cos^2\psi & \frac{1}{2}\sin 2\psi & 0 \\ \sin 2\psi & -\sin 2\psi & \cos 2\psi & 0 \\ 0 & 0 & 0 & 1 \end{bmatrix} \times$$

$$\times \begin{bmatrix} \cos^2\Theta & 0 & 0 & 0 \\ 0 & 1 & 0 & 0 \\ 0 & 0 & \cos\Theta & 0 \\ 0 & 0 & 0 & \cos\Theta \end{bmatrix}\begin{bmatrix} \cos^2\psi' & \sin^2\psi' & -\frac{1}{2}\sin 2\psi' & 0 \\ \sin^2\psi' & \cos^2\psi' & \frac{1}{2}\sin 2\psi' & 0 \\ \sin 2\psi' & -\sin 2\psi' & \cos 2\psi' & 0 \\ 0 & 0 & 0 & 1 \end{bmatrix}, \qquad (4.179)$$

where we have used Eqs. (4.148) and (4.178) for the matrices $\mathbf{L}$ and $\mathbf{R}$. Performing the multiplication of the last two matrices, we find

$$\mathbf{K}(\nu' \to \nu, \boldsymbol{\Omega}' \to \boldsymbol{\Omega}) = \frac{3}{8\pi}\,\delta(\nu - \nu')\begin{bmatrix} \cos^2\psi & \sin^2\psi & -\frac{1}{2}\sin 2\psi & 0 \\ \sin^2\psi & \cos^2\psi & \frac{1}{2}\sin 2\psi & 0 \\ \sin 2\psi & -\sin 2\psi & \cos 2\psi & 0 \\ 0 & 0 & 0 & 1 \end{bmatrix} \times$$

$$\times \begin{bmatrix} \cos^2\Theta\,\cos^2\psi' & \cos^2\Theta\,\sin^2\psi' & -\frac{1}{2}\cos^2\Theta\,\sin 2\psi' & 0 \\ \sin^2\psi' & \cos^2\psi' & \frac{1}{2}\sin 2\psi' & 0 \\ \cos\Theta\,\sin 2\psi' & -\cos\Theta\,\sin 2\psi' & \cos\Theta\,\cos 2\psi' & 0 \\ 0 & 0 & 0 & \cos\Theta \end{bmatrix}. \qquad (4.180)$$

Performing the matrix multiplication indicated in Eq. (4.180), we obtain

$$\mathbf{K}(\nu' \to \nu, \Omega' \to \Omega) = \frac{3}{8\pi}\, \delta(\nu - \nu') \times$$

$$\times \begin{bmatrix} (l,\,l)^2 & (r,\,l)^2 & (l,\,l)(r,\,l) & 0 \\ (l,\,r)^2 & (r,\,r)^2 & (l,\,r)(r,\,r) & 0 \\ 2(l,\,l)(l,\,r) & 2(r,\,r)(r,\,l) & (l,\,l)(r,\,r)+(r,\,l)(l,\,r) & 0 \\ 0 & 0 & 0 & (l,\,l)(r,\,r)-(l,\,r)(r,\,l) \end{bmatrix}, \quad (4.181)$$

where, following Chandrasekhar (1960b), we have defined

$$(r,\,r) = \sin \psi \sin \psi' \cos \Theta - \cos \psi \cos \psi', \tag{4.182}$$

$$(l,\,r) = -(\sin \psi \cos \psi' \cos \Theta + \cos \psi \sin \psi'), \tag{4.183}$$

$$(r,\,l) = \sin \psi' \cos \psi \cos \Theta + \cos \psi' \sin \psi, \tag{4.184}$$

$$(l,\,l) = \sin \psi \sin \psi' - \cos \psi \cos \psi' \cos \Theta. \tag{4.185}$$

The right hand sides of Eqs. (4.182) through (4.185) are identical, except for an overall minus sign in Eq. (4.183), to the left hand sides of Eqs. (4.174) through (4.177). Thus the spherical trigonometry relations, Eqs. (4.174) through (4.177), allow us to write

$$(r,\,r) = \cos(\varphi' - \varphi), \tag{4.186}$$

$$(l,\,r) = -\cos \theta' \sin(\varphi' - \varphi), \tag{4.187}$$

$$(r,\,l) = \cos \theta \sin(\varphi' - \varphi), \tag{4.188}$$

$$(l,\,l) = \sin \theta \sin \theta' + \cos \theta \cos \theta' \cos(\varphi' - \varphi). \tag{4.189}$$

The ten nonzero matrix elements k_{ij} in Eq. (4.181) are then given by

$$k_{11} = (l,\,l)^2 = \tfrac{1}{2}\left[2(1-\mu^2)(1-\mu'^2)+\mu^2\mu'^2\right]$$

$$+2\mu\mu'(1-\mu^2)^{1/2}(1-\mu'^2)^{1/2}\cos(\varphi'-\varphi)+\tfrac{1}{2}\mu^2\mu'^2\cos 2(\varphi'-\varphi), \tag{4.190}$$

$$k_{12} = (r,\,l)^2 = \tfrac{1}{2}\,\mu^2[1-\cos 2(\varphi'-\varphi)], \tag{4.191}$$

$$k_{13} = (l,\,l)(r,\,l) = \mu(1-\mu^2)^{1/2}(1-\mu'^2)^{1/2}\sin(\varphi'-\varphi)+\tfrac{1}{2}\,\mu^2\mu'\sin 2(\varphi'-\varphi), \tag{4.192}$$

$$k_{21} = (l,\,r)^2 = \tfrac{1}{2}\,\mu'^2[1-\cos 2(\varphi'-\varphi)], \tag{4.193}$$

$$k_{22} = (r,\,r)^2 = \tfrac{1}{2}\left[1+\cos 2(\varphi'-\varphi)\right], \tag{4.194}$$

$$k_{23} = (l,\,r)(r,\,r) = -\tfrac{1}{2}\,\mu' \sin 2(\varphi'-\varphi), \tag{4.195}$$

$$k_{31} = 2(l,\,l)(l,\,r) = -2\mu'(1-\mu^2)^{1/2}(1-\mu'^2)^{1/2}\sin(\varphi'-\varphi)-\mu\mu'^2 \sin 2(\varphi'-\varphi), \tag{4.196}$$

$$k_{32} = 2(r, r)(r, l) = \mu \sin 2(\varphi' - \varphi), \tag{4.197}$$

$$k_{33} = (l, l)(r, r) + (r, l)(l, r) = (1 - \mu^2)^{1/2}(1 - \mu'^2)^{1/2}\cos(\varphi' - \varphi) + \mu\mu' \cos 2(\varphi' - \varphi), \tag{4.198}$$

$$k_{44} = (l, l)(r, r) - (l, r)(r, l) = \mu\mu' + (1 - \mu^2)^{1/2}(1 - \mu'^2)^{1/2}\cos(\varphi' - \varphi). \tag{4.199}$$

Here we have defined

$$\mu \equiv \cos \theta; \quad \mu' \equiv \cos \theta'. \tag{4.200}$$

With these results we can write our final form for the matrix kernel $\mathbf{K}(\nu' \rightarrow \nu, \mathbf{\Omega}' \rightarrow \mathbf{\Omega})$ as

$$\mathbf{K}(\nu' \rightarrow \nu, \mathbf{\Omega}' \rightarrow \mathbf{\Omega}) = \mathbf{Q}[\mathbf{K}^{(0)}(\mu'; \mu) + (1 - \mu^2)^{1/2}(1 - \mu'^2)^{1/2}\mathbf{K}^{(1)}(\mu', \varphi'; \mu, \varphi)$$

$$+ \mathbf{K}^{(2)}(\mu', \varphi'; \mu, \varphi)], \tag{4.201}$$

where the matrix $\mathbf{Q}$ is the four by four diagonal matrix given by Eq. (4.152) and the three $\mathbf{K}^{(i)}$ matrices are given by

$$\mathbf{K}^{(0)}(\mu'; \mu) = \frac{3}{16\pi} \delta(\nu - \nu') \begin{bmatrix} 2(1 - \mu^2)(1 - \mu'^2) + \mu^2\mu'^2 & \mu^2 & 0 & 0 \\ \mu'^2 & 1 & 0 & 0 \\ 0 & 0 & 0 & 0 \\ 0 & 0 & 0 & \mu\mu' \end{bmatrix}, \tag{4.202}$$

$$\mathbf{K}^{(1)}(\mu', \varphi'; \mu, \varphi) = \frac{3}{16\pi} \delta(\nu - \nu') \begin{bmatrix} 4\mu\mu' \cos(\varphi' - \varphi) & 0 & 2\mu \sin(\varphi' - \varphi) & 0 \\ 0 & 0 & 0 & 0 \\ -2\mu' \sin(\varphi' - \varphi) & 0 & \cos(\varphi' - \varphi) & 0 \\ 0 & 0 & 0 & \cos(\varphi' - \varphi) \end{bmatrix}, \tag{4.203}$$

$$\mathbf{K}^{(2)}(\mu', \varphi'; \mu, \varphi) = \frac{3}{16\pi} \delta(\nu - \nu') \times$$

$$\times \begin{bmatrix} \mu^2\mu'^2 \cos 2(\varphi' - \varphi) & -\mu^2 \cos 2(\varphi' - \varphi) & \mu^2\mu' \sin 2(\varphi' - \varphi) & 0 \\ -\mu'^2 \cos 2(\varphi' - \varphi) & \cos 2(\varphi' - \varphi) & -\mu' \sin 2(\varphi' - \varphi) & 0 \\ -\mu\mu'^2 \sin 2(\varphi' - \varphi) & \mu \sin 2(\varphi' - \varphi) & \mu\mu' \cos 2(\varphi' - \varphi) & 0 \\ 0 & 0 & 0 & 0 \end{bmatrix}. \tag{4.204}$$

It should be noted that the $\mathbf{K}^{(i)}(\mu', \varphi'; \mu, \varphi)$ satisfy the symmetry relation

$$[\mathbf{K}^{(i)}(\mu, \varphi; \mu', \varphi')]^T = \mathbf{K}^{(i)}(\mu', \varphi'; \mu, \varphi), \tag{4.205}$$

in accord with the more general symmetry statement given by Eq. (4.164). It should also be noted that the structure of the $\mathbf{K}^{(i)}$ matrices is such that the equation of transfer for the Stokes parameter V is decoupled from the equations of transfer for the parameters I_l, I_r, and U. Thus in the case of Thomson scattering the complete solution of the equation of transfer including polarization effects is obtained in general geometry by solving a scalar equation of transfer for V and a three by three matrix equation of transfer for I_l, I_r, and U.

The situation simplifies even further if one is interested in systems with plane (or spherical) symmetry. In this case the $\mathbf{\Omega}$ dependences of the parameters I_l, I_r, U, and V describing the (partially) polarized light are fully described by μ dependences alone (there are no φ dependences). Thus the terms in the equation of transfer involving the $\mathbf{K}^{(1)}$ and $\mathbf{K}^{(2)}$ matrices, because all of the elements of these two matrices are periodic in φ' with period π or 2π, give a zero contribution to the inscattering term. We then find for a planar system with Thomson scattering that the four equations of transfer contained in Eq. (4.158) become

$$\frac{1}{c}\frac{\partial I_l(\mu, \nu)}{\partial t}+\mu\frac{\partial I_l(\nu, \mu)}{\partial z}+\sigma(\nu)\,I_l(\nu, \mu) = \frac{1}{2}\,S(\nu)$$

$$+\frac{3\sigma_s(\nu)}{8}\int_{-1}^{1} d\mu'\,\{[2(1-\mu^2)(1-\mu'^2)+\mu^2\mu'^2]I_l(\nu, \mu')+\mu^2 I_r(\nu, \mu')\},\qquad(4.206)$$

$$\frac{1}{c}\frac{\partial I_r(\nu, \mu)}{\partial t}+\mu\frac{\partial I_r(\nu, \mu)}{\partial z}+\sigma(\nu)\,I_r(\nu, \mu) = \frac{1}{2}\,S(\nu)$$

$$+\frac{3\sigma_s(\nu)}{8}\int_{-1}^{1} d\mu'\,[\mu'^2 I_l(\nu, \mu')+I_r(\nu, \mu')],\qquad(4.207)$$

$$\frac{1}{c}\frac{\partial U(\nu, \mu)}{\partial t}+\mu\frac{\partial U(\nu, \mu)}{\partial z}+\sigma(\nu)\,U(\nu, \mu) = 0,\qquad(4.208)$$

$$\frac{1}{c}\frac{\partial V(\nu, \mu)}{\partial t}+\mu\frac{\partial V(\nu, \mu)}{\partial z}+\sigma(\nu)\,V(\nu, \mu) = \frac{3\sigma_s(\nu)}{4}\int_{-1}^{1} d\mu'\,\mu\mu' V(\nu, \mu').\qquad(4.209)$$

In writing these equations we have performed the integrations over the Dirac delta functions in frequency in the inscattering terms, and have also (in the inscattering terms) integrated over φ', which simply gives a factor of 2π. Further, we have taken the source vector **S** as describing unpolarized light, and hence $S_l = S_r = \frac{1}{2}S$ and $S_U = S_V = 0$.

Clearly Eqs. (4.208) and (4.209) have as their solution

$$U = V = 0,\qquad(4.210)$$

and thus in systems with plane symmetry one need solve only two (coupled) equations of transfer, given by Eqs. (4.206) and (4.207), for I_l and I_r to obtain a complete solution to the radiative transfer problem including polarization effects. [We assume that the boundary and initial conditions are consistent with Eq. (4.210).] As noted in the last section, Eq. (4.210) is not restricted to the case of Thomson scattering, but is true for any scattering law in problems with plane (or spherical) symmetry [see discussion in the second paragraph following Eq. (4.164)]. Equivalently, we can write Eqs. (4.206) and (4.207) in terms of the total intensity $I \equiv I_l+I_r$, and the difference function $Q \equiv I_l-I_r$. We find, by first adding Eqs. (4.206) and (4.207) and then subtracting Eq. (4.207) from Eq. (4.206) respectively, the results

$$\frac{1}{c}\frac{\partial I(\nu, \mu)}{\partial t}+\mu\frac{\partial I(\nu, \mu)}{\partial z}+\sigma(\nu)\,I(\nu, \mu) = S(\nu)$$

$$+\frac{\sigma_s(\nu)}{2}\left[\int_{-1}^{1} d\mu'\,I(\nu, \mu')+\frac{1}{2}P_2(\mu)\int_{-1}^{1} d\mu'\,P_2(\mu')\,I(\nu, \mu')\right.$$

$$\left.-\frac{1}{2}P_2(\mu)\int_{-1}^{1} d\mu'\,Q(\nu, \mu')+\frac{1}{2}P_2(\mu)\int_{-1}^{1} d\mu'\,P_2(\mu')\,Q(\nu, \mu')\right],\qquad(4.211)$$

The Equations of Radiation Hydrodynamics

$$\frac{1}{c}\frac{\partial Q(\nu, \mu)}{\partial t}+\mu\,\frac{\partial Q(\nu, \mu)}{\partial z}+\sigma(\nu)\,Q(\nu, \mu) = \frac{\sigma_s(\nu)}{4}\,[1-P_2(\mu)]\times$$

$$\times\left[\int_{-1}^{1} d\mu'Q(\nu, \mu')-\int_{-1}^{1} d\mu'P_2(\mu')\,Q(\nu, \mu')-\int_{-1}^{1} d\mu'P_2(\mu')\,I(\nu, \mu')\right]. \quad (4.212)$$

In writing equations (4.211) and (4.212) we have introduced the Legendre polynomial $P_2(\mu) = \frac{3}{2}\mu^2-\frac{1}{2}$.

The usual equation of transfer for Thomson scattering, that neglecting polarization effects, is simply Eq. (4.211) with Q set equal to zero, i.e.,

$$\frac{1}{c}\frac{\partial I(\nu, \mu)}{\partial t}+\mu\frac{\partial I(\nu, \mu)}{\partial z}+\sigma(\nu)\,I(\nu, \mu) = S(\nu)$$

$$+\frac{\sigma_s(\nu)}{2}\left[\int_{-1}^{1} d\mu'I(\nu, \mu')+\frac{1}{2}\,P_2(\mu)\int_{-1}^{1} d\mu'P_2(\mu')\,I(\nu, \mu')\right], \quad (4.213)$$

or equivalently,

$$\frac{1}{c}\frac{\partial I(\nu,\mu)}{\partial t}+\mu\frac{\partial I(\nu, \mu)}{\partial z}+\sigma(\nu)\,I(\nu, \mu) = S(\nu)+\frac{3\sigma_s(\nu)}{16\pi}\int_{4\pi} d\Omega'[1+(\mathbf{\Omega}\cdot\mathbf{\Omega}')^2]\,I(\nu, \mu'). \quad (4.214)$$

However, $Q = 0$ will not in general satisfy Eq. (4.212), and thus an equation of transfer for the total intensity alone must in general be in error. Even if one is ultimately interested in only the specific intensity, one must nevertheless formulate the equation of transfer with proper account taken of the state of polarization in order to correctly calculate the total intensity I.

A question of practical importance is how large is the error associated with the neglect of the state of polarization in the formulation of the equation of transfer. To partially answer this question, let us consider diffusion-like approximations (namely the Eddington and asymptotic theories of Sections 2 through 4 of Chapter III) to the equation of transfer describing Thomson scattering, with and without a proper accounting of polarization effects. In addition, we shall find it useful to discuss the scalar equation of transfer (that which neglects polarization) with an isotropic scattering kernel, i.e.,

$$K(\nu' \to \nu, \mathbf{\Omega}'\cdot\mathbf{\Omega}) = \frac{1}{4\pi}\,\delta(\nu-\nu'). \quad (4.215)$$

Specifically, then, we shall consider in plane geometry diffusion-like approximations to three equations of transfer. These are, in order of increasing complexity:

A. Isotropic scattering neglecting polarization

$$\frac{1}{c}\frac{\partial I(\mu)}{\partial t}+\mu\,\frac{\partial I(\mu)}{\partial z}+\sigma I = \frac{\sigma\tilde{\omega}}{2}\int_{-1}^{1} d\mu'I(\mu')+S. \quad (4.216)$$

B. Thomson scattering neglecting polarization

$$\frac{1}{c}\frac{\partial I(\mu)}{\partial t}+\mu\frac{\partial I(\mu)}{\partial z}+\sigma I = S+\frac{\sigma\tilde\omega}{2}\left[\int_{-1}^{1}d\mu' I(\mu')+\frac{1}{2}P_2(\mu)\int_{-1}^{1}d\mu' P_2(\mu')I(\mu')\right]. \tag{4.217}$$

C. Thomson scattering with polarization

$$\frac{1}{c}\frac{\partial I(\mu)}{\partial t}+\mu\frac{\partial I(\mu)}{\partial z}+\sigma I(\mu) = S+\frac{\sigma\tilde\omega}{2}\left[\int_{-1}^{1}d\mu' I(\mu')+\frac{1}{2}P_2(\mu)\int_{-1}^{1}d\mu' P_2(\mu')I(\mu')\right.$$

$$\left.-\frac{1}{2}P_2(\mu)\int_{-1}^{1}d\mu' Q(\mu')+\frac{1}{2}P_2(\mu)\int_{-1}^{1}d\mu' P_2(\mu')Q(\mu')\right], \tag{4.218}$$

$$\frac{1}{c}\frac{\partial Q(\mu)}{\partial t}+\mu\frac{\partial Q(\mu)}{\partial z}+\sigma Q(\mu) = \frac{\sigma\tilde\omega}{4}[1-P_2(\mu)]\times$$

$$\times\left[\int_{-1}^{1}d\mu' Q(\mu')-\int_{-1}^{1}d\mu' P_2(\mu')Q(\mu')-\int_{-1}^{1}d\mu' P_2(\mu')I(\mu')\right]. \tag{4.219}$$

Here we have defined $\tilde\omega \equiv \sigma_s/\sigma = \sigma_s/(\sigma_s+\sigma_a)$. We have also dropped the frequency argument ν in all terms since it appears only as a parameter in all equations of transfer we are considering. (This is because scattering events in all three cases we are considering do not change the frequency of the photons. Hence photons of different frequencies diffuse independently, and the corresponding equations of transfer are uncoupled in frequency.)

We first consider Eq. (4.216), the scalar transport description with isotropic scattering. As a preliminary to obtaining diffusion-like descriptions, let us form the first two angular moments of this equation. Integrating Eq. (4.216) over all solid angle $(0 \le \varphi \le 2\pi, -1 \le \mu \le 1)$ we obtain

$$\frac{1}{c}\frac{\partial I_0}{\partial t}+\frac{\partial I_1}{\partial z}+\sigma(1-\tilde\omega) I_0 = 4\pi S, \tag{4.220}$$

where we have defined

$$I_0 \equiv \int_{4\pi} d\Omega I(\mu) = 2\pi\int_{-1}^{1}d\mu I(\mu), \tag{4.221}$$

$$I_1 \equiv \int_{4\pi} d\Omega \mu I(\mu) = 2\pi\int_{-1}^{1}d\mu\mu I(\mu). \tag{4.222}$$

Equation (4.220) is simply the continuity equation expressing conservation of photons at the frequency under consideration. Similarly, multiplication of Eq. (4.216) by μ prior to an integration over all solid angle yields

$$\frac{1}{c}\frac{\partial I_1}{\partial t}+\frac{\partial(\sigma D I_0)}{\partial z}+\sigma I_1 = 0, \tag{4.223}$$

where D is defined as

$$D \equiv \frac{\int\limits_{-1}^{1} d\mu\,\mu^2\, I(\mu)}{\sigma \int\limits_{-1}^{1} d\mu\, I(\mu)}. \tag{4.224}$$

Equations (4.220) and (4.223) are exact, and represent two equations for the two unknowns I_0 and I_1, if the diffusion coefficient D is known. To obtain the diffusion-like results of Chapter III, one must neglect the $\partial I_1/\partial t$ term in Eq. (4.223) [see the discussion in the paragraph containing Eq. (3.16)]. For the present purposes, however, we can just as well carry this term.

Calculation of the first two angular moments of the scalar equation of transfer describing Thomson scattering, Eq. (4.217), also leads to Eqs. (4.220) and (4.223) with D again defined by Eq. (4.224). Similarly, the equation of transfer for $I(\mu)$ with polarization effects included, Eq. (4.218), also has the same first two angular moments. It is an interesting fact that all three descriptions of scattering give formally the same two equations for the unknowns I_0 (essentially the radiative energy density) and I_1 (essentially the radiative flux). In particular, in the case of Thomson scattering with polarization effects included, the equations for I_0 and I_1 contain no reference to the function $Q \equiv I_l - I_r$ which is a measure of the degree of departure of the radiation field from natural (unpolarized) light. Of course, D as defined by Eq. (4.224) is a homogeneous functional of the specific intensity, and this intensity will depend upon the details of the scattering description. However, since the right hand side of Eq. (4.224) only involves integrals of $I(\mu)$, and the process of integration generally smooths out details of the integrand, one might expect the diffusion coefficient D to be a relatively weak functional of $I(\mu)$. This observation is the underlying basis of any diffusion-like description of radiative transfer.

In passing, we note that the first two angular moments of the equation for $Q(\mu)$, Eq. (4.219), are given by

$$\frac{1}{c}\frac{\partial Q_0}{\partial t} + \frac{\partial Q_1}{\partial z} + \sigma Q_0 = \frac{\sigma\tilde{\omega}}{4}\left[3(1-D')Q_0 + (1-3D)I_0\right], \tag{4.225}$$

$$\frac{1}{c}\frac{\partial Q_1}{\partial t} + \frac{\partial(\sigma D' Q_0)}{\partial z} + \sigma Q_1 = 0, \tag{4.226}$$

where we have defined

$$D' \equiv \frac{\int\limits_{-1}^{1} d\mu\,\mu^2\, Q(\mu)}{\sigma \int\limits_{-1}^{1} d\mu\, Q(\mu)}, \tag{4.227}$$

and the Q_n for $n = 0$ and 1 are the first two angular moments of $Q(\mu)$, defined analogously to I_0 and I_1 [see Eqs. (4.221) and (4.222)]. These equations do not enter in any direct way in our subsequent considerations, and are given here only for completeness. They must be considered, of course, if one is interested in the degree of polarization of the radiation field in addition to the radiative energy flow.

Returning now to Eqs. (4.220) and (4.223), we note that although these equations are formally exact relations involving I_0 and I_1, it remains to specify the diffusion coefficient D to close this set of equations. In diffusion-like approximations, one *a priori* assumes a form for $I(\mu)$ in Eq. (4.224) and uses the resulting value for D in Eq. (4.223) which, coupled with Eq. (4.220), gives a closed set of two equations for the two unknowns I_0 and I_1. We wish to show that in the diffusion-like theories described in Sections 2 through 4 of Chapter III, the diffusion coefficient D is either totally independent or a weak functional of the description of scattering.

We first discuss the Eddington, or classical diffusion, approximation introduced in Section 2 of Chapter III. The Eddington approximation corresponds to the assumption that $I(\mu)$ is given by [see Eq. (3.1)]

$$I(\mu) = \frac{1}{4\pi}(I_0 + 3\mu I_1). \tag{4.228}$$

Use of Eq. (4.228) in Eq. (4.224) leads to the result

$$D = \frac{1}{3\sigma}. \tag{4.229}$$

The important point here is that Eq. (4.229) is valid, in the Eddington approximation, for all three scattering descriptions under consideration.

We next consider asymptotic diffusion theory as introduced in Section 3 of Chapter III. In this case the diffusion coefficient will depend upon the description of the scattering process. The question of importance here is how strong is this dependence. The diffusion coefficient in asymptotic diffusion theory is that corresponding to an asymptotic angular distribution for the specific intensity. By an asymptotic distribution we mean that which corresponds to a source free ($S = 0$), homogeneous (σ_s and σ_a spatially independent), time independent transport problem.

The analysis for isotropic scattering has already been given in Chapter III, but we shall repeat it here in a form specifically suited to the present purposes. The asymptotic solution we seek is a solution of the equation, from Eq. (4.216),

$$\mu\,\frac{\partial I(z, \mu)}{\partial z} + \sigma I(z, \mu) = \frac{\sigma\bar{\omega}}{2}\int_{-1}^{1} d\mu'\, I(z, \mu'), \tag{4.230}$$

where we have explicitly indicated the spatial as well as the angular dependence of the specific intensity. We look for a separable solution of the form

$$I(z, \mu) = \psi(\mu)e^{-\alpha\sigma z}, \tag{4.231}$$

where $\psi(\mu)$ and α are to be determined. Use of Eq. (4.231) in Eq. (4.230) leads to

$$(1 - \mu\alpha)\psi(\mu) = \frac{\bar{\omega}}{2}\int_{-1}^{1} d\mu'\, \psi(\mu')\,. \tag{4.232}$$

The Equations of Radiation Hydrodynamics

Multiplication of Eq. (4.232) by $(1-\mu\alpha)^{-1}$ and subsequent integration over all μ yields

$$1 = \frac{\tilde{\omega}}{\alpha}\, Q_0(1/\alpha) \tag{4.233}$$

as the equation for α. Here $Q_0(\xi)$ is the zeroth order Legendre function of the second kind defined by (Abramowitz and Stegun, 1964)

$$Q_n(\xi) \equiv \frac{1}{2}\int_{-1}^{1} dt\, \frac{P_n(t)}{\xi - t}. \tag{4.234}$$

The $Q_n(\xi)$ introduced here are not to be confused with the angular moments of $Q(z, v, \mu, t)$ introduced earlier. The associated angular dependence is trivially obtained from Eq. (4.232) as, aside from an arbitrary normalization constant,

$$\psi(\mu) = (1-\mu\alpha)^{-1}. \tag{4.235}$$

Now it is easily shown from Eq. (4.232) that the expression for the diffusion coefficient, Eq. (4.224), is equivalent to

$$D = \frac{1-\tilde{\omega}}{\sigma\alpha^2}. \tag{4.236}$$

For the other two descriptions of scattering to be considered shortly, Eq. (4.236), with the value of α appropriate to the description under consideration, can also be shown to be equivalent to Eq. (4.224).

We now consider Thomson scattering neglecting polarization. From Eq. (4.217), the equation we need consider is

$$\mu\,\frac{\partial I(z, \mu)}{\partial z}+\sigma I(z, \mu) = \frac{\sigma\tilde{\omega}}{2}\left[\int_{-1}^{1} d\mu'\,I(z, \mu')+\frac{1}{2}\,P_2(\mu)\int_{-1}^{1} d\mu'\,P_2(\mu')\,I(z, \mu')\right]. \tag{4.237}$$

Use once again of the assumed form given by Eq. (4.231) yields

$$(1-\mu\alpha)\,\psi(\mu) = \frac{\tilde{\omega}}{2}\left[h_0(\alpha)+\frac{1}{2}\,P_2(\mu)\,h_2(\alpha)\right], \tag{4.238}$$

where we have defined

$$h_n(\alpha) = \int_{-1}^{1} d\mu\, P_n(\mu)\,\psi(\mu). \tag{4.239}$$

Equation (4.238) leads to, upon multiplication by $(1-\mu\alpha)^{-1}$ and integration over all μ,

$$h_0(\alpha) = \frac{\tilde{\omega}}{\alpha}\left[h_0(\alpha)\,Q_0(1/\alpha)+\frac{1}{2}\,h_2(\alpha)\,Q_2(1/\alpha)\right]. \tag{4.240}$$

Further, forming the first two angular moments of Eq. (4.238) we find

$$(1-\tilde{\omega})\,h_0(\alpha) = \alpha h_1(\alpha), \tag{4.241}$$

$$3h_1(\alpha) = \alpha[2h_2(\alpha)+h_0(\alpha)]. \tag{4.242}$$

Equations (4.240) through (4.242) represent three homogeneous equations for h_0, h_1, and h_2. The corresponding coefficient determinant must vanish for a nontrivial solution to exist, and this gives the condition on α. We defer giving an explicit equation for α until we have considered the third and final description of scattering, namely Thomson scattering with a proper accounting of polarization effects.

From Eqs. (4.218) and (4.219) we see that the equations we must consider in this case are

$$\mu\,\frac{\partial I(z,\,\mu)}{\partial z}+\sigma I(z,\,\mu) = \frac{\sigma\tilde{\omega}}{2}\left[\int_{-1}^{1} d\mu' I(z,\,\mu')+\frac{1}{2}\,P_2(\mu)\int_{-1}^{1} d\mu' P_2(\mu')\,I(z,\,\mu')\right.$$

$$\left.-\frac{1}{2}\,P_2(\mu)\int_{-1}^{1} d\mu' Q(z,\,\mu')+\frac{1}{2}\,P_2(\mu)\int_{-1}^{1} d\mu' P_2(\mu')\,Q(z,\,\mu')\right], \qquad (4.243)$$

$$\mu\,\frac{\partial Q(z,\,\mu)}{\partial z}+\sigma Q(z,\,\mu) = \frac{\sigma\tilde{\omega}}{4}\,[1-P_2(\mu)]\times$$

$$\times\left[\int_{-1}^{1} d\mu' Q(z,\,\mu')-\int_{-1}^{1} d\mu' P_2(\mu')\,Q(z,\,\mu')-\int_{-1}^{1} d\mu' P_2(\mu')\,I(z,\,\mu')\right].$$

$$(4.244)$$

In this case, in addition to Eq. (4.231), we use the assumed form

$$Q(z,\,\mu) = \varphi(\mu)\,e^{-\sigma\alpha z}, \qquad (4.245)$$

where α is the same α appearing in Eq. (4.231) and $\varphi(\mu)$ is to be determined. Use of these two forms in Eqs. (4.243) and (4.244) yields

$$(1-\mu\alpha)\,\psi(\mu) = \frac{\tilde{\omega}}{2}\left\{h_0(\alpha)+\frac{1}{2}\,P_2(\mu)\,[h_2(\alpha)-g_0(\alpha)+g_2(\alpha)]\right\}, \qquad (4.246)$$

$$(1-\mu\alpha)\,\varphi(\mu) = \frac{\tilde{\omega}}{4}\,[1-P_2(\mu)]\,[g_0(\alpha)-g_2(\alpha)-h_2(\alpha)], \qquad (4.247)$$

with $h_n(\alpha)$ given by Eq. (4.239) and

$$g_n(\alpha) = \int_{-1}^{1} d\mu P_n(\mu)\,\varphi(\mu). \qquad (4.248)$$

Equations (4.246) and (4.247) lead to, upon multiplication by $(1-\mu\alpha)^{-1}$ and integration over μ,

$$h_0(\alpha) = \frac{\tilde{\omega}}{\alpha}\left\{h_0(\alpha)\,Q_0(1/\alpha)+\frac{1}{2}\,[h_2(\alpha)-g_0(\alpha)+g_2(\alpha)]\,Q_2(1/\alpha)\right\}, \qquad (4.249)$$

$$g_0(\alpha) = \frac{\tilde{\omega}}{2\alpha}\,[g_0(\alpha)-g_2(\alpha)-h_2(\alpha)]\,[Q_0(1/\alpha)-Q_2(1/\alpha)]. \qquad (4.250)$$

Further, forming the first two angular moments of Eqs. (4.246) and (4.247) we find

$$(1-\tilde{\omega})\,h_0(\alpha) = \alpha h_1(\alpha), \tag{4.251}$$

$$3h_1(\alpha) = \alpha[2h_2(\alpha)+h_0(\alpha)], \tag{4.252}$$

$$2[g_0(\alpha)-\alpha g_1(\alpha)] = \tilde{\omega}[g_0(\alpha)-g_2(\alpha)-h_2(\alpha)], \tag{4.253}$$

$$3g_1(\alpha) = \alpha[2g_2(\alpha)+g_0(\alpha)]. \tag{4.254}$$

The requirement that Eqs. (4.249) through (4.254) have a nonzero solution for the g_n and h_n yields a transcendental equation for α.

After some algebra one can show that all of our results for the parameter α can be summarized by the equation

$$1 = \frac{\tilde{\omega}}{\alpha}\left\{Q_0(1/\alpha)+\frac{F(\alpha)}{4}\left[\frac{3(1-\tilde{\omega})}{\alpha^2}\right]Q_2(1/\alpha)\right\}. \tag{4.255}$$

In this equation, $F(\alpha) = 0$ for isotropic scattering, $F(\alpha) = 1$ for Thomson scattering neglecting polarization, and

$$\frac{1}{F(\alpha)} = 1-\frac{3\tilde{\omega}}{4\alpha^2}\left\{1+\frac{(1-\alpha^2)}{\alpha}[Q_2(1/\alpha)-Q_0(1/\alpha)]\right\} \tag{4.256}$$

for Thomson scattering with polarization. Once Eq. (4.255) has been solved for α, the diffusion coefficient D follows from Eq. (4.236). Table 4.1 gives the results of numerical

TABLE 4.1. *The Asymptotic Diffusion Coefficient*

$\tilde{\omega}$	$\sigma D^{(1)}$	$\sigma D^{(2)}$	$\sigma D^{(3)}$
0.0	1.0000	1.0000	1.0000
0.2	0.8001	0.8009	0.8011
0.4	0.6176	0.6271	0.6315
0.6	0.4859	0.4972	0.5065
0.8	0.3963	0.4026	0.4110
1.0	0.3333	0.3333	0.3333

[1] Isotopic scattering neglecting polarization.
[2] Thomson scattering neglecting polarization.
[3] Thomson scattering with polarization.

calculations of the diffusion coefficient for representative values of $\tilde{\omega}$. It can be seen that σD is a relatively weak functional of the scattering description. Further, the error introduced through the neglect of polarization and the error due to the neglect of the anisotropy of Thomson scattering are of comparable magnitude. This suggests that if polarization effects are neglected, as is generally the case in radiation hydrodynamic calculations, it is not inconsistent to neglect the anisotropy of Thomson scattering as well. This observation has very practical implications since the equation of transfer is significantly easier to solve, either analytically or numerically, if the scattering kernel can be assumed isotropic, as given by Eq. (4.215).

These conclusions only follow rigorously, of course, for that class of problems which is well described by one of the diffusion-like descriptions we have discussed (Eddington or asymptotic). For most problems, however, these diffusion-like theories are reasonably accurate and do describe the energy flow due to radiative processes in a semi-quantitative sense. It should also be recalled that in Section 4 of Chapter III we discussed a diffusion-like theory which combines certain features of both the Eddington and asymptotic theories. The analysis of the asymptotic theory we have just given carries over in a straightforward way to that combined theory. More to the point, the conclusion is the same. That is, we conclude that for problems well described by the diffusion-like theory discussed in Section 4 of Chapter III, a proper accounting of the effects of polarization in Thomson scattering is relatively unimportant from the point of view of radiative energy flow.

V

Radiative Transfer in Refractive
and Dispersive Media

---

1. Introduction

In Chapter II the equation of radiative transfer was derived under the assumption that photons travel in straight lines at speed c (the vacuum speed of light) with no change in frequency between collisions. It is well known, however, that a photon can travel at less than the vacuum speed of light in matter with a refractive index other than unity. It is also known that if the refractive index is a (continuous) function of position, a photon will not stream in a straight line between collisions but will undergo (continuous) refraction which leads to curved streaming paths. In addition, if the refractive index is a (continuous) function of time, a photon will (continuously) change its frequency as it streams between collisions. The assumption made in Chapter II is that the matter has a refractive index of unity at all points in space and at all times.

In the present chapter we relax this restriction, and show how one can incorporate refractive and dispersive effects into the streaming terms of the equation of radiative transfer. The next section discusses the "equations of motion" for a photon, which quantitatively describe the changes in the direction of propagation and the frequency of a photon as it streams in a refractive and dispersive medium. Section 3 then shows how these equations of motion can be incorporated into the equation of radiative transfer.

2. Hamilton's Equations for Photons

We begin by assuming that the equations (i.e., Maxwell's equations) describing waves of the form

$$\xi(\mathbf{r}, t) = \xi_0 \, e^{i(\mathbf{k}\cdot\mathbf{r} - \omega t)} \tag{5.1}$$

in a homogeneous, time independent medium have been solved to yield a relationship between the circular frequency $\omega(\omega = 2\pi\nu)$ and the wave vector $\mathbf{k}$. This relationship is generally referred to as the dispersion relation and written

$$D(\omega, \mathbf{k}) \equiv D(\omega, k_x, k_y, k_z) = 0. \tag{5.2}$$

Equation (5.2) can in principle be solved for ω as a function of $\mathbf{k}$ as

$$\omega_\alpha(\mathbf{k}) \equiv \omega_{\alpha r}(\mathbf{k}) + i\omega_{\alpha i}(\mathbf{k}), \tag{5.3}$$

where $i \equiv \sqrt{-1}$. In writing Eq. (5.3) we have treated $\mathbf{k}$ as a real vector and separated real and imaginary parts of $\omega_\alpha(\mathbf{k})$, denoting these by the subscripts r and i, respectively. The subscript α denotes a particular solution, corresponding to a particular type of wave, since in general the solution of Eq. (5.2) for ω as a function of $\mathbf{k}$ will yield more than one solution. Now, if the material properties depend upon position and time, then the dispersion relation will also depend upon these variables, and we write

$$\omega_\alpha \equiv \omega_\alpha(\mathbf{r},\ \mathbf{k},\ t) = \omega_{\alpha r}(\mathbf{r},\ \mathbf{k},\ t) + i\omega_{\alpha i}(\mathbf{r},\ \mathbf{k},\ t) \tag{5.4}$$

in place of Eq. (5.3).

We now assume that the dependence of ω_α on $\mathbf{r}$ and t is weak. In particular we assume that the change in ω_α over a distance of one wavelength and in a time of one period corresponding to the oscillations described by Eq. (5.1) is small compared to ω_α itself. Under these conditions it is reasonable to follow the motion of a wave packet (a wave train of finite extent) through matter (Harris, 1965). That is, under these conditions a wave packet will remain recognizable as a wave packet as it passes through matter. We further assume that the wave packet is sufficiently monochromatic so that one can assign a reasonably definite circular frequency ω to it. This implies an associated wave train for the packet which is long compared to the wave length. At the same time, however, we assume that the wave train is sufficiently short so that one can assign a reasonably definite spatial position $\mathbf{r}$ to the wave packet. (The term "reasonably definite" as used here implies an uncertainty which is for all practical purposes negligibly small.)

Under such conditions it is known (Weinberg, 1962; Stix, 1962; Harris, 1965) that the motion of a wave packet is the same as that of a particle of momentum $\mathbf{k}$ whose Hamiltonian is $\omega_{\alpha r}(\mathbf{k}, \mathbf{r}, t)$. The equations of motion are then Hamilton's equations

$$\dot{\mathbf{r}} = \mathbf{V}_{gr} = \nabla_k \omega, \tag{5.5}$$

$$\dot{\mathbf{k}} = -\nabla_r \omega, \tag{5.6}$$

where $\dot{\mathbf{r}}$ denotes the total time derivative of the position of the centre of the wave packet in physical space, and $\dot{\mathbf{k}}$ denotes the total time derivative of the position of the wave packet in wave vector space. In writing Eqs. (5.5) and (5.6) we have dropped the subscript α on $\omega_{\alpha r}$ since we have sufficiently emphasized that more than one type of wave may exist, and henceforth we shall deal only with waves of species α. We have also dropped the subscript r on $\omega_{\alpha r}$ for notational simplicity. In all subsequent work we shall use ω, but actually mean $\omega_{\alpha r}$. We shall further refer to the wave packets as photons.

Equation (5.5) is just the usual definition of the group velocity $\mathbf{V}_{gr}$, and Eq. (5.6) completes the specification of the dynamics of the system. We note that in general the direction of the group velocity $\mathbf{V}_{gr}$ does not coincide with the direction of the wave vector $\mathbf{k}$. These two directions coincide only if the medium is isotropic (has no preferred direction), in which case ω depends upon the magnitude of $\mathbf{k}$, denoted by k, but not its direction. For then we

The Equations of Radiation Hydrodynamics

have

$$\dot{\mathbf{r}} = \mathbf{V}_{\mathrm{gr}} = \nabla_k \omega = \frac{\partial \omega}{\partial k}\, \nabla_k k = V_{\mathrm{gr}} \boldsymbol{\Omega}, \tag{5.7}$$

where we have defined the group speed V_{gr} as

$$V_{\mathrm{gr}} \equiv \frac{\partial \omega}{\partial k}, \tag{5.8}$$

and a unit vector $\boldsymbol{\Omega}$ in the direction of propagation as

$$\boldsymbol{\Omega} \equiv \mathbf{k}/k. \tag{5.9}$$

In radiative transfer work, the phase space variables are conventionally taken as $\mathbf{r}$, ν (or equivalently ω since $\omega = 2\pi\nu$), and $\boldsymbol{\Omega}$ rather than $\mathbf{r}$ and $\mathbf{k}$. Accordingly, let us obtain the equations of motion for the variables $\mathbf{r}$, ω, and $\boldsymbol{\Omega}$. For isotropic matter, which we assume, Eq. (5.7) is the relevant equation of motion for the spatial position of the photon. We need derive expressions for $\dot{\omega}$ and $\dot{\boldsymbol{\Omega}}$. We first consider $\dot{\omega}$. Since the dot implies a total time derivative, and ω is an implicit function of time through its dependences upon $\mathbf{r}$ and $\mathbf{k}$, as well as an explicit function of time, we have

$$\dot{\omega} = \frac{\partial \omega}{\partial t} + \dot{\mathbf{r}} \cdot \nabla_r \omega + \dot{\mathbf{k}} \cdot \nabla_k \omega. \tag{5.10}$$

Use of the equations of motion, Eqs. (5.5) and (5.6), in Eq. (5.10) yields the simple result

$$\dot{\omega} = \frac{\partial \omega}{\partial t}. \tag{5.11}$$

Introducing the refractive index $n(\mathbf{r}, k, t)$ as

$$n \equiv ck/\omega, \tag{5.12}$$

we have

$$\dot{\omega} = \frac{\partial}{\partial t}(ck/n) = -\frac{ck}{n^2}\frac{\partial n}{\partial t} = -\frac{\omega}{n}\frac{\partial n}{\partial t}. \tag{5.13}$$

The partial derivatives with respect to time in Eq. (5.13) are taken with k held constant, i.e., the refractive index n is considered as a function of $\mathbf{r}$, k, and t. Let us now consider the refractive index to be a function of $\mathbf{r}$, ω, and t, and compute the partial time derivative of n with ω and $\mathbf{r}$ held constant. The total differential dn, considering $n \equiv n(\mathbf{r}, \omega, t)$, is given by

$$dn = dt\left(\frac{\partial n}{\partial t}\right)_{\mathbf{r},\,\omega} + d\omega\left(\frac{\partial n}{\partial \omega}\right)_{\mathbf{r},\,t} + d\mathbf{r}\cdot(\nabla_r \omega)_{\omega,\,t}. \tag{5.14}$$

From Eq. (5.14) we obtain

$$\left(\frac{\partial n}{\partial t}\right)_{\mathbf{r},\,k} = \left(\frac{\partial n}{\partial t}\right)_{\mathbf{r},\,\omega} + \left(\frac{\partial \omega}{\partial t}\right)_{\mathbf{r},\,k}\left(\frac{\partial n}{\partial \omega}\right)_{\mathbf{r},\,t}. \tag{5.15}$$

146

From Eq. (5.12) we find

$$\left(\frac{\partial \omega}{\partial t}\right)_{r,\,k} = -\frac{ck}{n^2}\left(\frac{\partial n}{\partial t}\right)_{r,\,k} = -\frac{\omega}{n}\left(\frac{\partial n}{\partial t}\right)_{r,\,k}, \tag{5.16}$$

and use of this result in Eq. (5.15) yields

$$\left(\frac{\partial n}{\partial t}\right)_{r,\,\omega} = \frac{1}{n}\left(\frac{\partial n}{\partial t}\right)_{r,\,k}\left(\frac{\partial[\omega n]}{\partial \omega}\right)_{r,\,t} = \frac{c}{n}\left(\frac{\partial n}{\partial t}\right)_{r,\,k}\left(\frac{\partial k}{\partial \omega}\right)_{r,\,t}. \tag{5.17}$$

Recognizing that $\partial k/\partial \omega$ in Eq. (5.17) is just the reciprocal of the group speed [see Eq. (5.8)], we obtain our final result

$$\left(\frac{\partial n}{\partial t}\right)_{r,\,k} = \frac{nV_{gr}}{c}\left(\frac{\partial n}{\partial t}\right)_{r,\,\omega}. \tag{5.18}$$

Use of Eq. (5.18) in Eq. (5.13) gives an alternate expression for $\dot{\omega}$ as

$$\dot{\omega} = -\frac{\omega V_{gr}}{c}\frac{\partial n}{\partial t}, \tag{5.19}$$

where here the time derivative is taken with ω [rather than k as in Eq. (5.13)] held constant. To obtain an expression for $\dot{\Omega}$ we note that $\mathbf{k} = k\Omega$ and hence

$$\dot{\mathbf{k}} = \dot{k}\Omega + k\dot{\Omega} = -\nabla_r \omega, \tag{5.20}$$

where the last equality is just a restatement of Eq. (5.6). Thus we have

$$k\dot{\Omega} = -\nabla_r \omega - \dot{k}\Omega. \tag{5.21}$$

Dotting Eq. (5.6) with the vector $\mathbf{k}$ we obtain

$$-\mathbf{k}\cdot\nabla_r \omega = \mathbf{k}\cdot\dot{\mathbf{k}} = \frac{1}{2}\frac{d}{dt}(\mathbf{k}\cdot\mathbf{k}) = \frac{1}{2}\frac{dk^2}{dt} = k\dot{k}, \tag{5.22}$$

or, since $\Omega \equiv \mathbf{k}/k$, we have

$$\dot{k} = -\Omega\cdot\nabla_r \omega. \tag{5.23}$$

Use of Eq. (5.23) in Eq. (5.21) gives

$$k\dot{\Omega} = -\nabla_r \omega + \Omega(\Omega\cdot\nabla_r \omega). \tag{5.24}$$

Introducing the refractive index as defined by Eq. (5.12) into Eq. (5.24), we find

$$\dot{\Omega} = \frac{c}{n^2}[\nabla_r n - \Omega(\Omega\cdot\nabla_r n)]. \tag{5.25}$$

As before, the refractive index in Eq. (5.25) is considered as a function of r, k, and t, and the gradient operations are with k held constant. Analogous to the algebra contained in Eqs. (5.14) through (5.18), we can obtain the result

$$(\nabla_r n)_{k,\,t} = \frac{nV_{gr}}{c}(\nabla_r n)_{\omega,\,t}, \tag{5.26}$$

147

The Equations of Radiation Hydrodynamics

and Eq. (5.25) can be written as

$$\dot{\Omega} = \frac{V_{gr}}{n} [\nabla_r n - \Omega(\Omega \cdot \nabla_r n)], \qquad (5.27)$$

where all gradients in Eq. (5.27) are with ω held constant.

- Equations (5.7), (5.19), and (5.27) are the equations of motion for the phase space variables r, ω, and Ω for photons in an isotropic medium. If we introduce s as a path length variable, then for any variable A (scalar or vector) we have

$$\dot{A} = \frac{dA}{ds} \dot{s}. \qquad (5.28)$$

Now, $\dot{s}$ is just the group speed and hence

$$\dot{A} = \frac{dA}{ds} V_{gr}. \qquad (5.29)$$

Thus Eqs. (5.7), (5.19), and (5.27) are equivalent to

$$\frac{d\mathbf{r}}{ds} = \Omega, \qquad (5.30)$$

$$\frac{d\omega}{ds} = -\frac{\omega}{c} \frac{\partial n}{\partial t}, \qquad (5.31)$$

$$\frac{d\Omega}{ds} = \frac{1}{n} [\nabla_r n - \Omega(\Omega \cdot \nabla_r n)]. \qquad (5.32)$$

The equation of transfer discussed in Chapter II is based on the simple dispersion relation

$$D(\omega, k) = \omega - ck = 0. \qquad (5.33)$$

This gives

$$\omega = ck; \quad n = ck/\omega = 1. \qquad (5.34)$$

Equations (5.30) through (5.32) then yield

$$\frac{d\mathbf{r}}{ds} = \Omega, \qquad (5.35)$$

$$\frac{d\omega}{ds} = \frac{d\Omega}{ds} = 0; \qquad (5.36)$$

i.e., a photon streams with no change in either frequency or direction of propagation between collisions.

In the next section we formulate the equation of transfer for a general dispersion relation. To this end, we record here the equations of motion for r, v, Ω (rather than r, ω, Ω) since the equation of transfer is conventionally written in terms of v rather than ω. Using the relation

$\omega = 2\pi\nu$ we have, from Eqs. (5.30) through (5.32),

$$\frac{d\mathbf{r}}{ds} = \mathbf{\Omega}, \tag{5.37}$$

$$\frac{d\nu}{ds} = -\frac{\nu}{c}\frac{\partial n}{\partial t}, \tag{5.38}$$

$$\frac{d\mathbf{\Omega}}{ds} = \frac{1}{n}[\nabla_r n - \mathbf{\Omega}(\mathbf{\Omega}\cdot\nabla_r n)], \tag{5.39}$$

where the refractive index n in these equations is to be considered as a function of $\mathbf{r}$, ν, and t.

3. The Equation of Transfer for a General Dispersion Relation

Our task now is to incorporate the equations of motion for photons, Eqs. (5.37) through (5.39), into an equation of transfer. We define the distribution function $f(\mathbf{r}, \nu, \mathbf{\Omega}, t)$ such that $f\,d\mathbf{r}\,d\nu\,d\mathbf{\Omega}$ is the number of photons at time t in a phase space volume $d\mathbf{r}\,d\nu\,d\mathbf{\Omega}$ centered at $\mathbf{r}$, ν, $\mathbf{\Omega}$. If we write $d\mathbf{\Omega} = d\mu\,d\varphi$, where $\mu = \cos\theta$ is the cosine of a polar angle defining $\mathbf{\Omega}$ and φ is the corresponding azimuthal angle, then $f\,d\mathbf{r}\,d\nu\,d\mu\,d\varphi$ is the number of photons in a differential volume of phase space. Since this number of photons is independent of time as the photons stream without interaction, we must have

$$\frac{d}{dt}(f\,d\mathbf{r}\,d\nu\,d\mu\,d\varphi) = 0, \tag{5.40}$$

where the time derivative is a total time derivative as introduced below Eq. (2.19) in Chapter II. In general, of course, the right hand side of Eq. (5.40) is nonzero due to absorption, scattering, and emission of photons. For simplicity, however, we shall neglect these terms in the present analysis and concern ourselves only with the so-called streaming term in the equation of transfer. These other terms present no difficulty and are handled just as in Chapter II. Following the analysis of Section 3 of Chapter II, in particular the development from Eq. (2.21) to Eq. (2.26), we find that Eq. (5.40) can be written

$$\frac{\partial f}{\partial t}+\frac{\partial(\dot{x}f)}{\partial x}+\frac{\partial(\dot{y}f)}{\partial y}+\frac{\partial(\dot{z}f)}{\partial z}+\frac{\partial(\dot{\mu}f)}{\partial\mu}+\frac{\partial(\dot{\varphi}f)}{\partial\varphi}+\frac{\partial(\dot{\nu}f)}{\partial\nu} = 0. \tag{5.41}$$

Equation (5.41) can be put in a more compact form by using vector notation for the spatial derivative terms. This is

$$\frac{\partial f}{\partial t}+\nabla_r\cdot(\dot{\mathbf{r}}f)+\frac{\partial}{\partial\mu}(\dot{\mu}f)+\frac{\partial}{\partial\varphi}(\dot{\varphi}f)+\frac{\partial}{\partial\nu}(\dot{\nu}f) = 0. \tag{5.42}$$

Let us now show that the angular derivatives in Eq. (5.42) can be regarded as a divergence term. We consider a vector $\mathbf{A}$ in a spherical coordinate system l, θ, φ with only components in the $\mathbf{e}_\theta$ and $\mathbf{e}_\varphi$ directions, i.e.,

$$\mathbf{A} = A_\theta\mathbf{e}_\theta + A_\varphi\mathbf{e}_\varphi, \tag{5.43}$$

where e_θ and e_φ are unit vectors. We define the two dimensional divergence $\nabla_\Omega \cdot (\ \ )$ as

$$\nabla_\Omega \cdot \mathbf{A} \equiv \frac{1}{\sin\theta} \frac{\partial}{\partial\theta} (\sin\theta\, A_\theta) + \frac{1}{\sin\theta} \frac{\partial A_\varphi}{\partial\varphi}. \qquad (5.44)$$

We apply this divergence operator to $\dot{\Omega}f$, where f is any function of θ and φ. Identifying the vector Ω with e_l, it is clear that $\dot{\Omega}$ has no e_l component since Ω is a unit vector at all times.

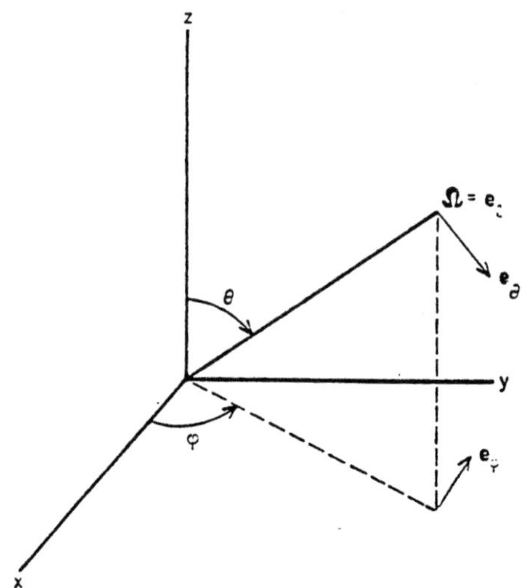

FIG. 5.1. The unit vectors e_l, e_θ, and e_φ.

Quantitatively we have (Fig. 5.1)

$$\dot{\Omega} = \dot{\theta}\, e_\theta + \sin\theta\,\dot{\varphi}\, e_\varphi, \qquad (5.45)$$

and thus

$$\nabla_\Omega \cdot (\dot{\Omega}f) = \frac{1}{\sin\theta} \frac{\partial}{\partial\theta} (\sin\theta\,\dot{\theta}f) + \frac{1}{\sin\theta} \frac{\partial}{\partial\varphi} (\sin\theta\,\dot{\varphi}) = \frac{\partial}{\partial\mu} (\dot{\mu}f) + \frac{\partial}{\partial\varphi} (\dot{\varphi}f). \qquad (5.46)$$

Use of Eq. (5.46) in Eq. (5.42) allows us to write

$$\frac{\partial f}{\partial t} + \nabla_r \cdot (\dot{r}f) + \nabla_\Omega \cdot (\dot{\Omega}f) + \frac{\partial}{\partial v} (\dot{v}f) = 0. \qquad (5.47)$$

We note that the usual spherical divergence operator $\nabla_l \cdot (\ \ )$ operating on a vector $\mathbf{A}$

$$\mathbf{A} = A_l e_l + A_\theta e_\theta + A_\varphi e_\varphi \qquad (5.48)$$

is given by

$$\nabla_l \cdot \mathbf{A} = \frac{1}{l^2} \frac{\partial}{\partial l} (l^2 A_l) + \frac{1}{l\sin\theta} \frac{\partial}{\partial\theta} (\sin\theta\, A_\theta) + \frac{1}{l\sin\theta} \frac{\partial A_\varphi}{\partial\varphi}. \qquad (5.49)$$

Since $\dot{\Omega}f$ has no e_l component, $\nabla_\Omega \cdot (\ \ )$ in Eq. (5.47) can be considered as $\nabla_l \cdot (\ \ )$ with l

set equal to one [compare Eqs. (5.44) and (5.49)], i.e.,

$$\nabla_\Omega \cdot (\dot{\Omega} f) = \nabla_l \cdot (\dot{\Omega} f)\big|_{l=1}. \tag{5.50}$$

This identity is important because it shows that the two dimensional divergence in Eq. (5.47) can be considered as a special case of the usual three dimensional divergence and hence the usual vector identities can be employed.

Use of Eq. (5.29) allows us to write Eq. (5.47) in terms of path length derivatives as

$$\frac{\partial f}{\partial t} + \nabla_r \cdot \left(\frac{d\mathbf{r}}{ds} V_{gr} f\right) + \nabla_\Omega \cdot \left(\frac{d\Omega}{ds} V_{gr} f\right) + \frac{\partial}{\partial \nu}\left(\frac{d\nu}{ds} V_{gr} f\right) = 0. \tag{5.51}$$

If we introduce the specific intensity of radiation as

$$I \equiv h\nu V_{gr} f, \tag{5.52}$$

Eq. (5.51) can be rewritten as

$$\frac{\partial}{\partial t}\left(\frac{I}{V_{gr}}\right) + \nabla_r \cdot \left(\frac{d\mathbf{r}}{ds} I\right) + \nabla_\Omega \cdot \left(\frac{d\Omega}{ds} I\right) + \nu \frac{\partial}{\partial \nu}\left(\frac{1}{\nu}\frac{d\nu}{ds} I\right) = 0. \tag{5.53}$$

Noting that $d\mathbf{r}/ds = \Omega$ [see Eq. (5.37)], we see that $d\mathbf{r}/ds$ in Eq. (5.53) is unaffected by the spatial divergence operator, and Eq. (5.53) can also be written

$$\frac{\partial}{\partial t}\left(\frac{I}{V_{gr}}\right) + \frac{d\mathbf{r}}{ds}\cdot \nabla_r I + \nabla_\Omega \cdot \left(\frac{d\Omega}{ds} I\right) + \nu \frac{\partial}{\partial \nu}\left(\frac{1}{\nu}\frac{d\nu}{ds} I\right) = 0. \tag{5.54}$$

Considering the angular divergence term in Eq. (5.54), we have from the chain rule of differentiation

$$\nabla_\Omega \cdot \left(\frac{d\Omega}{ds} I\right) = \frac{d\Omega}{ds}\cdot \nabla_\Omega I + I\nabla_\Omega \cdot \frac{d\Omega}{ds}. \tag{5.55}$$

Use of Eq. (5.39) for $d\Omega/ds$ in Eq. (5.55) yields

$$\nabla_\Omega \cdot \left(\frac{d\Omega}{ds} I\right) = \frac{d\Omega}{ds}\cdot \nabla_\Omega I - \frac{I}{n}[(\Omega \cdot \nabla_r n)\nabla_\Omega \cdot \Omega + \Omega \cdot \nabla_\Omega(\Omega \cdot \nabla_r n)]. \tag{5.56}$$

Equation (5.56) simplifies greatly by noting

$$\nabla_\Omega \cdot \Omega = \nabla_l \cdot \Omega\big|_{l=1} = \nabla_l \cdot e_l\big|_{l=1} = \frac{1}{l^2}\frac{\partial}{\partial l}(l^2)\big|_{l=1} = 2, \tag{5.57}$$

and

$$\Omega \cdot \nabla_\Omega(\Omega \cdot \nabla_r n) = \Omega \cdot \nabla_l(\Omega \cdot \nabla_r n)\big|_{l=1} = 0. \tag{5.58}$$

Equation (5.58) follows from the fact that $\Omega = e_l$ and the gradient of $\Omega \cdot \nabla_r n$ has no e_l component ($\Omega \cdot \nabla_r n$ is not a function of l, only θ and φ). Using these two results, we find that Eq. (5.56) becomes

$$\nabla_\Omega \cdot \left(\frac{d\Omega}{ds} I\right) = \frac{d\Omega}{ds}\cdot \nabla_\Omega I - \frac{2I}{n}(\Omega \cdot \nabla_r n), \tag{5.59}$$

The Equations of Radiation Hydrodynamics

and hence Eq. (5.54) can then be written (once again using the fact that $\boldsymbol{\Omega} = d\mathbf{r}/ds$)

$$\frac{\partial}{\partial t}\left(\frac{I}{V_{\text{gr}}}\right)+n^2\left[\frac{d\mathbf{r}}{ds}\cdot\nabla_r\left(\frac{I}{n^2}\right)+\frac{d\boldsymbol{\Omega}}{ds}\cdot\nabla_\Omega\left(\frac{I}{n^2}\right)\right]+v\frac{\partial}{\partial v}\left(\frac{1}{v}\frac{dv}{ds}I\right)=0. \tag{5.60}$$

We now consider the time and frequency derivative terms in Eq. (5.60). Since I/n^2 appears as the natural dependent variable in the spatial gradient term in Eq. (5.60), we write

$$\frac{\partial}{\partial t}\left(\frac{I}{V_{\text{gr}}}\right)+v\frac{\partial}{\partial v}\left(\frac{1}{v}\frac{dv}{ds}I\right)=\frac{n^2}{V_{\text{gr}}}\frac{\partial}{\partial t}\left(\frac{I}{n^2}\right)+n^2\frac{dv}{ds}\frac{\partial}{\partial v}\left(\frac{I}{n^2}\right)$$

$$+\frac{I}{n^2}\frac{\partial}{\partial t}\left(\frac{n^2}{V_{\text{gr}}}\right)-\frac{v}{c}\frac{I}{n^2}\frac{\partial}{\partial t}\left(n^2\frac{\partial n}{\partial v}\right). \tag{5.61}$$

Equation (5.61) is the result of algebraic manipulation and the use of Eq. (5.38) for dv/ds. It is easily shown that the definitions of the group speed and index of refraction, Eqs. (5.8) and (5.12), give

$$\omega\frac{\partial n}{\partial \omega}=v\frac{\partial n}{\partial v}=\frac{c}{V_{\text{gr}}}-n. \tag{5.62}$$

Use of Eq. (5.62) in Eq. (5.61) yields

$$\frac{\partial}{\partial t}\left(\frac{I}{V_{\text{gr}}}\right)+v\frac{\partial}{\partial v}\left(\frac{1}{v}\frac{dv}{ds}I\right)=\frac{n^2}{V_{\text{gr}}}\frac{\partial}{\partial t}\left(\frac{I}{n^2}\right)+n^2\frac{dv}{ds}\frac{\partial}{\partial v}\left(\frac{I}{n^2}\right)+\frac{3}{c}I\frac{\partial n}{\partial t}, \tag{5.63}$$

and use of this result in the equation of transfer, Eq. (5.60), gives

$$\frac{1}{V_{\text{gr}}}\frac{\partial}{\partial t}\left(\frac{I}{n^2}\right)+\frac{d\mathbf{r}}{ds}\cdot\nabla_r\left(\frac{I}{n^2}\right)+\frac{d\boldsymbol{\Omega}}{ds}\cdot\nabla_\Omega\left(\frac{I}{n^2}\right)+\frac{dv}{ds}\frac{\partial}{\partial v}\left(\frac{I}{n^2}\right)+\frac{3}{c}\frac{\partial n}{\partial t}\left(\frac{I}{n^2}\right)=0. \tag{5.64}$$

We can eliminate the group velocity in Eq. (5.64) in favor of the refractive index by noting that $1/V_{\text{gr}}=\partial k/\partial\omega=c^{-1}\partial(\omega n)/\partial\omega=c^{-1}\partial(vn)/\partial v$. This gives

$$\frac{1}{c}\frac{\partial(vn)}{\partial v}\frac{\partial}{\partial t}\left(\frac{I}{n^2}\right)+\frac{d\mathbf{r}}{ds}\cdot\nabla_r\left(\frac{I}{n^2}\right)+\frac{d\boldsymbol{\Omega}}{ds}\cdot\nabla_\Omega\left(\frac{I}{n^2}\right)+\frac{dv}{ds}\frac{\partial}{\partial v}\left(\frac{I}{n^2}\right)+\frac{3}{c}\frac{\partial n}{\partial t}\left(\frac{I}{n^2}\right)=0. \tag{5.65}$$

Finally, explicit use of the equations of motion, Eqs. (5.37) through (5.39), gives the result

$$\frac{1}{c}\frac{\partial(vn)}{\partial v_i}\cdot\frac{\partial}{\partial t}\left(\frac{I}{n^2}\right)+\boldsymbol{\Omega}\cdot\nabla_r\left(\frac{I}{n^2}\right)+\frac{1}{n}\nabla_r n\cdot\nabla_\Omega\left(\frac{I}{n^2}\right)-\frac{v}{c}\frac{\partial n}{\partial t}\frac{\partial}{\partial v}\left(\frac{I}{n^2}\right)+\frac{3}{c}\frac{\partial n}{\partial t}\left(\frac{I}{n^2}\right)=0, \tag{5.66}$$

where we have again used the fact the vector $\boldsymbol{\Omega}$ is orthogonal to $\nabla_\Omega\cdot(\ \ )$. Combining the last two terms of Eq. (5.66), we can rewrite this result in a more compact form as

$$\frac{1}{c}\frac{\partial(vn)}{\partial v}\frac{\partial}{\partial t}\left(\frac{I}{n^2}\right)+\boldsymbol{\Omega}\cdot\nabla_r\left(\frac{I}{n^2}\right)+\frac{1}{n}\nabla_r n\cdot\nabla_\Omega\left(\frac{I}{n^2}\right)-\frac{v^4}{c}\frac{\partial n}{\partial t}\frac{\partial}{\partial v}\left(\frac{I}{n^2 v^3}\right)=0. \tag{5.67}$$

If the refractive index n is unity, Eq. (5.67) reduces to the usual streaming term as discussed in Chapter II [see, for example, Eq. (2.15)]. It is clear that refractive and dispersive effects severely complicate the equation of transfer.

Our result can be cast in a somewhat different form by introducing the total path length derivative $d(\ )/ds$ along the photon flight path, similar to the derivative discussed in Chapter II [see Eq. (2.20)]. We have, since I/n^2 is a function of t, $\mathbf{r}$, Ω, and v,

$$\frac{d}{ds}\left(\frac{I}{n^2}\right) = \frac{dt}{ds}\frac{\partial}{\partial t}\left(\frac{I}{n^2}\right) + \frac{d\mathbf{r}}{ds}\cdot\nabla_r\left(\frac{I}{n^2}\right) + \frac{d\Omega}{ds}\cdot\nabla_\Omega\left(\frac{I}{n^2}\right) + \frac{dv}{ds}\frac{\partial}{\partial v}\left(\frac{I}{n^2}\right). \tag{5.68}$$

Recalling that $ds/dt = V_{gr}$, we see that the right hand side of Eq. (5.68) is identical to the first four terms on the left hand side of Eq. (5.64). Thus Eq. (5.64) can be rewritten

$$\frac{d}{ds}\left(\frac{I}{n^2}\right) + \frac{3}{c}\frac{\partial n}{\partial t}\left(\frac{I}{n^2}\right) = 0. \tag{5.69}$$

If the refractive index is independent of time, then $\partial n/\partial t = 0$ and Eq. (5.69) reduces to

$$\frac{d}{ds}\left(\frac{I}{n^2}\right) = 0, \tag{5.70}$$

which shows that the fundamental quantity which remains unchanged between collisions is I/n^2, the specific intensity divided by the square of the refractive index.

VI

The Effects of Fluid Motion on the Equation of Transfer

1. Introduction

It was pointed out in Section 10 of Chapter II that for large wave packets the equation of transfer is only valid if the interaction centers of the matter (i.e., the atoms, ions, and electrons) are randomly distributed. This implies no inherent preferred direction in the matter and hence σ_a, σ_a', S, and B, which describe the absorption and source of photons, should be independent of $\mathbf{\Omega}$, the flight direction of a photon. Further, σ_s, the scattering kernel, should only depend upon the scattering angle, whose cosine is $\mathbf{\Omega \cdot \Omega'}$, rather than upon the directions $\mathbf{\Omega}$ and $\mathbf{\Omega'}$ separately. However, the fact that in radiation hydrodynamic problems the material is in general in motion changes the situation. As seen by an inertial frame observer, this motion does introduce a preferred direction in the matter, namely the direction of motion of the fluid. This in turn, in the relativistic limit, introduces an $\mathbf{\Omega}$ dependence in σ_a, σ_a', S, and B, and separate $\mathbf{\Omega}$ and $\mathbf{\Omega'}$ dependences in σ_s.

Taking this into account, the equation of transfer derived in Chapter II, namely Eq. (2.167) or equivalently Eq. (2.170), should be written

$$\frac{1}{c}\frac{\partial I(v,\mathbf{\Omega})}{\partial t}+\mathbf{\Omega \cdot \nabla} I(v,\mathbf{\Omega}) = S(v,\mathbf{\Omega})[1+c^2 I(v,\mathbf{\Omega})/2hv^3]-\sigma_a(v,\mathbf{\Omega})\,I(v,\mathbf{\Omega})$$

$$+\int_0^\infty dv' \int_{4\pi} d\mathbf{\Omega}'\,\frac{v}{v'}\,\sigma_s(v' \to v,\mathbf{\Omega}' \to \mathbf{\Omega})\,I(v',\mathbf{\Omega}')\,[1+c^2 I(v,\mathbf{\Omega})/2hv^3]$$

$$-\int_0^\infty dv' \int_{4\pi} d\mathbf{\Omega}'\sigma_s(v \to v',\mathbf{\Omega} \to \mathbf{\Omega}')\,I(v,\mathbf{\Omega})[1+c^2 I(v',\mathbf{\Omega}')/2hv'^3], \tag{6.1}$$

or equivalently, defining σ_a' and B as in Eqs. (2.168) and (2.169),

$$\frac{1}{c}\frac{\partial I(v,\mathbf{\Omega})}{\partial t}+\mathbf{\Omega \cdot \nabla} I(v,\mathbf{\Omega}) = \sigma_a'(v,\mathbf{\Omega})[B(v,\mathbf{\Omega})-I(v,\mathbf{\Omega})]$$

$$+\int_0^\infty dv' \int_{4\pi} d\mathbf{\Omega}'\,\frac{v}{v'}\,\sigma_s(v' \to v,\mathbf{\Omega}' \to \mathbf{\Omega})\,I(v',\mathbf{\Omega}')[1+c^2 I(v,\mathbf{\Omega})/2hv^3]$$

$$-\int_0^\infty dv' \int_{4\pi} d\Omega' \sigma_s(v \rightarrow v', \Omega \rightarrow \Omega')\, I(v, \Omega)\, [1 + c^2 I(v', \Omega')/2hv'^3]. \qquad (6.2)$$

It should be emphasized that these additional angular dependences are not inherent properties of the material, but arise only from the relative motion between the fluid and the observer. Hence these angular dependences can be computed from the special theory of relativity. This subject is treated in the next section, using the results of the Appendix. These angular effects turn out to be, as one might expect, of the order of u/c, where u is the macroscopic speed of the fluid and c is the vacuum speed of light.

2. Application of the Lorentz Transformation

We consider an observer in an inertial frame of reference observing radiative transfer in a moving fluid. At a particular space point $\mathbf{r}$ and time t the fluid has a macroscopic velocity $\mathbf{u}(\mathbf{r}, t)$ as seen by this observer. We call the frame of reference in which the fluid is at rest the zero frame, and the frame of the observer the unadorned frame. The transformation velocity of the Appendix, i.e., the velocity $\mathbf{v}$ of the unadorned frame with respect to the zero frame, is

$$\mathbf{v} = -\mathbf{u}(\mathbf{r}, t). \qquad (6.3)$$

Since the observer is in an inertial frame of reference, Eq. (A.1) or (A.2) of the Appendix with the zero subscripts dropped, or Eq. (6.1) or (6.2), is the relevant transport equation. We assume that the source functions S and B, the absorption coefficients σ_a and σ_a', and the scattering kernel σ_s are known in the zero frame (the fluid rest frame), and use the Lorentz transformations of the Appendix to obtain these functions in the unadorned frame. As argued in Chapter II, in the zero frame $S, B, \sigma_a,$ and σ_a' are independent of Ω, and σ_s depends only upon $\Omega \cdot \Omega'$ rather than upon Ω and Ω' separately.

It must be noted that the zero frame, defined as the frame for which the fluid is at rest, is not in general an inertial frame since the fluid velocity is a function of both space and time. Hence, Eq. (A.1) or (A.2), valid only in an inertial frame, is not a proper description of radiative transfer in the fluid rest frame. More to the point, the fact that the fluid rest frame is not an inertial frame implies that the Lorentz transformations of the Appendix are, in general, not valid. However, certain of these transformations can be used in the present context. That is, even though the fluid rest frame is not an inertial frame, one can envision an inertial frame which instantaneously, at time t and space point $\mathbf{r}$, coincides with the fluid rest frame. Since the source terms (by source terms here we mean all terms except the streaming terms; i.e., the emission source, absorption, and scattering terms) in the equation of transfer are well defined at a single time t and space point $\mathbf{r}$ (i.e., t and $\mathbf{r}$ are only parameters in the source terms; no operators involving time or space appear in these terms), the transformations of the Appendix can indeed be used to relate the source terms in the fluid rest frame to those in the unadorned (observer) frame. The streaming terms, on the other hand, involve derivatives with respect to space and time. To define these derivatives via a Lorentz transformation the fluid rest frame must be an inertial frame for an arbitrarily small, but nonzero, interval of space and time. Since in general such an interval does not exist for a fluid in nonuniform motion, the transformations developed in the Appendix are not valid for the streaming terms. Fortunately, we need concern ourselves here only with the source terms since we wish to write the equation of transfer in the unadorned frame and in this frame, since it is inertial, the streaming terms are already known.

We first consider the terms S, B, σ_a, and σ_a'. Since the fluid is isotropic, all of these functions are independent of Ω and hence depend only upon frequency in the fluid rest (zero) frame. From Eqs. (A.80), (A.81), (A.83), (A.88) of the Appendix and Eq. (6.3) we find

$$S(\nu, \Omega) = \frac{1}{(\varLambda E)^2} S_0(\nu_0), \tag{6.4}$$

where

$$\nu_0 = \varLambda E\nu, \tag{6.5}$$
$$\varLambda = (1-u^2/c^2)^{-1/2}, \tag{6.6}$$
$$E = 1-\Omega\cdot\mathbf{u}/c. \tag{6.7}$$

For $u/c \ll 1$, it is sensible to expand $S(\nu, \Omega)$ in powers of u/c. Correct to first order we find

$$S(\nu, \Omega) = S_0(\nu)+(\Omega\cdot\mathbf{u}/c)[2S_0(\nu)-\nu dS_0(\nu)/d\nu]+O(u^2/c^2). \tag{6.8}$$

Similarly, we find from Eqs. (A.89) through (A.91)

$$B(\nu, \Omega) = \frac{1}{(\varLambda E)^3} B_0(\nu_0), \tag{6.9}$$

$$B(\nu, \Omega) = B_0(\nu)+(\Omega\cdot\mathbf{u}/c)[3B_0(\nu)-\nu dB_0(\nu)/d\nu]+O(u^2/c^2), \tag{6.10}$$
$$\sigma_a(\nu, \Omega) = \varLambda E\sigma_{a0}(\nu_0), \tag{6.11}$$
$$\sigma_a(\nu, \Omega) = \sigma_{a0}(\nu)-(\Omega\cdot\mathbf{u}/c)[\sigma_{a0}(\nu)+\nu d\sigma_{a0}(\nu)/d\nu]+O(u^2/c^2), \tag{6.12}$$
$$\sigma_a'(\nu, \Omega) = \varLambda E\sigma_{a0}'(\nu_0), \tag{6.13}$$
$$\sigma_a'(\nu, \Omega) = \sigma_{a0}'(\nu)-(\Omega\cdot\mathbf{u}/c)[\sigma_{a0}'(\nu)+\nu d\sigma_{a0}'(\nu)/d\nu]+O(u^2/c^2). \tag{6.14}$$

In addition, from Eq. (A.92) of the Appendix we can deduce the scattering kernel in the unadorned frame. We have

$$\sigma_s(\nu \rightarrow \nu', \Omega \rightarrow \Omega') = \frac{E}{E'} \sigma_{s0}(\nu_0 \rightarrow \nu_0', \Omega_0\cdot\Omega_0'), \tag{6.15}$$

where E' is defined by Eq. (6.7) with a prime placed on E and Ω. In writing Eq. (6.15) we have explicitly shown that in the zero frame the scattering kernel depends only upon the scattering angle. For small $\mathbf{u}/c$, Eq. (6.15) yields

$$\begin{aligned}
\sigma_s(\nu \rightarrow \nu', \Omega \rightarrow \Omega') = {}&\sigma_{s0}(\nu \rightarrow \nu', \Omega\cdot\Omega') \\
&-(\Omega\cdot\mathbf{u}/c)[\sigma_{s0}+\nu\partial\sigma_{s0}/\partial\nu+(1-\Omega\cdot\Omega')\partial\sigma_{s0}/\partial(\Omega\cdot\Omega')] \\
&-(\Omega'\cdot\mathbf{u}/c)[-\sigma_{s0}+\nu'\partial\sigma_{s0}/\partial\nu'+(1-\Omega\cdot\Omega')\partial\sigma_{s0}/\partial(\Omega\cdot\Omega')] \\
&+O(u^2/c^2).
\end{aligned} \tag{6.16}$$

For notational simplicity, we have omitted the arguments ν, ν', and $\Omega\cdot\Omega'$ in all σ_{s0} terms multiplying $\mathbf{u}/c$ in Eq. (6.16).

VII

The Interaction of the Radiation Field with Matter

1. Introduction

In the first six chapters we have discussed at some length the equation of radiative transfer in various forms. For the most part all of these results followed from the simple notion of conservation of photons and various geometric arguments. The underlying physics of radiative transfer is contained in the absorption coefficient $\sigma_a(\nu)$, the scattering kernel $\sigma_s(\nu' \to \nu, \Omega' \cdot \Omega)$, and the spontaneous emission source $S(\nu)$. A complete discussion of these three items involves detailed analysis from the fields of both statistical mechanics and quantum mechanics, and would require a volume in itself. Our intent in this chapter is to give only the briefest introduction to this subject. We will quote results, without derivation, following the outline given by Campbell (1969) and give several references where the reader can find more detail on the general subject of photon–matter interaction.

Following a few general remarks in the next section, we introduce in Section 3 the so-called Einstein coefficients which are frequently used in any discussion of absorption and emission of photons by matter. Section 4 concentrates on the absorption coefficient $\sigma_a(\nu)$, and the last section of this chapter briefly discusses the various forms of the scattering kernel $\sigma_s(\nu' \to \nu, \Omega' \cdot \Omega)$ used in radiative transfer work.

2. General Remarks

The calculation of the absorption and scattering coefficients and the source function involves two conceptually distinct steps. In the first place, given the temperature, density, and atomic composition of a plasma, one requires a quantitative statement concerning the population of the various ionic species present. In addition, for each ionic species one needs the population of each quantum energy state. This is a problem in statistical mechanics. Excellent introductions to this field are given by Rushbrooke (1949) and ter Haar (1954). Standard references in statistical mechanics are the books by Fowler (1936), Tolman (1938), Fowler and Guggenheim (1939), and Mayer and Mayer (1940).

Secondly, given these populations one requires the probability that a photon will induce a transition from one population state of the plasma to another. This requires a study of atomic and molecular processes together with the quantum theory of radiation (quantum electrodynamics). In quantum electrodynamics, the atoms (or molecules) and the radiation field are considered to be a mutually interacting quantum system. The interaction between an atom and the radiation field is represented quantitatively by an operator, called the interaction operator. The interaction causes a perturbation that induces transitions between the possible states of the collective system of the matter (plasma) and the radiation. The probability that a particular event (transition) will occur is proportional to the square of the relevant matrix element of the interaction operator. This transition probability, together with the information from statistical mechanics concerning the initial populations of the various states of the matter, gives the absorption and scattering coefficients as well as the source function. Leighton (1959) gives a very nice introduction to quantum electrodynamics, and the standard reference in this area is Heitler (1954). A useful summary of radiation theory is given by Condon and Shortley (1957). References on atomic and molecular physics are Richtmyer *et al.* (1955a), Leighton (1959), and Eisberg (1961). More advanced treatments are given by Slater (1960) and Bates (1962).

The mechanisms of absorption of radiation by matter are bound–bound (line) absorption, bound–free (photoelectric) absorption, and free–free absorption. In line absorption, an electron in a bound state is excited to another bound state of higher energy by the absorption of a quantum of radiation (photon). The frequency of the absorption line is given by Bohr's relation, $h\nu_{nm} = E_n - E_m$, where E_n and E_m are the higher and lower energy states, respectively. In photoelectric absorption, the electron is ejected from the atom (or ion) and goes into one of the continuum of free energy states. Photoelectric absorption occurs whenever the energy of the incident photon is greater than the binding energies of the electrons of the atom. In free–free absorption, an electron in a free state makes a transition to another free state of higher energy with the absorption of a photon. Figure 7.1 shows schematically the arrangement of energy levels and possible transitions.

Similar processes occur when the material is composed of molecules. Molecular processes are somewhat more complicated because of the extra degrees of freedom (vibrational and rotational) that are present when atoms join together to form molecules. Two important references on molecular spectroscopy are Herzberg (1950) and Penner (1959). Two absorption processes that occur at high (gamma ray) energies are electron–positron pair production and nuclear absorption. These processes are seldom of interest in usual radiative transfer problems.

Each absorption process has an inverse that results in the emission of radiation. The emission processes which correspond to bound–bound, bound–free, and free–free absorption are line emission, electron capture, and bremsstrahlung. The inverse of pair production is electron–positron annihilation. The inverse of nuclear absorption is the nuclear emission of a photon due to de-excitation of an excited nuclear state. We have shown in Chapter II [see Eq. (2.168)] that if local thermodynamic equilibrium (LTE) is assumed, the absorption coefficient and source function are simply related through the Planck function. Throughout our discussion we will make the LTE assumption, and hence a discussion of the absorption coefficient suffices as a discussion of the source function (emission) as well.

The scattering of electromagnetic radiation by an atom (or ion) can be considered as the absorption of a primary photon and the nearly simultaneous emission of a secondary photon

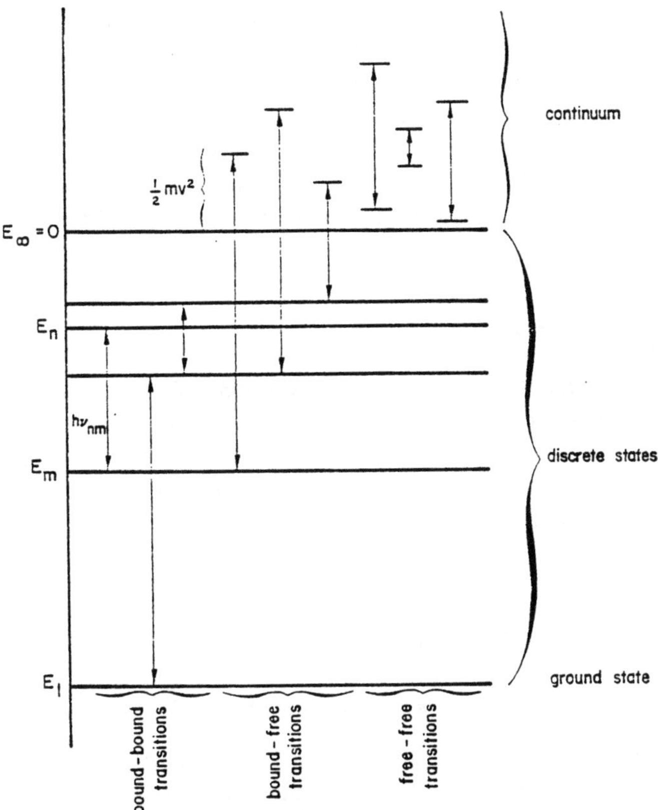

Fig. 7.1. Atomic energy levels and transitions.

in generally a different direction. If the frequencies of the primary and secondary photons are the same, the process is called coherent scattering. In some cases the scattered photon will not have the same frequency as the primary photon. During the scattering process the atom may become excited or partially de-excited by collision or by spontaneous emission. The scattering process results in a change of frequency and is called Raman scattering. Two cases can be distinguished in coherent scattering. When the frequency of the primary photon is much different from the frequency of a line (i.e., a resonance frequency), the scattering is called Rayleigh (low energy) or Thomson (high energy) scattering. When the frequency of the incident photon is the same (or nearly the same) as that of a line, the process is called resonance scattering (or resonance fluorescence).

More important than scattering from atoms is scattering from free electrons, called Compton scattering. In the low energy limit ($h\nu \ll m_0c^2 = 0.51$ MeV), Compton scattering reduces to classical Thomson scattering. The scattering coefficient for bound electrons (coherent and Raman scattering) is small compared to the photoelectric absorption coefficient, and roughly of the same general form as the Compton scattering coefficient. It is usually sufficient, therefore, to assume that all electrons are free when evaluating the total effect of scattering by a material. In some instances, however, it is important to take into account the motion (say Maxwellian at the local temperature) of these free electrons. Scattering from a given velocity distribution of free electrons is referred to as Compton and inverse Compton scattering. Because scattering from moving free electrons, and the low temperature (Compton) and low temperature and photon energy (Thomson) limits, are the most important scattering interactions in radiative transfer work, we will devote all of Chapter VIII to a discussion of Compton and inverse Compton scattering. For the remainder of this chapter, we concentrate on a brief description of the photon absorption coefficient, and the scattering kernels for electrons at rest.

3. The Einstein Coefficients

The fundamental quantities describing the absorption and emission of photons by matter are the Einstein coefficients A_{nm} and B_{nm}. These coefficients are defined in the following way: Consider an atom with energy levels E_n ($n = 1, 2, \ldots$; these levels may be discrete or lie in the continuum) in the presence of a radiation field. Transitions between any two states with energies E_n and E_m, where $E_n > E_m$, are associated with the emission and absorption of frequency ν_{nm}, where

$$h\nu_{nm} = E_n - E_m. \tag{7.1}$$

The Einstein coefficient B_{mn} is defined such that

$$P_{mn} = B_{mn}I(\nu_{nm})\,d\Omega. \tag{7.2}$$

Here P_{mn} is the probability per unit time that an atom in state m, exposed to radiation of frequency ν_{nm}, will absorb a photon from the radiation field that prior to the absorption was contained in a solid angle element $d\Omega$, and $I(\nu_{nm})$ is the specific intensity of radiation at frequency ν_{nm}. Likewise, the coefficients A_{nm} and B_{nm} are defined such that P_{nm}, the probability per unit time that an excited atom in state n will emit a photon with frequency ν_{nm} in the directions within $d\Omega$, is given by

$$P_{nm} = [A_{nm} + B_{nm}I(\nu_{nm})]d\Omega. \tag{7.3}$$

It should be remarked that in writing Eqs. (7.2) and (7.3) we have assumed the electrons to be nondegenerate. That is, we have assumed that all of the quantum states corresponding to the final energy state are essentially empty so that the Pauli exclusion principle for the electrons does not reduce the probability of any transition.

Equation (7.3) shows that the probability of transition to a lower energy is increased by the presence of an external radiation field. This is induced, or stimulated, emission as discussed earlier in Section 9 of Chapter II. The coefficient A_{nm} describes the rate at which the atom undergoes spontaneous emission in the absence of a radiation field. The coefficients B_{nm} and B_{mn} give the rate at which the radiation field induces transitions, both upward and downward in energy, between the various states of the atom. In terms of these probabilities, the energy emitted per unit volume in a time dt into an element of solid angle $d\Omega$ is given by

$$dW_e = \sum_{n,m} N_n h\nu_{nm}[A_{nm} + B_{nm}I(\nu_{nm})]\,d\Omega\,dt, \tag{7.4}$$

and the corresponding energy absorbed is

$$dW_a = \sum_{n,m} N_m h\nu_{nm} B_{mn} I(\nu_{nm})\, d\Omega\, dt. \tag{7.5}$$

Here N_n and N_m are the number of atoms per unit volume in the energy states E_n and E_m, and the sums extend over all pairs of energy states consistent with Eq. (7.1), i.e., over all transitions involving photons of frequency ν_{nm}.

By making a photon balance as in Chapter II, we find that the equation of transfer for radiation of frequency ν_{nm} is, in the absence of scattering,

$$\frac{1}{c}\frac{\partial I(\nu_{nm})}{\partial t} + \mathbf{\Omega}\cdot\nabla I(\nu_{nm}) = h\nu_{nm}\sum_{n,m} N_n[A_{nm} + B_{nm}I(\nu_{nm})]$$

$$- h\nu_{nm}\sum_{n,m} N_m B_{mn} I(\nu_{nm}). \tag{7.6}$$

Comparison of Eq. (7.6) with the equation of transfer obtained in Chapter II, Eq. (2.167), shows that the absorption coefficient $\sigma_a(\nu_{nm})$ and the source function $S(\nu_{nm})$ can be expressed in terms of the Einstein coefficients and level populations as

$$\sigma_a(\nu_{nm}) = h\nu_{nm}\sum_{n,m} N_m B_{mn}, \tag{7.7}$$

$$S(\nu_{nm}) = h\nu_{nm}\sum_{n,m} N_n A_{nm}. \tag{7.8}$$

Further, for Eqs. (7.6) and (2.167) to be consistent, we must have

$$\frac{\displaystyle\sum_{n,m} N_n B_{nm}}{\displaystyle\sum_{n,m} N_n A_{nm}} = \frac{c^2}{2h(\nu_{nm})^3}, \tag{7.9}$$

which is a general relationship among the Einstein coefficients.

Equation (7.6) can be algebraically rearranged to give

$$\frac{1}{c}\frac{\partial I(\nu_{nm})}{\partial t} + \mathbf{\Omega}\cdot\nabla I(\nu_{nm}) = h\nu_{nm}\sum_{n,m} N_n A_{nm}$$

$$- h\nu_{nm}\left[\sum_{n,m} N_m B_{mn}\right]\left[1 - \frac{\displaystyle\sum_{n,m} N_n B_{nm}}{\displaystyle\sum_{n,m} N_m B_{mn}}\right]. \tag{7.10}$$

Comparison of Eq. (7.10) with Eq. (2.170) gives the relationship

$$\sigma_a'(\nu_{nm}) = h\nu_{nm}\left[\sum_{n,m} N_m B_{mn}\right]\left[1 - \frac{\displaystyle\sum_{n,m} N_n B_{nm}}{\displaystyle\sum_{n,m} N_m B_{mn}}\right]$$

$$= \sigma_a(\nu_{nm})\left[1 - \frac{\displaystyle\sum_{n,m} N_n B_{nm}}{\displaystyle\sum_{n,m} N_m B_{mn}}\right], \tag{7.11}$$

which shows, in terms of the Einstein coefficients and level populations, the effective reduction in the absorption coefficient due to induced emission. We will shortly show that if the matter is in thermodynamic equilibrium, the ratio of the two summations in Eq. (7.11) is less than unity, and hence $\sigma_a'(\nu_{nm})$ is positive. However, if the matter is far from thermodynamic equilibrium, it is possible for this ratio of summations to be greater than unity (population inversion), and then the effective absorption coefficient $\sigma_a'(\nu_{nm})$ is negative. In this case the intensity of a beam of light, upon traveling through matter, will be enhanced due to "absorption". This is the principle upon which the laser works.

The relationship Eq. (7.9) among the Einstein coefficients can be sharpened and another relationship derived by using the principle of detailed balance. This principle states that at complete thermodynamic equilibrium the probability of a transition from state n to state m must be precisely equal to the probability of a transition from state m to state n. With the aid of Eqs. (7.2) and (7.3), we then have the equality

$$N_m B_{mn} I(\nu_{nm}) = N_n [A_{nm} + B_{nm} I(\nu_{nm})], \tag{7.12}$$

where here the N_m and N_n refer to the populations of the states m and n at thermodynamic equilibrium. Further, $I(\nu_{nm})$ in this equation is the thermodynamic equilibrium distribution for the radiation field, namely the Planck function. Solving Eq. (7.12) for $I(\nu_{nm})$ we find

$$I(\nu_{nm}) = \frac{A_{nm}/B_{nm}}{\left[\dfrac{N_m B_{mn}}{N_n B_{nm}} - 1\right]}. \tag{7.13}$$

Setting $I(\nu_{nm})$ equal to the Planck function [see Eq. (2.172)], we have

$$\frac{A_{nm}/B_{nm}}{\left[\dfrac{N_m B_{mn}}{N_n B_{nm}} - 1\right]} = \frac{2h(\nu_{nm})^3/c^2}{e^{h\nu_{nm}/kT} - 1}. \tag{7.14}$$

Equation (7.14) gives the following relationships between the Einstein coefficients:

$$\frac{A_{nm}}{B_{nm}} = \frac{2h(\nu_{nm})^3}{c^2}, \tag{7.15}$$

$$\frac{N_m B_{mn}}{N_n B_{nm}} = \exp\left(h\nu_{nm}/kT\right). \tag{7.16}$$

Equation (7.15) is a refinement of our previous result, Eq. (7.9). Equation (7.16), a new relationship, can be used to obtain a general relationship between B_{nm} and B_{mn} [as it now stands, Eq. (7.16) is only valid at thermodynamic equilibrium]. If the electrons are nondegenerate, as we have assumed, it is well known that in thermodynamic equilibrium the energy levels are populated according to the usual Boltzmann distribution (see, for example, Richtmyer *et al.*, 1955b)

$$\frac{N_m}{N_n} = \frac{g_m}{g_n} e^{(E_n - E_m)/kT} = \frac{g_m}{g_n} e^{h\nu_{nm}/kT}. \tag{7.17}$$

The quantity g_n in this equation is the statistical weight or degeneracy of the level n. The degeneracy is the number of different states with the same energy E_n. A degenerate level would be split into g_n different states if the degeneracy were removed by an external perturbation (e.g., a magnetic field). Use of Eq. (7.17) in Eq. (7.16) gives

$$g_m B_{mn} = g_n B_{nm} \tag{7.18}$$

as the second relationship among the Einstein coefficients.

We emphasize that the Einstein coefficients A_{nm} and B_{nm} are fundamental measures of the interaction between an atom (or ion) and a radiation field. They have the same values whether the matter is in thermal equilibrium or not. We have used the case of thermodynamic equilibrium as a simple way to derive Eqs. (7.15) and (7.18), but these relationships do not depend upon the equilibrium condition. We only need to calculate one set of coefficients, e.g. the A_{nm}, and use Eqs. (7.15) and (7.18) to determine the others, B_{nm} and B_{mn}.

We note that if we assume the matter to be in local thermodynamic equilibrium, Eq. (7.16) is valid and can be written

$$N_n B_{nm} = N_m B_{mn} e^{-h\nu_{nm}/kT}. \tag{7.19}$$

Equation (7.19) allows Eq. (7.11) to be written as

$$\sigma_a'(\nu_{nm}) = h\nu_{nm} \left[\sum_{n,\,m} N_m B_{mn} \right] (1 - e^{-h\nu_{nm}/kT})$$
$$= \sigma_a(\nu_{nm})(1 - e^{-h\nu_{nm}/kT}). \tag{7.20}$$

Thus the effective absorption coefficient, under the assumption of local thermodynamic equilibrium, is a product of two factors, the true absorption coefficient

$$\sigma_a(\nu_{nm}) = h\nu_{nm} \sum_{n,\,m} N_m B_{mn}, \tag{7.21}$$

and the correction for induced emission

$$\frac{\sigma_a'(\nu_{nm})}{\sigma_a(\nu_{nm})} = 1 - e^{-h\nu_{nm}/kT}. \tag{7.22}$$

When the absorption coefficient is expressed in this way, stimulated emission is considered as negative absorption. The effect of stimulated emission is to reduce the true absorption coefficient by the factor $[1 - \exp(-h\nu_{nm}/kT)]$. We have obtained this same result in Chapter II [see Eq. (2.173)].

Finally, we note that the Einstein coefficient A_{nm} can be given a useful physical interpretation. We consider N_n atoms in the energy state E_n in the absence of a radiation field. The probability that any one atom will decay in a time dt to an energy state E_m is $4\pi A_{nm} dt$ [the factor 4π arises upon integration of Eq. (7.3) over all solid angle]. Therefore, the total number of decays in a time dt is

$$|dN_n| = N_n \sum_m 4\pi A_{nm}\, dt, \tag{7.23}$$

or, since dN_n is negative,

$$\frac{dN_n}{dt} = -N_n \sum_m 4\pi A_{nm}. \tag{7.24}$$

The Equations of Radiation Hydrodynamics

Here the summation is over all states m such that $E_m < E_n$. If the original number of atoms at zero time is denoted by $N_n^{(0)}$, the solution to Eq. (7.24) is simply

$$N_n = N_n^{(0)} \exp\left[-4\pi \sum_m A_{nm} t\right].$$ (7.25)

We can define the mean life of the state E_n as the time required for a large number of excited atoms to decay to $1/e$ of their original number. The mean life τ_n is given in terms of the transition coefficients by

$$\tau_n = \left[4\pi \sum_m A_{nm}\right]^{-1}.$$ (7.26)

Thus we see that the coefficients A_{nm} are simply related to the mean life of the energy state E_n against radiative decay.

4. The Absorption Coefficient

We first consider the bound–free (photoelectric) absorption coefficient. From our discussion of the last section it is clear that a necessary ingredient to obtain this absorption coefficient is the population level of each quantum state of the matter. It is conventional in a discussion of these population levels to separate the calculation into two conceptually distinct parts, namely (1) the populations of the various ionic species and the free electron population, and (2) the population of the various energy states for a given ion. If local thermodynamic equilibrium is assumed, the Saha equation concerns itself with the ionic species and free electrons, and quantum statistics gives the population of the energy states for a given ion.

We consider a simple gas consisting of atoms of a single element with nuclear charge Z and assume, as is usually the case, that the temperature is high enough so that no molecules are present. We denote by N the number of atoms per unit volume, and let N_m denote the number of ions per unit volume with a charge me ($m = 0$ is a neutral atom; $m = Z$ is a bare nucleus), where $-e$ is the charge of an electron. The density of free electrons is denoted by N_e. For simplicity, we assume the gas to be sufficiently rarefied, or at a high enough temperature, so that the electrons are nondegenerate. (We shall give shortly a quantitative statement as to when the electrons can be taken as nondegenerate.) Conservation of the number of nuclei implies the condition

$$\sum_{m=0}^{Z} N_m = N,$$ (7.27)

and conservation of charge gives

$$\sum_{m=1}^{Z} m N_m = N_e.$$ (7.28)

The Saha equation relating the various ionic densities N_m is (Zel'dovich and Raizer, 1966)

$$\frac{N_{m+1} N_e}{N_m} = 2 \frac{u_{m+1}}{u_m} \left(\frac{2\pi m_0 kT}{h^2}\right)^{3/2} \exp\left(-I_{m+1}/kT\right), \qquad m = 0, 1, \ldots, Z-1, \quad (7.29)$$

where m_0 is the mass of an electron. Here the I_m denote the successive ionization potentials.

164

That is, I_1 is the energy required to remove the first electron from a neutral atom, I_2 is the energy required to remove an electron from a singly ionized atom, etc. The quantity u_m is the excitation partition function for an atom with m electrons removed, and is given by

$$u_m = \sum_{k=1}^{\infty} g_{mk} \exp\left(-w_{mk}/kT\right). \tag{7.30}$$

Here the index k denotes the kth energy level of the ion under consideration. In particular, g_{mk} is the degeneracy factor for the kth level, and w_{mk} denotes the excitation energy of the kth state, i.e., the energy of the kth state in excess of the ground state $k = 1$ (thus $w_{m1} = 0$). Equations (7.27) through (7.29) represent $Z+2$ nonlinear equations for the $Z+2$ unknowns $N_0, N_1, \ldots, N_Z$, and N_e.

The summation in Eq. (7.30) defining the partition function is over all discrete energy states k for the ion under consideration (characterized by the index m). An isolated atom (or ion) has an infinite number of discrete energy levels which converge to the bottom of the continuum. Thus the partition function formally contains an infinite number of terms and is divergent. This divergence never occurs in practice, however, since an atom is never isolated but is part of a gas of finite density. The effect of the other atoms present is to reduce the infinite number of discrete states to a finite number. That is, when the dimensions of the orbit of an electron are comparable to the spacing between atoms (as will always occur if the index k is large enough), such an orbit is distorted and the electron acts more like a free electron than one bound to a specific atom. The effect of these considerations on Eq. (7.30) is to truncate the sum with $k = k^*$ rather than letting it go to infinity. The value of k^* is not precise, but should correspond roughly to that index associated with a Bohr orbit whose major semi-axis is equal to the mean spacing between atoms.

Once the densities of the various ionic species have been determined from Eqs. (7.27) through (7.29), we require information concerning the population of the kth discrete level for each ion in order to compute the bound–free absorption coefficient. This can in principle be obtained as follows. One solves the Schroedinger equation for all the energy levels corresponding to a system of $Z+1$ particles, namely the nucleus and Z electrons. The complete solution can be divided into $Z+1$ categories, those with 0, 1, 2, $\ldots$, Z electrons in the continuum (free electrons), respectively. Each one of these categories corresponds to a certain state of ionization which we have previously characterized by an index m. Within each class, i.e., for a fixed value of m, we label the energy states by an index k denoting the excitation state of the bound electrons, and by a (continuous) index k' denoting the translational states of the free electrons and the ion. Thus for a given ion with index m, we have energy states $E_{mk}(k')$. We have previously indicated that for nondegenerate electrons, the probability of a state with energy E being occupied at thermal equilibrium is proportional to $\exp\left(-E/kT\right)$. Quantum statistics gives an absolute number, rather than a proportionality, for the occupation number of a given state. According to Fermi–Dirac statistics (Chandrasekhar, 1957) the probability $f_{mk}(k')$ that a state with energy $E_{mk}(k')$ is occupied at temperature T is

$$f_{mk}(k') = \cfrac{1}{\exp\left(\cfrac{E_{mk}(k')-\mu}{kT}\right)+1}, \tag{7.31}$$

where μ is referred to as the chemical potential or Fermi energy. If we write $\mu = -\alpha kT$, the quantity α is given by the relationship (Cox, 1965a)

$$N_e = \frac{4\pi}{h^3}(2m_0kT)^{3/2}\int_0^\infty dx \frac{\sqrt{x}}{\exp(x+\alpha)+1}. \tag{7.32}$$

For large α, Eq. (7.32) gives

$$e^\alpha = \frac{2}{N_e}\left(\frac{2\pi m_0 kT}{h^2}\right)^{3/2}. \tag{7.33}$$

The condition for Eq. (7.31) to reduce to nondegenerate Boltzmann statistics, namely $f_{mk}(k')$ proportional to $\exp(-E_{mk}(k')/kT)$, is that α be much greater than unity. From Eq. (7.33) we see this is the case at low free electron densities or high temperatures.

If we define N_{mk} as the density of the kth discrete state for an atom with m electrons removed, Eq. (7.31) gives

$$N_{mk} = N\sum_{k'}\frac{g_{mk}(k')}{\exp\left(\alpha+\dfrac{E_{mk}(k')}{kT}\right)+1}, \tag{7.34}$$

where N is the atomic density of the system (nuclei per unit volume). Here $g_{mk}(k')$ is the degeneracy factor corresponding to the energy $E_{mk}(k')$, and the sum over k' reflects the fact that we are interested in the various discrete states regardless of the translational configuration of the free electrons and ion. [The sum in Eq. (7.34) is actually an integral, as we shall show shortly.] An alternate way of expressing Eq. (7.34), consistent with our breakup of the level population problem into first finding the relative proportions of the various ions present and then asking for the probability of a given discrete level of a given ion being populated, is

$$p_{mk} = \frac{N}{N_m}\sum_{k'}\frac{g_{mk}(k')}{\exp\left(\alpha+\dfrac{E_{mk}(k')}{kT}\right)+1}. \tag{7.35}$$

Here N_m is the density of atoms with m electrons removed, and p_{mk} is the probability for this type of ion that the kth discrete energy level is populated.

We note that this expression for p_{mk} and the entire discussion of the Saha equation is really redundant since Eq. (7.31) in principle gives all the information required concerning the populations of the various energy levels for all states of ionization, as well as the population of the continuum (the free electron density). In fact, the Saha equation follows quite simply from Eq. (7.34). It is instructive to show this connection between quantum statistics and the Saha equation. This derivation of the Saha equation also shows how one handles the summation over k' in Eq. (7.34).

If we sum Eq. (7.34) over all k, i.e., over all discrete states for an atom with m electrons removed, we obtain N_m, the ionic density for this state of ionization. If we assume $\alpha \gg 1$ (nondegenerate electrons), we have

$$N_m = Ne^{-\alpha}\sum_{k=1}\sum_{k'}g_{mk}(k')\exp(-E_{mk}(k')/kT). \tag{7.36}$$

Now, the energy $E_{mk}(k')$ is the sum of the ground state ($k = 1$) energy E_{m1}, the excitation energy for the kth state w_{mk}, the ion translational energy $p_i^2/2m_i$, and the free electron translational energies $p_j^2/2m_0$ ($j = 1, 2, \ldots, m$ since these are m free electrons per atom). Here the subscript i refers to the ion, m is the mass, and p is the momentum. Thus

$$E_{mk}(k') = E_{m1} + w_{mk} + \frac{p_i^2}{2m_i} + \sum_{j=1}^{m} \frac{p_j^2}{2m_0}. \tag{7.37}$$

The degeneracy factor $g_{mk}(k')$ is the product of g_{mk}, the degeneracy factor for the kth discrete level of the mth ion, and the degeneracy factors for the translational energies of the m free electrons and the ion. The degeneracy factor for one free electron is twice the available phase space divided by h^3, the basic unit in phase space. The factor of two arises from the two spin states of a free electron. Thus the degeneracy factor for the jth free electron is $8\pi p_j^2 dp_j / N_e h^3$. One has a similar expression for the degeneracy factor corresponding to the translational energy of the ion (except no factor of two is required). Thus we have

$$g_{mk}(k') = g_{mk} \left(\frac{4\pi p_i^2 \, dp_i}{N h^3} \right) \prod_{j=1}^{m} \left(\frac{8\pi p_j^2 \, dp_j}{N_e h^3} \right). \tag{7.38}$$

Using Eqs. (7.37) and (7.38) in Eq. (7.36), writing the result for the index $m+1$ as well, and dividing the $m+1$ result by the m result, we find (the sum over k' becomes a multiple integral)

$$\frac{N_{m+1}}{N_m} = \left[\frac{\exp(-E_{m+1,1}/kT) \sum_{k=1} g_{m+1,k} \exp(-w_{m+1,k}/kT)}{\exp(-E_{m1}/kT) \sum_{k=1} g_{mk} \exp(-w_{mk}/kT)} \right] \times$$

$$\times \frac{8\pi}{N_e h^3} \int_0^\infty dp\, p^2 \exp(-p^2/2m_0 kT). \tag{7.39}$$

In writing Eq. (7.39) we have cancelled the contribution from the translational energy of the two ionic species m and $m+1$ since they only differ in the mass of the ion and these two masses are nearly equal. Defining

$$I_{m+1} = E_{m+1,1} - E_{m1}, \tag{7.40}$$

and carrying out the integral over momentum, we find

$$\frac{N_{m+1} N_e}{N_m} = 2 \frac{u_{m+1}}{u_m} \left(\frac{2\pi m_0 kT}{h^2} \right)^{3/2} \exp(-I_{m+1}/kT), \tag{7.41}$$

where u_m is the partition function given by Eq. (7.30). Equation (7.41) is just the Saha equation [see Eq. (7.29)] with Eq. (7.40) defining the mth ionization potential.

A more complete treatment of energy levels, occupation numbers, and the general problem of absorption coefficients is given by Mayer (1947a) and Cox (1965b). A number of important papers were presented at both of the Air Force Weapons Laboratory Conferences on Opacities (1964, 1965).

The Equations of Radiation Hydrodynamics

Unfortunately, only for the hydrogen atom, or for a hydrogen-like atom (i.e., an atom with nuclear charge Ze and one associated electron) can one in practice carry out the calculations associated with the above discussion of the level population problem. Only in this case can one obtain a complete solution to the Schroedinger equation for the energy levels and the associated degeneracies. Further, only for a hydrogen-like atom can one obtain relatively complete results for the transition probability. In this case we have the result (Kramers, 1923; Gaunt, 1930)

$$\mu_{bf}^{(k)}(\nu) = \begin{cases} \dfrac{64\pi^4 e^{10} m_0 Z^4}{3\sqrt{3}\,h^6 c\nu^3 k^5}\, g_{bf}^{(k)}(\nu), & h\nu \geq |E_k| \\[2mm] 0, & h\nu \leq |E_k|. \end{cases} \tag{7.42}$$

Here $\mu_{bf}^{(k)}(\nu)$ is the bound–free absorption cross section, expressed in cm^2/electron, for photo-ionization from the kth energy level. (We have dropped the index m since it is redundant in this case. A hydrogen-like atom has only two ionic states, and only one of these has a bound electron.) For a hydrogen-like atom with nuclear charge Ze, these energy levels are

$$E_k = -I_H \left(\frac{Z}{k}\right)^2, \qquad k = 1, 2, \ldots, \tag{7.43}$$

where I_H is the ionization potential for the hydrogen atom, given by

$$I_H = \frac{2\pi^2 m_0 e^4}{h^2} = 13.6 \text{ eV}. \tag{7.44}$$

The corresponding degeneracy factor is $g_k = 2k^2$ (the two arises from the two spin states of the electron, and the k^2 arises from a sum over azimuthal quantum numbers $0 \leq l \leq k-1$, and magnetic quantum numbers $|m| \leq l$). The factor $g_{bf}^{(k)}(\nu)$ is the bound–free Gaunt factor for the kth discrete state, and is given approximately by (Menzel and Pekeris, 1935)

$$g_{bf}^{(k)}(\nu) = 1 - 0.1728 \left(\frac{h\nu}{I_H Z^2}\right)^{1/3} \left[\frac{2}{k^2}\left(\frac{I_H Z^2}{h\nu}\right) - 1\right]. \tag{7.45}$$

The Gaunt factor is a quantum correction to the semi-classical result. Often the Gaunt factor gives a small correction and can be taken as unity.

For complex atoms (those with more than one electron) one must necessarily use approximate techniques. We describe briefly one such approximation here, the "average ion" approach of Mayer (1947b). The basic idea is to make use of the known hydrogen-like results as much as possible. That is, we consider an individual electron in a complex atom to be an isolated electron subjected to an effective nuclear charge Z^*e. We then use Eq. (7.42) for the absorption cross section for the various bound states, coupled with Eq. (7.31) for the level populations, to obtain the bound–free absorption coefficient.

The effective atomic number Z_j^* for an electron in the jth energy state is given in terms of the Slater screening constants (Slater, 1930) σ_{jk} by

$$Z_j^* = Z - \sum_{k \neq j} n_k \sigma_{jk} - n_j \left(1 - \frac{1}{g_j}\right) \sigma_{jj}, \tag{7.46}$$

where $g_j = 2j^2$ is the degeneracy of the jth level in a hydrogen-like atom. The n_k in Eq. (7.46) are the number of bound electrons in the kth level (the occupation numbers), which at this point are unknown. The screening constants σ_{jk} never exceed unity and give the effectiveness of an electron in the kth level screening the nuclear charge as seen by an electron in

TABLE 7.1. *The Slater Screening Constants σ_{ij}*

i \ j	1	2	3	4	5
1	0.6250	0.9383	0.9811	0.987	0.994
2	0.2346	0.6895	0.8932	0.94	0.97
3	0.1090	0.3970	0.7018	0.85	0.92
4	0.0617	0.2350	0.4781	0.705	0.83
5	0.0398	0.1552	0.3312	0.531	0.72
6	0.0277	0.1093	0.2388	0.400	0.854
7	0.0204	0.0808	0.1782	0.3102	0.459
8	0.0156	0.0625	0.1378	0.2425	0.371
9	0.0123	0.0494	0.1106	0.1936	0.299
10	0.0100	0.0400	0.0900	0.1584	0.245

i \ j	6	7	8	9	10
1	0.997	0.999	1.000	1.000	1.000
2	0.984	0.990	0.993	0.995	1.00
3	0.955	0.97	0.98	0.99	1.00
4	0.90	0.95	0.97	0.98	0.99
5	0.83	0.90	0.95	0.97	0.98
6	0.735	0.83	0.90	0.95	0.97
7	0.610	0.745	0.83	0.90	0.95
8	0.506	0.635	0.750	0.83	0.90
9	0.431	0.544	0.656	0.760	0.83
10	0.353	0.466	0.576	0.67	0.765

the jth level. Values of the screening constants are given in Table 7.1. The energy levels for a hydrogen-like atom with atomic number Z_j^* are given by [see Eq. (7.43)]

$$E_j = -I_H \left(\frac{Z_j^*}{j} \right)^2, \qquad j = 1, 2, \ldots. \tag{7.47}$$

The energy levels given by Eq. (7.47) must be corrected for the interaction of the bound electrons with free electrons in the neighborhood of the ion. The average ion has a net positive charge $n_f e$, where n_f is the number of free electrons per atom, and obviously given by

$$n_f = Z - \sum_j n_j. \tag{7.48}$$

We let R be the radius of the spherical region surrounding the ion which contains zero net charge; that is, n_f free electrons. If N denotes the atomic density (nuclei per unit volume),

169

then $1/N$ is the volume associated with one ion and its associated free electrons. Thus

$$\frac{4}{3}\pi R^3 = \frac{1}{N},$$ (7.49)

which serves to define the radius of the so-called ion sphere. Any free electron outside this ion sphere has no interaction with the ion.

Now, the electrostatic potential due to the free electrons within the ion sphere is

$$\varphi(r) = -\frac{n_f e}{2R}\left(3 - \frac{r^2}{R^2}\right).$$ (7.50)

In order to obtain the contribution to the energy levels, the potential $\varphi(r)$ must be averaged over the charge distribution $-e\psi_j\psi_j^*$ of the bound state described by the wave function ψ_j. With this correction, the energy levels become (Mayer, 1947b)

$$E_j = -I_H\left(\frac{Z_j^*}{j}\right)^2 + \frac{n_f e^2}{2R}\left[3 - \frac{\overline{(r^2)}_j}{R^2}\right], \qquad j = 1, 2, \ldots,$$ (7.51)

where

$$\overline{(r^2)}_j = \int d\mathbf{r}\, r^2 \psi_j(\mathbf{r})\,\psi_j^*(\mathbf{r})$$ (7.52)

is the mean square radius for the electron orbital ψ_j. The mean square radius for a hydrogen-like atom is given by Pauling and Wilson (1935) as

$$\overline{(r^2)}_{jl} = a_0^2 \frac{j^4}{(Z_j^*)^2}\left\{1 + \frac{3}{2}\left[1 - \frac{l(l+1) - \frac{1}{3}}{j^2}\right]\right\},$$ (7.53)

where a_0 is the first Bohr radius for the hydrogen atom. Averaging Eq. (7.53) over angular momentum states $0 \le l \le j-1$, weighting the lth state with its degeneracy $2(2l+1)$, we find

$$\overline{(r^2)}_j = \frac{a_0^2 j^4}{(Z_j^*)^2}\left(\frac{7}{4} + \frac{5}{4j^2}\right).$$ (7.54)

[Mayer (1947b) obtained a slightly different result, namely $\overline{(r^2)}_j = a_0^2 j^4 (2 + j^{-2})/(Z_j^*)^2$, by using a weight function of unity, rather than $4l+2$, in averaging.]

One further correction to the energy levels is needed. A free electron in the ion sphere has a certain interaction energy with the ion and with other free electrons. For this reason, an electron in the ion sphere with zero kinetic energy will have a negative total energy. We therefore add a small correction to Eq. (7.49) so that any electron in the ion sphere has positive total energy and is therefore a free electron. With this correction the energy levels become (Mayer, 1947b)

$$E_j = -I_H\left(\frac{Z_j^*}{j}\right)^2 + \frac{n_f e^2}{2R}\left[3 - \frac{\overline{(r^2)}_j}{R^2}\right] + \frac{3}{5}\frac{n_f e^2}{2R},$$ (7.55)

or, combining terms,

$$E_j = -I_H\left(\frac{Z_j^*}{j}\right)^2 + \frac{n_f e^2}{2R}\left[\frac{18}{5} - \frac{\overline{(r^2)}_j}{R^2}\right], \qquad j = 1, 2, \ldots.$$ (7.56)

The energy levels can be obtained from Eq. (7.56) if the occupation numbers n_j are known. The effective atomic numbers Z_j^* are given by Eq. (7.46) and Table 7.1; the number of free electrons per atom n_f is given by Eq. (7.48); the radius of the ion sphere R is given by Eq. (7.49); and the mean square radii are given by Eq. (7.54). It should be emphasized that we have considered a system consisting of only one atomic species of atomic number Z. The extension of this model to mixtures is given by Mayer (1947b). Mayer (1947c) also gives a more sophisticated model, the "ionic model", for treating complex atoms.

The occupation numbers n_j, needed to complete our discussion of the "average ion" model, follow directly from Eq. (7.31). Since the degeneracy of a hydrogen-like atom in state j is $2j^2$, we have

$$n_j = \frac{2j^2}{\exp\left(\alpha + \dfrac{E_j}{kT}\right) + 1}, \tag{7.57}$$

with α given by Eq. (7.32), where

$$N_e = Nn_f. \tag{7.58}$$

Equations (7.46), (7.48), (7.55), (7.57), and (7.58) represent a set of nonlinear equations for the unknowns Z_j^*, n_f, E_j, n_j, and N_e.

The bound–free cross section, expressed in cm²/electron, for photoionization from the jth energy level is [see Eq. (7.42)]

$$\mu_{bf}^{(j)}(\nu) = \begin{cases} \dfrac{64\pi^4 e^{10} m_0 (Z_j^*)^4}{3\sqrt{3}h^6 c\nu^3 j^5}\, g_{bf}^{(j)}(\nu), & h\nu \geq |E_j| \\[2ex] 0, & h\nu \leq |E_j|, \end{cases} \tag{7.59}$$

with Z_j^* given by Eq. (7.46) and E_j given by Eq. (7.59). The Gaunt factor in Eq. (7.59) is [see Eq. (7.45)]

$$g_{bf}^{(j)}(\nu) = 1 - 0.1728 \left[\frac{h\nu}{I_H(Z_j^*)^2}\right]^{1/3} \left\{\frac{2}{j^2}\left[\frac{I_H(Z_j^*)^2}{h\nu}\right] - 1\right\}. \tag{7.60}$$

We note that the cross section for an electron in the energy level E_j is zero unless $h\nu \geq |E_j|$. Only if the radiation has enough energy to remove an electron to the continuum can that electron contribute to photoelectric absorption. The total bound–free cross section, in cm²/atom, is given by

$$\mu_{bf}(\nu) = \sum_j n_j \mu_{bf}^{(j)}(\nu), \quad |E_j| \leq h\nu. \tag{7.61}$$

In Eq. (7.61) n_j is the number of electrons in level j as discussed earlier, and the sum extends only over those levels for which the ionization energy $|E_j|$ is less than the radiation energy $h\nu$. The cross sections for different levels are shown schematically in Fig. 7.2.

The bound–free, or photoelectric, absorption coefficient is the product of $\mu_{bf}(\nu)$, the cross section per atom, and N, the number of atoms per unit volume. Thus

$$\sigma_{bf}(\nu) = N\mu_{bf}(\nu). \tag{7.62}$$

The absorption coefficient is also conventionally expressed in the form

$$\sigma_{bf}(\nu) = \varrho D_k(\nu)\,(h\nu)^{-3}, \tag{7.63}$$

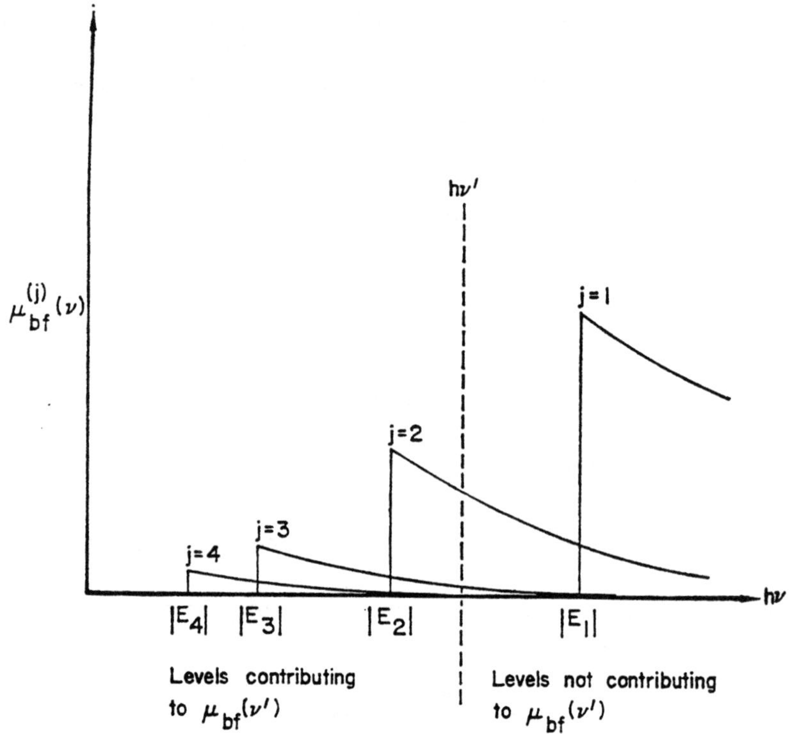

FIG. 7.2. Photoelectric cross sections for bound electrons.

where the index k is defined by

$$|E_k| \leq h\nu < |E_{k+1}|, \tag{7.64}$$

and

$$D_k(\nu) = \frac{N}{\varrho} \frac{64\pi^4 e^{10} m_0}{3\sqrt{3}\,ch^3} \sum_{j=k}^{J} n_j \frac{(Z_j^*)^4}{j^5} g_{bf}^{(j)}(\nu). \tag{7.65}$$

The sum in Eq. (7.65) starts at k, the first level for which $h\nu$ is greater than the ionization energy, and extends up to some maximum value J. Since the contribution of the higher levels falls off as ν^{-3}, it is usually sufficient to terminate the sum at $J = 5$ or 6. [Note that N/ϱ in Eq. (7.65) is just A/M, where A is Avogadro's number and M is the atomic weight of the material.]

It should be noticed that the absorption coefficient depends strongly on the temperature through n_j and Z_j^* because of ionization. The positions of the edges $|E_j|$ are temperature dependent as well. The form of the absorption coefficient is shown in Fig. 7.3. The K, L, M, ... edges represent the ionization energies for the K, L, M, ... shells. Actually, the L, M, ... edges have a fine structure, shown schematically in Fig. 7.4, due to the different binding energies of the angular momentum subshells. These corrections for the subshells are small and have been ignored in the expression for the energy levels, Eq. (7.47). These corrections arise from relativity and spin–orbit coupling.

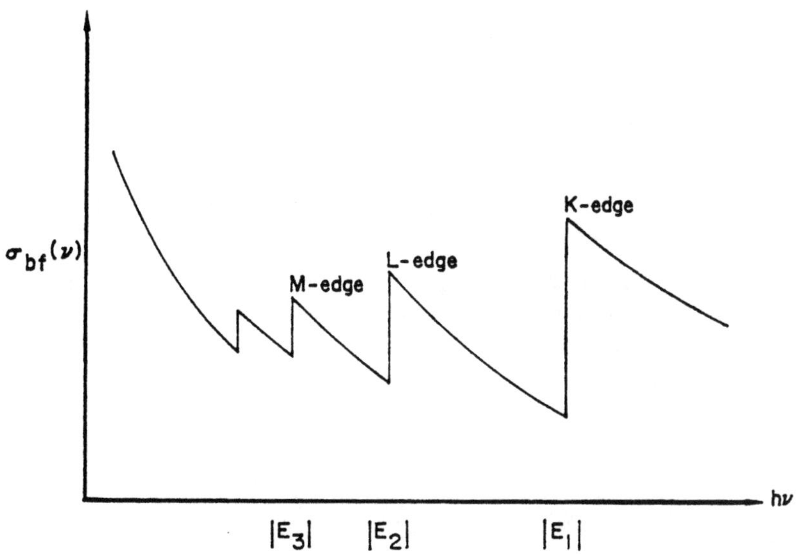

Fig. 7.3. Photoelectric absorption coefficient.

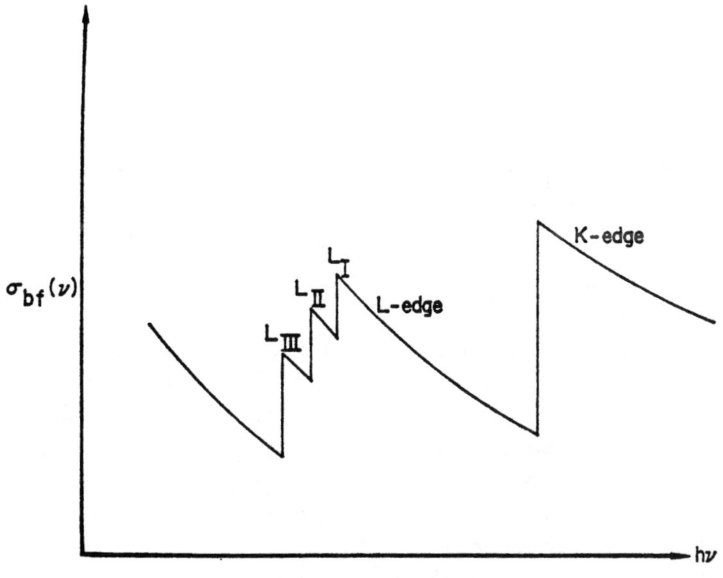

Fig. 7.4. Fine structure of the L-edge.

The Equations of Radiation Hydrodynamics

We now consider the free–free (inverse bremsstrahlung) absorption coefficient. For a hydrogen-like atom of atomic number Z, the cross section for free–free absorption, in cm^2/atom, is given by (Kramers, 1923; Gaunt, 1930)

$$\mu_{ff}(\nu) = \frac{4}{3}\left(\frac{2\pi}{3m_0 kT}\right)^{1/2}\frac{Z^2 e^6}{hcm_0\nu^3}N_e g_{ff}(\nu), \tag{7.66}$$

where N_e is the free electron density and $g_{ff}(\nu)$ is the free–free Gaunt factor given approximately by (Menzel and Pekeris, 1935)

$$g_{ff}(\nu) = 1+0.1728\left(\frac{h\nu}{I_H Z^2}\right)^{1/3}\left(1+2\frac{kT}{h\nu}\right). \tag{7.67}$$

One can apply this result to the "average ion" model by replacing Z in Eqs. (7.66) and (7.67) by n_f, the number of free electrons per atom, since $n_f e$ is the net positive charge per atom [see Eq. (7.48)]. The free–free absorption coefficient is obtained by multiplying Eq. (7.66) by N, the number of atoms per unit volume. Thus we have

$$\sigma_{ff}(\nu) = \varrho D_f(\nu)\,(h\nu)^{-3}, \tag{7.68}$$

where

$$D_f(\nu) = \frac{N}{\varrho}\frac{4}{3}\left(\frac{2\pi}{3m_0 kT}\right)^{1/2}\frac{n_f^2 h^2 e^6}{cm_0}g_{ff}(\nu), \tag{7.69}$$

and

$$g_{ff}(\nu) = 1+0.1728\left(\frac{h\nu}{I_H n_f^2}\right)^{1/3}\left(1+2\frac{kT}{h\nu}\right). \tag{7.70}$$

The total coefficient of continuous absorption, $\sigma_c(\nu)$, is thus given by

$$\sigma_c(\nu) = \sigma_{bf}(\nu)+\sigma_{ff}(\nu) = \varrho[D_k(\nu)+D_f(\nu)](h\nu)^{-3}. \tag{7.71}$$

It should be recalled that Eq. (7.71) gives the true continuous absorption coefficient, and it must be corrected for induced emission for use in the equation of transfer. Under the assumption of local thermodynamic equilibrium, we have [see Eqs. (2.173) and (7.22)]

$$\sigma_c'(\nu) = \varrho[D_k(\nu)+D_f(\nu)](1-e^{-h\nu/kT})(h\nu)^{-3}. \tag{7.72}$$

In contrast to bound electrons of a given level, free electrons contribute to the absorption coefficient at all frequencies. The relative importance of bound–free and free–free absorption depends on the state of ionization. At low temperatures, there are very few free electrons, and bound–free absorption dominates. At very high temperatures, most electrons are free, and free–free absorption is dominant. However, for heavy elements free–free absorption is usually much smaller than photoelectric absorption for the usual range of temperatures encountered in radiative transfer work.

The final item to be discussed in this section is bound–bound (line) absorption. The classical theory of line absorption (or emission) corresponding to radiation of frequency ν_0 treats a bound electron as a harmonic oscillator. The classical oscillator can emit and absorb radiation in a narrow spectral line. The classical theory of line absorption is described by Woolley and Stibbs (1953) and Ambartsumyan (1956). The classical cross section for line absorption, in cm^2/oscillator, is given by

$$\mu_{bb}(\nu) = \frac{e^2}{m_0 c} \frac{\gamma_0}{4\pi} \left[\frac{1}{(\nu - \nu_0)^2 + (\gamma_0/4\pi)^2} \right], \tag{7.73}$$

where ν_0 is the frequency of the center of the line, and

$$\gamma_0 = \frac{8\pi^2 e^2 \nu_0^2}{3 m_0 c^3} \tag{7.74}$$

determines the line width. The quantity γ_0 is called the radiation damping constant since it represents the reciprocal of the time during which a harmonic oscillator will reduce its energy by a factor of $e = 2.71828 \dots$ by the emission of radiation. In addition to the natural width, a line profile is broadened by collisional damping (energy loss due to interatomic collisions). Taking this into account, Eq. (7.73) remains the relevant expression for the cross section, but γ_0 is replaced by γ, defined as

$$\gamma = \gamma_0 + \frac{2}{\tau_c}. \tag{7.75}$$

Here τ_c is the average time between collisions which results in collisional damping of the oscillator, and is given by

$$\frac{1}{\tau_c} = \sigma_0 \bar{v} N, \tag{7.76}$$

where σ_0 is the cross section for collisional damping, in cm^2/atom, $\bar{v}$ is the mean speed of the atoms, and N is the atomic density.

Thus far our discussion has been based on classical physics. The quantum mechanical treatment of an atom gives a similar result except for a correction factor f called the oscillator strength. The cross section for line absorption by an atom is (Cox, 1965b)

$$\mu_{bb}(\nu) = \frac{e^2}{m_0 c} \frac{\gamma}{4\pi} f \left[\frac{1}{(\nu - \nu_0)^2 + (\gamma/4\pi)^2} \right], \tag{7.77}$$

with γ given by Eq. (7.75).

One can relate the oscillator strength f to the Einstein coefficients discussed in the last section in the following way. We compute the energy absorbed in this line in a time dt which, before absorption, was propagating in direction Ω within $d\Omega$. This is given by

$$dW_a = \frac{e^2}{m_0 c} \frac{\gamma}{4\pi} f \int_0^\infty d\nu \left[\frac{1}{(\nu - \nu_0)^2 + (\gamma/4\pi)^2} \right] I(\nu) \, d\Omega \, dt. \tag{7.78}$$

Assuming the line to be narrow, we can set $I(\nu) \approx I(\nu_0)$, and extend the bottom integration limit to $-\infty$. Performing the integral, we then find

$$dW_a = \frac{\pi e^2}{m_0 c} f I(\nu_0) \, d\Omega \, dt. \tag{7.79}$$

From Eq. (7.5) one sees that dW_a in terms of the Einstein coefficient B_{mn} is given by

$$dW_a = h\nu_{nm}B_{mn}I(\nu_{nm})\,d\Omega\,dt,\tag{7.80}$$

where $\nu_{nm} = \nu_0$, the frequency of the line. Equating Eqs. (7.79) and (7.80), we obtain

$$f = f_{mn} = \frac{m_0c}{\pi e^2}\,h\nu_{nm}B_{mn}.\tag{7.81}$$

Use of Eqs. (7.15) and (7.18), the two relationships among the Einstein coefficients, and Eq. (7.74), the definition of γ_0, allows Eq. (7.81) to be rewritten as

$$f_{mn} = \frac{4\pi}{3\gamma_0}\frac{g_n}{g_m}A_{nm}.\tag{7.82}$$

Here n and m are such that $E_n > E_m$. Often one introduces a negative oscillator strength f_{nm} for emission, defined as

$$f_{nm} = -\frac{g_m}{g_n}f_{mn}.\tag{7.83}$$

From Eq. (7.82) we then find

$$f_{nm} = -\frac{4\pi}{3\gamma_0}A_{nm}.\tag{7.84}$$

Oscillator strengths for hydrogen-like atoms have been calculated by Menzel and Pekeris (1935).

In addition to natural and collisional broadening, a spectral line can be smeared by the interatomic Stark effect and Doppler shifts. The Stark effect is the result of the electric fields of neighboring ions, and is discussed by Griem (1960). The cross section for the Doppler case, assuming a Maxwellian distribution of velocities for the atoms, is (Cox, 1965b)

$$\mu_{bb}(\nu) = \frac{\pi e^2}{m_0 c}\frac{f}{\nu_0}\left(\frac{Mc^2}{2\pi kT}\right)^{1/2}\exp\left[-\frac{Mc^2}{2kT}\left(\frac{\nu-\nu_0}{\nu_0}\right)^2\right],\tag{7.85}$$

where M is the mass of the atom.

The atomic cross section for line absorption is conveniently expressed in the form

$$\mu_{bb}(\nu) \equiv \mu_{mn}(\nu) = \left(\frac{\pi e^2}{m_0 c}\right)f_{mn}b(\nu),\tag{7.86}$$

where $b(\nu)$ is the shape factor or dispersion of the line. The shape factor is normalized to unity,

$$\int_0^\infty d\nu\,b(\nu) = 1,\tag{7.87}$$

so that the strength of the line does not depend upon the shape. The bound–bound, or line, absorption coefficient corresponding to a line of frequency ν_{nm} is given by

$$\sigma_{bb}(\nu) = \left(\frac{\pi e^2}{m_0 c}\right)N_m f_{mn}b(\nu),\tag{7.88}$$

where N_m is the density of atoms in the state corresponding to energy E_m. The shape factor can be either the natural and collisional (Lorentz) profile

$$b(\nu) = \frac{\gamma/4\pi}{(\nu-\nu_0)^2+(\gamma/4\pi)^2}\,, \tag{7.89}$$

the profile resulting from the Stark effect (Griem, 1960), the Doppler profile

$$b(\nu) = \frac{1}{\nu_0}\left(\frac{Mc^2}{2\pi kT}\right)^{1/2}\exp\left[-\frac{Mc^2}{2kT}\left(\frac{\nu-\nu_0}{\nu_0}\right)^2\right], \tag{7.90}$$

or, more generally, a combination of all three.

TABLE 7.2. *Oscillator Strengths for Hydrogen*

Final level j \ Initial level i	$i=1$ Lyman	$i=2$ Balmer	$i=3$ Paschen
$j=1$	—	-0.1041	-0.0088
2	0.4162	—	-0.2848
3	0.0791	0.6408	—
4	0.0290	0.1193	0.8420
5	0.0139	0.0447	0.1506
6	0.0078	0.0221	0.0559
7	0.0048	0.0127	0.0277
8	0.0032	0.0080	0.0160
9	0.0022	0.0054	0.0102
10	0.0016	0.0039	0.0070

Oscillator strengths for absorption in hydrogen are given in Table 7.2 (Menzel and Pekeris, 1935). Negative values indicate emission, and are related to absorption values by Eq. (7.83). From this table it can be seen that the principal lines are at the heads of the series. The higher transitions become weaker and closer together and blend into the bound–free absorption edge as shown schematically in Fig. 7.5. The effect of the large number of weak, closely spaced lines is to extend the photoelectric ionization edge. This provides a means of approximating the effect of these lines (Hunt and Sibulkin, 1967). The total absorption coefficient including bound–free, free–free, and bound–bound transitions for a typical heavy element is shown very schematically in Fig. 7.6. It should be noted that the absorption coefficient varies over many orders of magnitude. Bound–bound absorption effects at low frequencies are depressed because the higher levels are depopulated at typical temperatures occurring in radiative transfer work.

5. The Scattering Coefficient

The physics of the scattering interaction is contained in the scattering kernel, or differential scattering coefficient, $\sigma_s(\nu \to \nu', \xi)$, where $\xi = \mathbf{\Omega}\cdot\mathbf{\Omega}'$ is the cosine of the scattering angle (see Section 3 of Chapter I). Let us decompose $\sigma_s(\nu \to \nu', \xi)$ according to

$$\sigma_s(\nu \to \nu', \xi) = N\mu_s(\nu)\, K(\nu \to \nu', \xi), \tag{7.91}$$

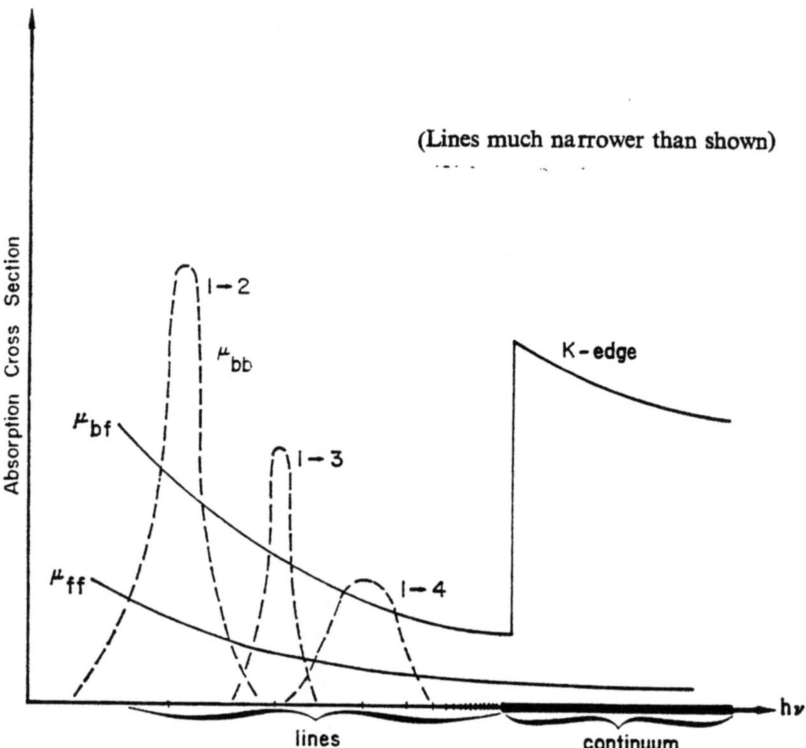

FIG. 7.5. Comparison of absorption processes.

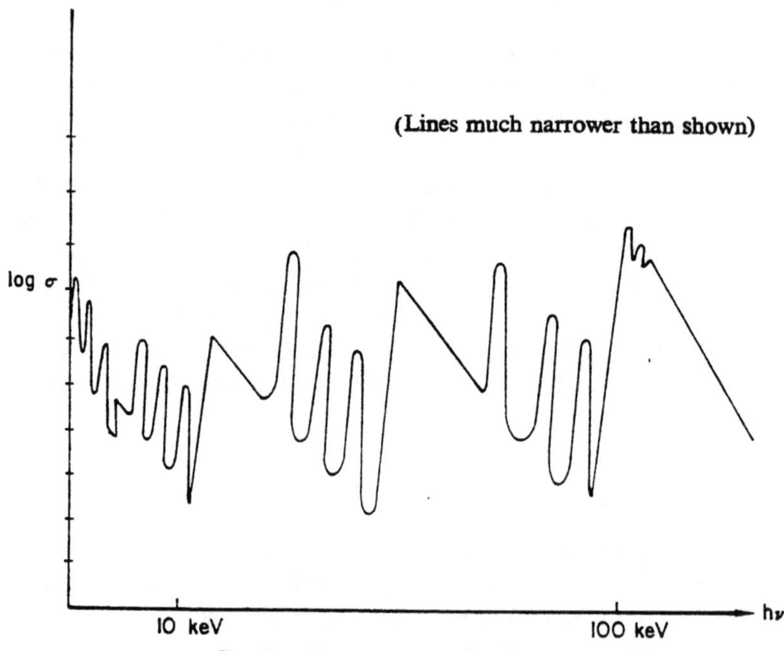

FIG. 7.6. Total absorption coefficient.

where $\mu_s(\nu)$ is the scattering cross section per scatterer, N is the number of scatterers per unit volume, and $K(\nu \to \nu', \xi)$ is a normalized scattering kernel, i.e.,

$$\int_0^\infty d\nu' \int_{4\pi} d\Omega' K(\nu \to \nu', \mathbf{\Omega} \cdot \mathbf{\Omega}') = 2\pi \int_0^\infty d\nu' \int_{-1}^1 d\xi K(\nu \to \nu', \xi) = 1. \tag{7.92}$$

If the scattering is coherent (no energy exchange), one has

$$K(\nu \to \nu', \xi) = K(\xi)\, \delta(\nu - \nu'), \tag{7.93}$$

where $\delta(z)$ is the Dirac delta function. The phase function $K(\xi)$ is normalized as

$$\int_{4\pi} d\Omega' K(\mathbf{\Omega} \cdot \mathbf{\Omega}') = 2\pi \int_{-1}^1 d\xi K(\xi) = 1. \tag{7.94}$$

If the scattering is isotropic (equal probability at all scattering angles), one has

$$K(\nu \to \nu', \xi) = \frac{1}{4\pi} K(\nu \to \nu'), \tag{7.95}$$

where the frequency redistribution function $K(\nu \to \nu')$ is normalized as

$$\int_0^\infty d\nu' K(\nu \to \nu') = 1. \tag{7.96}$$

The simplest scattering kernel is both coherent and isotropic, and in this case we have

$$K(\nu \to \nu', \xi) = \frac{1}{4\pi} \delta(\nu - \nu'). \tag{7.97}$$

Classically, scattering by atoms (or ions) is calculated assuming the scattering to be from bound electrons which are represented as harmonic oscillators at frequency ν_0. The cross section for scattering from such an oscillator, in cm^2/oscillator, is given by

$$\mu_s(\nu) = \frac{\pi e^2}{m_0 c} \frac{2\nu^2 \left(\dfrac{\nu}{\nu_0}\right)^2 \dfrac{\gamma}{2\pi^2}}{(\nu^2 - \nu_0^2)^2 + \nu^2 \left(\dfrac{\gamma}{2\pi}\right)^2} f_s, \tag{7.98}$$

where γ indicates the width of the resonance which occurs at ν_0, and is given by Eqs. (7.74) through (7.76). If one sets $f_s = 1$ in Eq. (7.98), $\mu_s(\nu)$ then describes the energy absorbed by a classical oscillator (Cox, 1965b). The factor f_s accounts for the ultimate disposition of the energy absorbed. In a rarefied gas the energy absorbed will, for the most part, be reradiated (i.e., scattered, since classically scattering is viewed as absorption followed by reradiation), and f_s will be very nearly equal to one. On the other hand, if the gas is very dense there will be very little reradiation since most of the energy absorbed by the oscillator will be transferred to other atoms and ions by collisions (i.e., the absorbed energy will be converted into heat). In this case, the process is primarily absorption and f_s will be small. An alternate way of viewing f_s is as a Gaunt factor or oscillator strength; that is, f_s can be viewed as a quantum correction factor to the classical result.

For ν in the vicinity of ν_0, Eq. (7.98) can be written

$$\mu_s(\nu) = \frac{e^2}{m_0 c} \frac{\gamma}{4\pi} \left[\frac{1}{(\nu - \nu_0)^2 + (\gamma/4\pi)^2} \right] f_s, \tag{7.99}$$

which describes resonance scattering (or resonance fluorescence). When the photon frequency ν is large compared to ν_0, Eq. (7.98) gives the Thomson result [in this case, neglecting quantum effects, f_s goes to unity and γ goes to γ_0 given by Eq. (7.74)]

$$\mu_s(\nu) = \frac{\pi e^2}{m_0 c} \frac{\gamma_0}{\nu_0^2 \pi^2} = \frac{8\pi e^4}{3 m_0^2 c^4} \cdot \tag{7.100}$$

This is the Thomson cross section which has a numerical value of 0.665×10^{-24} cm^2. In the other extreme, $\nu \ll \nu_0$, Eq. (7.98) gives

$$\mu_s(\nu) = \frac{\pi e^2}{m_0 c} \frac{\nu^4 \gamma}{\nu_0^6 \pi^2} f_s = \frac{8\pi e^4}{3 m_0^2 c^4} \left(\frac{\nu}{\nu_0} \right)^4 f_s. \tag{7.101}$$

This is generally referred to as the Rayleigh cross section, and can be seen to be just the Thomson result times the dimensionless quantity $(\nu/\nu_0)^4 f_s$. A similar expression, also called Rayleigh scattering, describes the classical scattering of light by a dielectric sphere (Chandrasekhar, 1960d). In this case one has

$$\mu_s(\nu) = \frac{128\pi^5 \alpha^2 \nu^4}{3 c^2} f_s, \tag{7.102}$$

where α is the polarizability of the sphere. The quantity f_s approaches unity as the wavelength of the light $\lambda = c/\nu$ becomes large compared to the radius of the sphere.

We now consider the scattering of photons from free, rather than bound, electrons. We shall discuss this case in slightly more detail since, as argued in Section 2 of this chapter, this is the most important scattering interaction in radiative transfer work. The classical cross section, in cm^2/electron, for scattering from free electrons is the Thomson result given by Eq. (7.100). (This result was derived as the high energy limit of the bound electron scattering cross section. In this limit, the electron acts as if it were free since the oscillator energy is small compared to the incident photon energy.) The corresponding normalized scattering kernel [see Eq. (7.92)] is

$$K(\nu \to \nu', \xi) = \frac{3}{16\pi} (1 + \xi^2) \, \delta(\nu - \nu'). \tag{7.103}$$

Thus the classical scattering coefficient for free electrons is given by [see Eqs. (7.91), (7.100), and (7.103)]

$$\sigma_s(\nu \to \nu', \xi) = N \frac{e^4}{2 m_0^2 c^4} (1 + \xi^2) \, \delta(\nu - \nu'), \tag{7.104}$$

where N is the free electron density. The two distinguishing features of Thomson scattering are: (1) it is coherent, and (2) it is symmetric in the forward and backward hemispheres of the scattering angle (it depends only upon the absolute value of ξ, the cosine of the scattering angle).

Equation (7.104) is the classical result of scattering from free electrons after averaging over polarization states. The more complete classical result which accounts for the polarization state of the light has been discussed in Chapter IV. In particular, Eq. (4.178) gives the basic normalized Thomson scattering kernel including the effects of polarization. The polarization averaged result, Eq. (7.103), also follows from Eq. (4.178) in conjunction with Eq. (4.130).

The quantum mechanical result for the scattering of photons by free electrons is given by the Klein–Nishina formula. As in the classical result, the scattering kernel actually depends upon the state of polarization of the incident photon. If one averages over polarization states, the scattering kernel is given by (Evans, 1955)

$$\sigma_s(\nu \to \nu', \xi) = N\frac{r_0^2}{2}\frac{1+\xi^2}{[1+\gamma(1-\xi)]^2} \times$$

$$\times \left\{1+\frac{\gamma^2(1-\xi)^2}{(1+\xi^2)[1+\gamma(1-\xi)]}\right\}\delta\left(\nu'-\frac{\nu}{1+\gamma(1-\xi)}\right), \qquad (7.105)$$

where N is the free electron density, r_0 is the classical electron radius

$$r_0 = \frac{e^2}{m_0 c^2}, \qquad (7.106)$$

and γ is a dimensionless frequency

$$\gamma = \frac{h\nu}{m_0 c^2}. \qquad (7.107)$$

The Dirac delta function in Eq. (7.105) states that, given an initial photon frequency, the scattering angle and final photon energy are correlated. This correlation results from simple conservation of energy and momentum in the scattering process. Another widely seen form of the Klein–Nishina formula follows from Eq. (7.105) by changing the delta function from one in ν' to one in ξ. Suppressing the algebraic detail, we find

$$\sigma_s(\nu \to \nu', \xi) = N\frac{r_0^2}{2}\frac{1}{\gamma\nu}\left[\frac{\gamma}{\gamma'}+\frac{\gamma'}{\gamma}+2\left(\frac{1}{\gamma}-\frac{1}{\gamma'}\right)+\left(\frac{1}{\gamma}-\frac{1}{\gamma'}\right)^2\right] \times$$

$$\times \delta\left(\xi-1+\frac{1}{\gamma'}-\frac{1}{\gamma}\right), \qquad (7.108)$$

where γ is given by Eq. (7.107) and

$$\gamma' = \frac{h\nu'}{m_0 c^2}. \qquad (7.109)$$

A third form for $\sigma_s(\nu \to \nu', \xi)$ follows very simply from Eq. (7.108). This is

$$\sigma_s(\nu \to \nu', \xi) = N\frac{r_0^2}{2}\frac{1}{\gamma\nu}[1+\xi^2+\gamma\gamma'(1-\xi)^2] \times$$

$$\times \delta\left(\xi-1+\frac{1}{\gamma'}-\frac{1}{\gamma}\right). \qquad (7.110)$$

For low incident energies, $\gamma \ll 1$, it is sensible to expand Eq. (7.105) in powers of γ. To second order, we have

$$\sigma_s(\nu \to \nu', \xi) = N \frac{r_0^2}{2} (1+\xi^2) \left[1 - 2\gamma(1-\xi) + \gamma^2 \frac{(1-\xi)^2(4+3\xi^2)}{1+\xi^2} \right] \times$$

$$\times \delta(\nu' - \nu[1 - \gamma(1-\xi) + \gamma^2(1-\xi)^2]). \qquad (7.111)$$

Setting $\gamma = 0$ in Eq. (7.111), we obtain

$$\sigma_s(\nu \to \nu', \xi) = N \frac{r_0^2}{2} (1+\xi^2)\, \delta(\nu'-\nu), \qquad (7.112)$$

which is just the classical Thomson scattering kernel [see Eq. (7.104)].

Integration of Eq. (7.105) over all Ω' and ν' gives the scattering coefficient $\sigma_s(\nu)$. We find

$$\sigma_s(\nu) = \int_{4\pi} d\Omega' \int_0^\infty d\nu' \sigma_s(\nu \to \nu', \Omega \cdot \Omega') = 2\pi \int_{-1}^{1} d\xi \int_0^\infty d\nu' \sigma_s(\nu \to \nu', \xi)$$

$$= \frac{3Na_0}{4} \left\{ \left(\frac{1+\gamma}{\gamma^3} \right) \left[\frac{2\gamma(1+\gamma)}{1+2\gamma} - \ln(1+2\gamma) \right] + \frac{1}{2\gamma} \ln(1+2\gamma) - \frac{1+3\gamma}{(1+2\gamma)^2} \right\}, \qquad (7.113)$$

where a_0 is the Thomson scattering cross section per electron

$$a_0 = \frac{8\pi}{3} r_0^2 = \frac{8\pi e^4}{3m_0^2 c^4}. \qquad (7.114)$$

Correct to second order in γ, Eq. (7.113) gives

$$\sigma_s(\nu) = Na_0 \left(1 - 2\gamma + \frac{26}{5}\gamma^2 + \ldots \right). \qquad (7.115)$$

For small γ, Eq. (7.115) clearly shows that the Compton scattering coefficient is smaller than the classical Thomson value. It can be shown from Eq. (7.113) that this inequality is true for all γ (i.e., all photon energies).

In certain radiative transfer problems it is important to account for the motion of the free electrons prior to scattering in the scattering interaction. Scattering from moving free electrons is generally referred to as Compton and inverse Compton scattering. We consider the details of this type of scattering in the next chapter.

VIII

Compton and Inverse Compton Scattering

---

1. Introduction

In this chapter we discuss the scattering kernel and the equation of transfer describing photon scattering from free electrons which are in motion. Particular emphasis will be placed on the special case of scattering from a Maxwellian gas of free electrons at some temperature T.

We first show how the Klein–Nishina formula, in conjunction with the Lorentz transformation discussed in the Appendix, can be used to derive such a scattering kernel. Section 3 discusses a simplification of the equation of transfer describing Compton and inverse Compton scattering known as the Fokker–Planck method. The end result of this method is the conversion of the integral scattering operator in frequency to a second order differential operator. Finally, Sections 4 and 5 discuss some recent work in which the frequency variable in the equation of transfer is described, within the Fokker–Planck treatment, by an eigenfunction expansion.

2. The Scattering Kernel

The Klein–Nishina formula describes Compton scattering from free electrons at rest, and exhibits the characteristic that photons cannot gain energy upon scattering. If the interaction is between a photon and a moving electron, however, the electron can impart some or all of its energy to the photon and increase the photon's frequency upon scattering. Such an event is often referred to as inverse Compton scattering. In this section we derive the scattering kernel in the case of photon scattering from a relativistic Maxwellian gas of free, nondegenerate, electrons. The elements of the derivation, however, are applicable to the more general case of photon scattering from an arbitrary distribution of moving particles.

Before proceeding analytically, it is worth while to briefly discuss the nature of the scattering kernel expected in this case. As pointed out by Dirac (1925), the kernel will have three rather distinct characteristics. In the first place, a photon will, upon scattering, have its wavelength increased due to the usual Compton shift associated with scattering from an electron at rest. Secondly, it will undergo broadening due to the classical Doppler effect of scattering from a distribution of moving electrons. Finally, there will be a reduction in the wavelength upon scattering due to the relativistic effect that the photon density will appear more intense to an electron moving toward the photon than away from it. This last effect, the blue shift, is needed to "balance" the Compton red shift, for, as pointed out by Milne (1924), if black body radiation at a certain temperature scatters from a Maxwellian gas of free electrons at the same temperature, the scattered radiation must have the same distribution in wavelength as the incident radiation.

Since the Maxwellian distribution is the thermodynamic equilibrium distribution for the electrons, the scattering kernel $\sigma_s(v' \to v, \, \boldsymbol{\Omega}' \cdot \boldsymbol{\Omega})$ must also satisfy the detailed balance condition. This condition states that in complete thermodynamic equilibrium the number of photons which scatter from $dv' \, d\Omega'$ about $v', \boldsymbol{\Omega}'$ to $dv \, d\Omega$ about $v, \boldsymbol{\Omega}$ must equal the number scattered from $dv \, d\Omega$ to $dv' \, d\Omega'$. Quantitatively, this condition takes the form

$$[1 + c^2 B(v)/2hv^3] \, \sigma_s(v' \to v, \boldsymbol{\Omega}' \cdot \boldsymbol{\Omega}) B(v')/hv' = [1 + c^2 B(v')/2hv'^3] \, \sigma_s(v \to v', \boldsymbol{\Omega} \cdot \boldsymbol{\Omega}') B(v)/hv, \tag{8.1}$$

where $B(v)$ is the Planck distribution given by Eq. (1.8). Equation (8.1) relates the scattering kernel, at a given scattering angle, to the kernel with the frequency variables v and v' interchanged, at the same angle. Explicit use of Eq. (1.8) in Eq. (8.1) yields

$$\sigma_s(v' \to v, \boldsymbol{\Omega}' \cdot \boldsymbol{\Omega}) W(v')/hv' = \sigma_s(v \to v', \boldsymbol{\Omega} \cdot \boldsymbol{\Omega}') W(v)/hv, \tag{8.2}$$

where $W(v)$ is the Wien approximation to the Planck function, i.e., aside from a normalization constant,

$$W(v) = v^3 e^{-hv/kT}. \tag{8.3}$$

This result can be interpreted as the detailed balance condition in the absence of induced scattering and shows that the neglect of these induced terms in the scattering description leads to the Wien law, rather than the Planck function, as the equilibrium distribution of the scattering operator.

To compute the scattering kernel we have just discussed, we consider a frame of reference in which a group of electrons is at rest. We call this the e frame and subscript all quantities in this frame with an e. We take these electrons to have a density N_e in this frame. If the unadorned frame moves with velocity $-\mathbf{v}$ with respect to the e frame (so that, as observed from the unadorned frame, the electrons have velocity $\mathbf{v}$) we have, from Eq. (A.92) of the Appendix,

$$\sigma_s(v \to v', \boldsymbol{\Omega} \to \boldsymbol{\Omega}') = \frac{D}{D'} \, \sigma_{se}(v_e \to v'_e, \boldsymbol{\Omega}_e \to \boldsymbol{\Omega}'_e), \tag{8.4}$$

where

$$D = 1 - \boldsymbol{\Omega} \cdot \mathbf{v}/c, \tag{8.5}$$

$$D' = 1 - \boldsymbol{\Omega}' \cdot \mathbf{v}/c. \tag{8.6}$$

[Note the change of sign in Eqs. (8.5) and (8.6) as compared to Eqs. (A.81) and (A.82). This is because here the unadorned frame moves with velocity $-\mathbf{v}$ with respect to the e frame. The e frame here is to be identified with the zero-frame in the Appendix.] In the e frame, the scattering kernel, $\sigma_{se}(v_e \to v'_e, \boldsymbol{\Omega}_e \to \boldsymbol{\Omega}'_e)$, is just the Klein–Nishina formula, Eq. (7.110), with $\xi = \boldsymbol{\Omega} \cdot \boldsymbol{\Omega}'$ and all quantities subscripted with an e. The independent variables, frequency and angle, transform as, from Eqs. (A.83) and (A.86),

$$v_e = \lambda Dv, \tag{8.7}$$

$$v_e' = \lambda Dv', \tag{8.8}$$

$$1 - \mathbf{\Omega}_e \cdot \mathbf{\Omega}_e' = (1 - \mathbf{\Omega} \cdot \mathbf{\Omega}')/\lambda^2 DD', \tag{8.9}$$

with D and D' given by Eqs. (8.5) and (8.6) and

$$\lambda = (1 - v^2/c^2)^{-1/2}. \tag{8.10}$$

Also, due to the Lorentz contraction, the electron density in the unadorned frame is given by $N = \lambda N_e$, with λ given above in Eq. (8.10). Combining all of these results, we obtain

$$\sigma_s(v \to v', \mathbf{\Omega} \to \mathbf{\Omega}') = \frac{Nr_0^2}{2\gamma v\lambda} \, \delta \left(\xi - 1 + \frac{\lambda D}{\gamma'} - \frac{\lambda D'}{\gamma} \right) \times$$

$$\times \left\{ 1 + \left[1 - \frac{(1-\xi)}{\lambda^2 DD'} \right]^2 + \frac{\gamma\gamma'(1-\xi)^2}{\lambda^2 DD'} \right\}, \tag{8.11}$$

where $\xi = \mathbf{\Omega} \cdot \mathbf{\Omega}'$.

Equation (4.16) is the scattering kernel corresponding to all electrons moving with a velocity $\mathbf{v}$. To account for the fact that the electrons have a velocity distribution as seen by the observer in the unadorned frame, we replace N in Eq. (8.11) by $Nf(v)\,d\mathbf{v}$, where $f(v)$ is an isotropic distribution function (and hence depends only upon v rather than $\mathbf{v}$) normalized according to

$$\int d\mathbf{v} f(v) = 4\pi \int_0^c dv v^2 f(v) = 1, \tag{8.12}$$

and integrate the resulting expression over all $\mathbf{v}$. This gives

$$\sigma_s(v \to v', \xi) = \frac{Nr_0^2}{2\gamma v} \int d\mathbf{v} f(v) \frac{1}{\lambda} \delta \left(\xi - 1 + \frac{\lambda D}{\gamma'} - \frac{\lambda D'}{\gamma} \right) \times$$

$$\times \left\{ 1 + \left[1 - \frac{(1-\xi)}{\lambda^2 DD'} \right]^2 + \frac{\gamma\gamma'(1-\xi)^2}{\lambda^2 DD'} \right\} \tag{8.13}$$

as the scattering kernel corresponding to Compton and inverse Compton scattering from an isotropic distribution of free electrons. In writing the left hand side of Eq. (8.13) we have anticipated the fact that the kernel will depend only upon $\xi = \mathbf{\Omega} \cdot \mathbf{\Omega}'$ rather than $\mathbf{\Omega}$ and $\mathbf{\Omega}'$ separately, since the electron distribution is isotropic.

The triple integral indicated in Eq. (8.13) can be reduced somewhat because of the appearance of the Dirac delta function. Let us sketch the details of this. We write

$$d\mathbf{v} = v^2 \, d\Omega_v, \tag{8.14}$$

and Eq. (8.13) becomes

$$\sigma_s(v \to v', \xi) = \frac{Nr_0^2}{2\gamma v} \int_0^c dv \, \frac{v^2}{\lambda} \, f(v) \int_{4\pi} d\Omega_v \, \delta \left(\xi - 1 + \frac{\lambda D}{\gamma'} - \frac{\lambda D'}{\gamma} \right) \times$$

$$\times \left\{ 1 + \left[1 - \frac{(1-\xi)}{\lambda^2 DD'} \right]^2 + \frac{\gamma\gamma'(1-\xi)^2}{\lambda^2 DD'} \right\}. \tag{8.15}$$

The Equations of Radiation Hydrodynamics

In performing the integral over Ω_v in Eq. (8.15), we choose a coordinate system such that Ω is the z axis and Ω' lies in the x–z plane. We describe Ω_v by θ_v and φ_v and Ω' by θ' and φ', where θ_v and θ' are polar angles measured with respect to the z axis and φ_v and φ' are azimuthal angles measured with respect to the x axis. Since Ω' lies in the x–z plane, $\varphi' = 0$. Further, θ' is the angle between Ω and Ω' and hence $\cos\theta' = \Omega\cdot\Omega' = \xi$. Equation (8.15) then becomes

$$\sigma_s(v \rightarrow v', \xi) = \frac{Nr_0^2}{\gamma v}\int_0^c dv\,\frac{v^2}{\lambda}f(v)\int_{-1}^1 d\mu_v\int_0^\pi d\varphi_v\delta\left(\xi-1+\frac{\lambda D}{\gamma'}-\frac{\lambda D'}{\gamma}\right)\times$$

$$\times\left\{1+\left[1-\frac{(1-\xi)}{\lambda^2 DD'}\right]^2+\frac{\gamma\gamma'(1-\xi)^2}{\lambda^2 DD'}\right\},\tag{8.16}$$

where the integral over φ_v from 0 to 2π has been written as twice the integral from 0 to π. This follows since the integrand is only a function of $\cos\varphi_v$. In this equation

$$\mu_v = \cos\theta_v,\tag{8.17}$$

$$D = 1-v\mu_v/c,\tag{8.18}$$

$$D' = 1-\frac{v}{c}\left[\mu_v\xi+\sqrt{1-\mu_v^2}\sqrt{1-\xi^2}\cos\varphi_v\right].\tag{8.19}$$

In writing Eq. (8.19) we have used a well known identity to expand $\mathbf{v}\cdot\Omega'$. Let us use the delta function in Eq. (8.16) to integrate over the azimuthal angle φ_v. Given ξ, γ, γ', and v, only for certain values of μ_v will the argument of the delta function vanish when φ_v runs from 0 to π. Hence the integral over φ_v gives a nonzero value only for certain values of μ_v. We denote these values of μ_v as belonging to the set A. If we define

$$g(v, \mu_v) = \lambda\delta(1-v\mu_v/c)[\xi-1+\lambda(1-v\mu_v/c)/\gamma'],\tag{8.20}$$

then, performing the integral over φ_v in Eq. (8.16) we find

$$\sigma_s(v \rightarrow v', \xi) = \frac{Ncr_0^2}{v}\int_0^c dv\,\frac{v}{\lambda^2}f(v)\times$$

$$\times\int_{\mu_v\in A} d\mu_v\frac{\left\{1+\left[1-\frac{(1-\xi)}{g(v,\mu_v)}\right]^2+\frac{\gamma\gamma'(1-\xi)^2}{g(v,\mu_v)}\right\}}{\sqrt{1-\mu_v^2}\sqrt{1-\xi^2}|\sin\bar\varphi_v|},\tag{8.21}$$

where $\bar\varphi_v$ is the angle which makes the argument of the delta function vanish. We have

$$\cos\bar\varphi_v = \frac{\dfrac{c}{v}\left\{1-\dfrac{\gamma}{\lambda}[\xi-1+\lambda(1-v\mu_v/c)/\gamma']\right\}-\mu_v\xi}{\sqrt{1-\mu_v^2}\sqrt{1-\xi^2}}.\tag{8.22}$$

Hence if we define

$$h(v, \mu_v) = \left\{(1-\mu_v^2)(1-\xi^2)-\left[\frac{c}{v}+\frac{c\gamma}{v\lambda}(1-\xi)-\frac{c\gamma}{v\gamma'}(1-v\mu_v/c)-\mu_v\xi\right]^2\right\}^{1/2},\tag{8.23}$$

Eq. (8.21) becomes

$$\sigma_s(\nu \to \nu', \xi) = \frac{Ncr_0^2}{\nu} \int\limits_0^c dv \, \frac{v}{\lambda^2} f(v) \int\limits_{\mu_v \in A} d\mu_v \, \frac{1}{h(v, \mu_v)} \times$$

$$\times \left\{ 1 + \left[1 - \frac{(1-\xi)}{g(v, \mu_v)} \right]^2 + \frac{\gamma\gamma'(1-\xi)^2}{g(v, \mu_v)} \right\}. \tag{8.24}$$

To complete the definition of the right hand side of Eq. (8.24) we need specify the set A. Since the original integral over μ_v in Eq. (8.16) is over the interval $(-1, 1)$, A must be a subset of this interval. Further, for the argument of the delta function to vanish as φ_v runs from 0 to π, we must have $-1 \le \cos \bar{\varphi}_v \le 1$. With reference to Eq. (8.22), we define the set B as all values of μ_v satisfying the inequality

$$\left| \frac{\frac{c}{v} \left\{ 1 - \frac{\gamma}{\lambda} [\xi - 1 + \lambda(1 - \mu_v/c)/\gamma'] \right\} - \mu_v \xi}{\sqrt{1 - \mu_v^2} \sqrt{1 - \xi^2}} \right| \le 1. \tag{8.25}$$

Then the set A is the intersection of the interval $(-1, 1)$ and the set B.

This is as far as it appears profitable to proceed analytically. Given ν, ν', and ξ, the scattering kernel is probably best computed from Eq. (8.24) by performing the double integral numerically. A computer program to do this has been written by Stone and Nelson (1966).

Of course, to evaluate the right hand side of Eq. (8.24) it is necessary to know $f(v)$, the relativistic Maxwellian distribution, and we now turn to this item. We know from elementary statistical mechanics that $\psi(\mathbf{p})$, the distribution function per unit momentum, is given by the Maxwell–Boltzmann distribution

$$\psi(\mathbf{p}) = Ce^{-E/kT}, \tag{8.26}$$

where C is a normalization constant. To go from $\psi(\mathbf{p})$ to $f(v)$, the distribution function per unit velocity, we need introduce the Jacobian of the transformation from $\mathbf{p}$ to $\mathbf{v}$, i.e.,

$$J(\mathbf{p}; \mathbf{v}) = \frac{p^2}{v^2} \frac{dp}{dv}. \tag{8.27}$$

Since

$$p = m_0 \lambda v, \tag{8.28}$$

and

$$E = m_0 c^2 \lambda, \tag{8.29}$$

we find

$$f(v) = C' \lambda^5 e^{-m_0 c^2 \lambda/kT}, \tag{8.30}$$

where C' is another constant and, as before, λ is the Lorentz factor given by Eq. (8.10). If we demand that $f(v)$ be normalized to have an integral of unity according to Eq. (8.12), the constant C' is easily evaluated. Our final result for a normalized relativistic Maxwellian distribution is then

$$f(v) = \frac{m_0 \lambda^5 e^{-m_0 c^2 \lambda/kT}}{4\pi ckT K_2(m_0 c^2/kT)}, \tag{8.31}$$

where $K_2(z)$ is the modified Bessel function of the second kind of order two, and m_0 is the rest mass of an electron.

Figures 8.1 through 8.3 show typical results for the scattering kernel describing the scattering of photons from a Maxwellian gas of free electrons. These results were obtained by evaluating Eq. (8.24), with $f(v)$ given by Eq. (8.31), numerically (with $N = 1$), and then

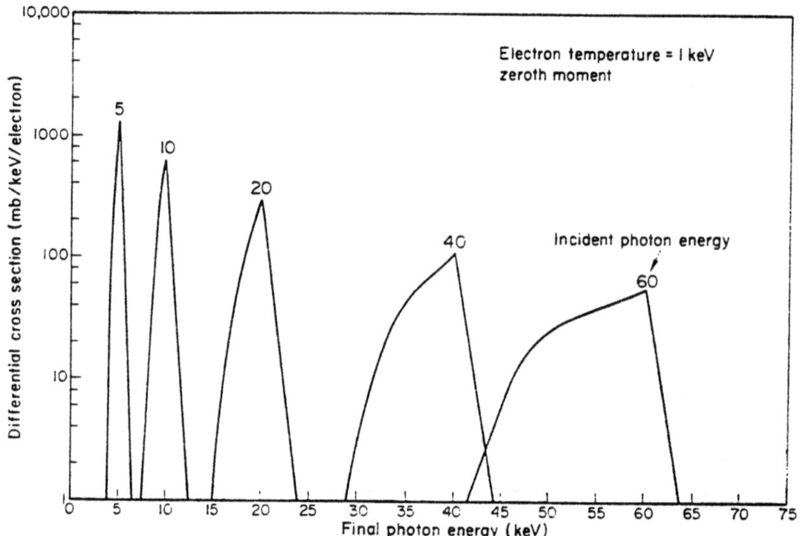

FIG. 8.1a. The differential scattering cross section: $n = 0$; $T = 1$.

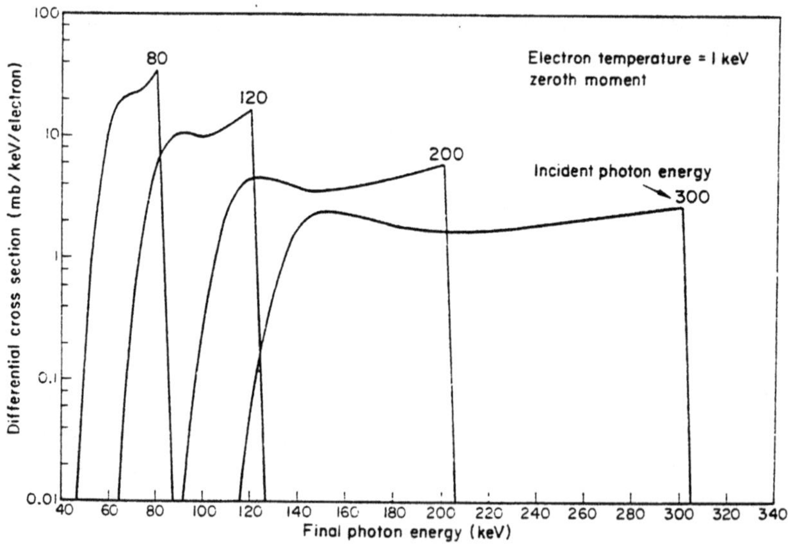

FIG. 8.1b. The differential scattering cross section: $n = 0$; $T = 1$.

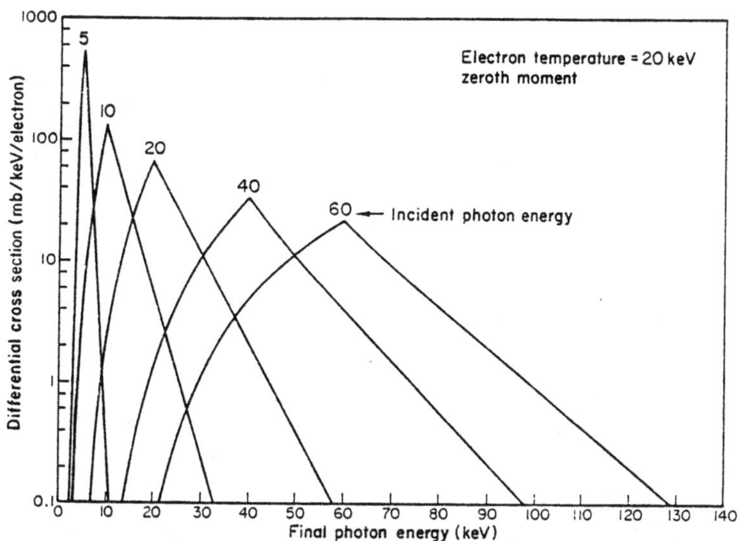

FIG. 8.2a. The differential scattering cross section: $n = 0$; $T = 20$.

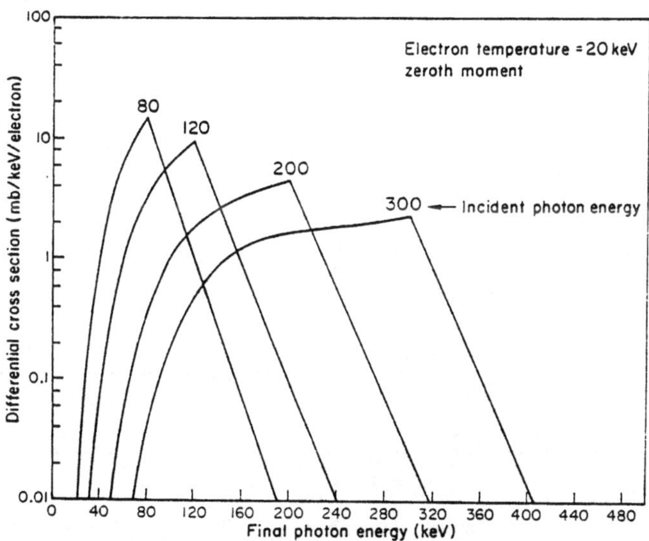

FIG. 8.2b. The differential scattering cross section: $n = 0$; $T = 20$.

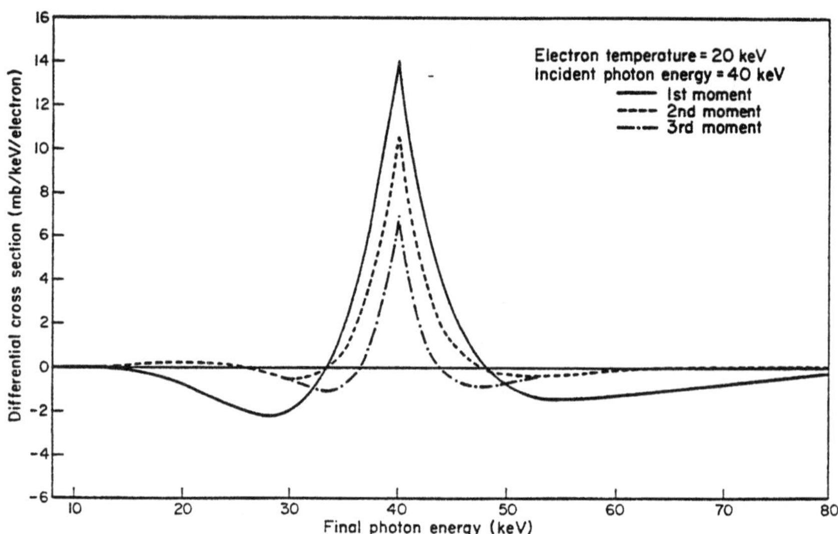

FIG. 8.3a. The differential scattering cross section: $n = 1, 2, 3$; $T = 20$.

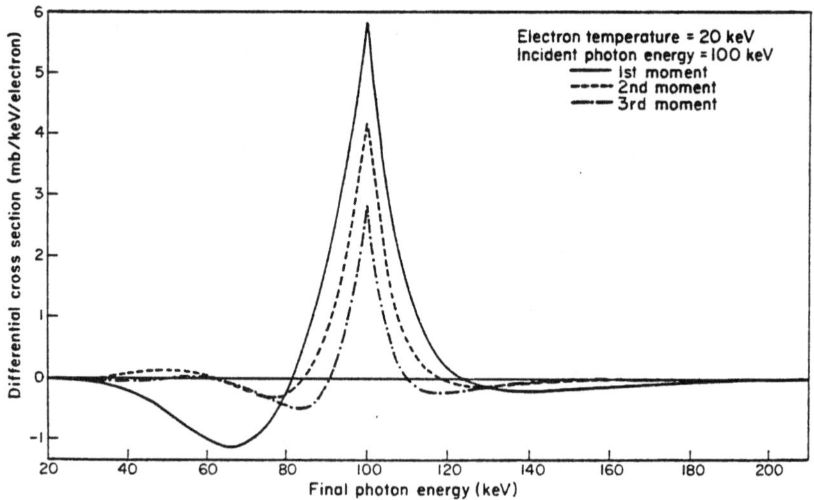

FIG. 8.3b. The differential scattering cross section: $n = 1, 2, 3$; $T = 20$.

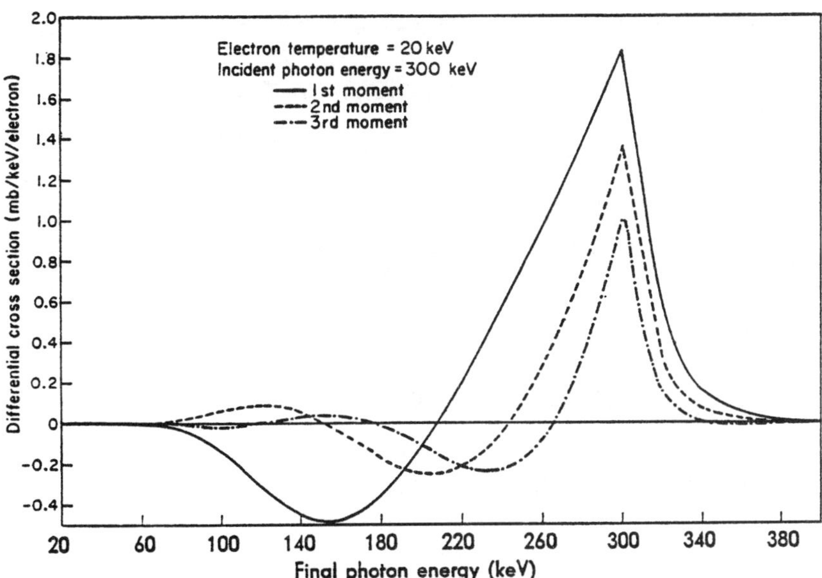

FIG. 8.3c. The differential scattering cross section: $n = 1, 2, 3$; $T = 20$.

computing the Legendre moments, i.e.,

$$\sigma_{sn}(v \to v') = 2\pi \int_{-1}^{1} d\xi P_n(\xi)\, \sigma_s(v \to v', \xi). \tag{8.32}$$

These figures show the four lowest Legendre moments ($n = 0, 1, 2, 3$), with all temperatures expressed in kiloelectron volts (keV), and the differential scattering cross sections, σ_{sn}, expressed in millibarns (mb) per keV ($1 \text{ mb} = 10^{-27} \text{ cm}^2$).

3. The Fokker–Planck Treatment of Scattering

The scattering kernel for Compton and inverse Compton effects just derived, while accurately describing the physics of photon scattering from a Maxwellian gas of free electrons, is rather complex in that it is defined in terms of a multiple integral [see Eq. (8.24)]. In addition, the use of this kernel in the equation of transfer introduces further integrals over frequency and angle [see Eq. (2.170)]. In this section we describe in some detail recent work concerning a simplification of this scattering description which leads to the elimination of all integrals defining the scattering kernel as well as the integral over frequency in the equation of transfer. The assumption required to achieve this result (the Fokker–Planck approximation) is that the electron temperatures and photon frequencies of interest are small compared to the rest energy of the electron.

In order to effect this simplification, we first expand $I(v', \Omega')$ in Eq. (2.170) in a Taylor series about $v' = v$, i.e.,

$$I(v', \Omega') = \sum_{n=0}^{\infty} \frac{1}{n!} \frac{\partial^n I(v, \Omega')}{\partial v^n} (v' - v)^n. \tag{8.33}$$

The Equations of Radiation Hydrodynamics

Inserting Eq. (8.33) into Eq. (2.170) and integrating term by term over v', we obtain

$$\frac{1}{c}\frac{\partial I(v, \Omega)}{\partial t}+\Omega\cdot\nabla I(v, \Omega) = \sigma_a'(v)[B(v)-I(v, \Omega)]$$

$$+\int_{4\pi} d\Omega' \sum_{n=0}^{\infty} N_n(v, \Omega\cdot\Omega')\, v^n\, \frac{\partial^n I(v, \Omega')}{\partial v^n}-\sigma_s(v)\, I(v, \Omega)$$

$$+\frac{c^2}{2hv^3}\, I(v, \Omega) \int_{4\pi} d\Omega' \sum_{n=0}^{\infty} M_n(v, \Omega\cdot\Omega')\, v^n\, \frac{\partial^n I(v, \Omega')}{\partial v^n}, \tag{8.34}$$

where we have defined

$$\sigma_s(v) = \int_0^{\infty} dv' \int_{4\pi} d\Omega' \sigma_s(v \to v', \Omega\cdot\Omega'), \tag{8.35}$$

$$N_n(v, \Omega\cdot\Omega') = \frac{1}{n!} \int_0^{\infty} dv'\, \frac{v}{v'} \left(\frac{v'-v}{v}\right)^n \sigma_s(v' \to v, \Omega'\cdot\Omega), \tag{8.36}$$

$$M_n(v, \Omega\cdot\Omega') = \frac{1}{n!} \int_0^{\infty} dv' \left(\frac{v'-v}{v}\right)^n \left[\frac{v}{v'}\sigma_s(v' \to v, \Omega'\cdot\Omega)- \left(\frac{v}{v'}\right)^3 \sigma_s(v \to v', \Omega\cdot\Omega')\right].$$

$$\tag{8.37}$$

This formal Taylor series expansion has converted the integral operator in frequency in Eq. (2.170) into an infinite order differential operator. The advantage of this procedure is that for small electron temperatures ($\alpha \equiv kT/m_0c^2 \ll 1$) and photon energies ($\gamma \equiv hv/m_0c^2 \ll 1$) this infinite order operator effectively truncates itself to one of finite order. In particular, to first order in α and γ, Fraser (1966) has given the results

$$\sigma_s(v) = \sigma_{\text{Th}}(1-2\gamma), \tag{8.38}$$

$$N_0(v, \Omega\cdot\Omega') = \frac{3}{16\pi}\, \sigma_{\text{Th}}[(1-\gamma+2\alpha)+(\Omega\cdot\Omega')(\gamma-4\alpha)$$

$$+(\Omega\cdot\Omega')^2(1-\gamma-6\alpha)+(\Omega\cdot\Omega')^3(\gamma+4\alpha)], \tag{8.39}$$

$$N_1(v, \Omega\cdot\Omega') = \frac{3}{16\pi}\, \sigma_{\text{Th}}[1-(\Omega\cdot\Omega')+(\Omega\cdot\Omega')^2-(\Omega\cdot\Omega')^3](\gamma-2\alpha), \tag{8.40}$$

$$N_2(v, \Omega\cdot\Omega') = \frac{3}{16\pi}\, \sigma_{\text{Th}}[1-(\Omega\cdot\Omega')+(\Omega\cdot\Omega')^2-(\Omega\cdot\Omega')^3](\alpha), \tag{8.41}$$

$$M_0(v, \Omega\cdot\Omega') = -\frac{3}{16\pi}\, \sigma_{\text{Th}}[1-(\Omega\cdot\Omega')+(\Omega\cdot\Omega')^2-(\Omega\cdot\Omega')^3]\,(2\gamma), \tag{8.42}$$

$$M_1(v, \Omega\cdot\Omega') = \frac{3}{16\pi}\, \sigma_{\text{Th}}[1-(\Omega\cdot\Omega')+(\Omega\cdot\Omega')^2-(\Omega\cdot\Omega')^3]\,(2\gamma). \tag{8.43}$$

Here σ_{Th} is the Thomson scattering coefficient given by

$$\sigma_{\text{Th}} = \frac{8\pi}{3}\, Nr_0^2. \tag{8.44}$$

All of the other $N_n(\nu)$ and $M_n(\nu)$ are of higher order in α and γ. Introducing these results into Eq. (8.34) and employing a somewhat more compact notation, we can write the equation of transfer with scattering described to first order in α and γ as

$$\frac{1}{c}\frac{\partial I(\nu,\boldsymbol{\Omega})}{\partial t}+\boldsymbol{\Omega}\cdot\nabla I(\nu,\boldsymbol{\Omega}) = \sigma_a'(\nu)[B(\nu)-I(\nu,\boldsymbol{\Omega})]-\sigma_{\mathrm{Th}}(1-2\gamma)\,I(\nu,\boldsymbol{\Omega})$$

$$+\sigma_{\mathrm{Th}}\int_{4\pi} d\Omega' \sum_{n=0}^{3}\left(\frac{2n+1}{4\pi}\right)P_n(\boldsymbol{\Omega}\cdot\boldsymbol{\Omega}')\,S_n I(\nu,\boldsymbol{\Omega}')$$

$$-\frac{3\sigma_{\mathrm{Th}}}{16\pi}\frac{c^2}{h\nu^3}\,\gamma I(\nu,\boldsymbol{\Omega})\left(1-\nu\frac{\partial}{\partial\nu}\right)\int_{4\pi} d\Omega'[1-(\boldsymbol{\Omega}\cdot\boldsymbol{\Omega}')+(\boldsymbol{\Omega}\cdot\boldsymbol{\Omega}')^2-(\boldsymbol{\Omega}\cdot\boldsymbol{\Omega}')^3]\,I(\nu,\boldsymbol{\Omega}').$$

$$(8.45)$$

Here $P_n(z)$ is the usual nth order Legendre polynomial and the scattering operators S_n are defined as

$$S_0 = \left[1-\gamma\left(1-\nu\frac{\partial}{\partial\nu}\right)-\alpha\left(2\nu\frac{\partial}{\partial\nu}-\nu^2\frac{\partial^2}{\partial\nu^2}\right)\right], \qquad (8.46)$$

$$S_1 = \frac{2}{5}\left[\gamma\left(1-\nu\frac{\partial}{\partial\nu}\right)-\alpha\left(1-2\nu\frac{\partial}{\partial\nu}+\nu^2\frac{\partial^2}{\partial\nu^2}\right)\right], \qquad (8.47)$$

$$S_2 = \frac{1}{10}\left[1-\gamma\left(1-\nu\frac{\partial}{\partial\nu}\right)-\alpha\left(6+2\nu\frac{\partial}{\partial\nu}-\nu^2\frac{\partial^2}{\partial\nu^2}\right)\right], \qquad (8.48)$$

$$S_3 = \frac{3}{70}\left[\gamma\left(1-\nu\frac{\partial}{\partial\nu}\right)+\alpha\left(4+2\nu\frac{\partial}{\partial\nu}-\nu^2\frac{\partial^2}{\partial\nu^2}\right)\right]. \qquad (8.49)$$

Equation (8.45) is the result of the formal expansion of the scattering operator to first order in α and γ and was first given by Fraser (1966). The usefulness of this expansion is that the integral operator of the exact description of scattering is replaced by a second order differential operator. It can be verified by direct substitution that the small α and γ expansion has not destroyed the equilibrium solution of the scattering operator. That is, with the induced (quadratic in I) scattering terms retained, the equilibrium solution of Eq. (8.45) is the Planck function, Eq. (1.8), and with the neglect of the induced terms the Wien distribution, Eq. (8.3), is the equilibrium solution.

We now proceed to effect a further simplification in the description of scattering without introducing any additional assumptions. The result of this simplification is, in fact, more accurate, in a qualitative but specific sense, than is Eq. (8.45). We shall elaborate on this point following the derivation of the simplified form of Eq. (8.45).

A straightforward way to effect this simplification is to consider Eq. (8.45) projected onto the basis elements of a spherical harmonic function space. We shall follow the vectorial method introduced in Section 5 of Chapter III. Since these spherical harmonic equations are in the present context only an intermediate result, we shall simply quote the results which are very similar to those of Chapter III [see Eqs. (3.168) and (3.169)]. If for simplicity we

The Equations of Radiation Hydrodynamics

momentarily neglect the induced scattering (nonlinear) terms in Eq. (8.45), the result is

$$\frac{1}{c}\frac{\partial\psi_0}{\partial t}+\nabla_U\cdot\nabla_r\psi_1+\sigma_a'(\psi_0-4\pi B)+\sigma_{\mathrm{Th}}(1-2\gamma-S_0)\,\psi_0 = 0, \quad (8.50)$$

$$\frac{3}{c}\frac{\partial\psi_1}{\partial t}+\nabla_U\cdot\nabla_r\psi_2+3[\sigma_a'+\sigma_{\mathrm{Th}}(1-2\gamma-S_1)]\,\psi_1+\mathbf{U}\cdot\nabla_r\psi_0 = 0, \quad (8.51)$$

$$\frac{5}{c}\frac{\partial\psi_2}{\partial t}+\nabla_U\cdot\nabla_r\psi_3+5[\sigma_a'+\sigma_{\mathrm{Th}}(1-2\gamma-S_2)]\,\psi_2+3\mathbf{U}\cdot\nabla_r\psi_1-U^2\nabla_U\cdot\nabla_r\psi_1 = 0, \quad (8.52)$$

$$\frac{7}{c}\frac{\partial\psi_3}{\partial t}+\nabla_U\cdot\nabla_r\psi_4+7[\sigma_a'+\sigma_{\mathrm{Th}}(1-2\gamma-S_3)]\,\psi_3+5\mathbf{U}\cdot\nabla_r\psi_2-U^2\nabla_U\cdot\nabla_r\psi_2 = 0, \quad (8.53)$$

$$\frac{2n+1}{c}\frac{\partial\psi_n}{\partial t}+\nabla_U\cdot\nabla_r\psi_{n+1}+(2n+1)[\sigma_a'+\sigma_{\mathrm{Th}}(1-2\gamma)]\,\psi_n$$
$$+(2n-1)\,\mathbf{U}\cdot\nabla_r\psi_{n-1}-U^2\nabla_U\cdot\nabla_r\psi_{n-1} = 0, \quad n\geq 4. \quad (8.54)$$

In these equations the vector $\mathbf{U}$ is in the direction $\mathbf{\Omega}$ and has an arbitrary magnitude U. The functions ψ_n are defined as

$$J_n = \frac{U^n}{2n+1}\sum_{m=-n}^{n} A_n^m I_n^m(\nu)\,Y_n^m(\mathbf{\Omega}), \quad (8.55)$$

where the $I_n^m(\nu)$ are the coefficients of an expansion of the specific intensity in surface harmonics according to

$$I(\nu,\mathbf{\Omega}) = \frac{1}{4\pi}\sum_{n=0}^{\infty}\sum_{m=-n}^{n} A_n^m I_n^m(\nu)\,Y_n^m(\mathbf{\Omega}). \quad (8.56)$$

Here the surface harmonics are defined in the usual way:

$$Y_n^m(\mathbf{\Omega}) = P_n^{|m|}(\cos\theta)\,e^{im\varphi}, \quad (8.57)$$

where the $P_n^m(z)$ are the associated Legendre functions and the constants A_n^m are normalization coefficients

$$A_n^m = 4\pi\left[\int_{4\pi} d\mathbf{\Omega}\,Y_n^m(\mathbf{\Omega})\,Y_n^{m*}(\mathbf{\Omega})\right]^{-1} = \frac{(2n+1)(n-|m|)!}{(n+|m|)!}, \quad (8.58)$$

with the superscript * on $Y_n^m(\mathbf{\Omega})$ indicating the complex conjugate. Due to the biorthogonality relationship between the surface harmonics and their complex conjugates, one has an explicit expression for $I_n^m(\nu)$ in terms of the specific intensity, i.e.,

$$I_n^m(\nu) = \int_{4\pi} d\mathbf{\Omega}\,Y_n^{m*}(\mathbf{\Omega})\,I(\nu,\mathbf{\Omega}). \quad (8.59)$$

Now, in Eq. (8.51) we replace $\sigma_{\mathrm{Th}}(1-2\gamma-S_1)$ by just σ_{Th}, since $S_1+2\gamma$ is of order α and γ and hence to lowest order $\sigma_{\mathrm{Th}}(1-2\gamma-S_1) = \sigma_{\mathrm{Th}}$. By similar arguments, we replace $\sigma_{\mathrm{Th}}(1-2\gamma-S_2)$ in Eq. (8.52) by $9\sigma_{\mathrm{Th}}/10$, and in Eqs. (8.53) and (8.54) we replace $\sigma_{\mathrm{Th}}(1-2\gamma-S_3)$ and $\sigma_{\mathrm{Th}}(1-2\gamma)$ in each case by σ_{Th}. We note, however, that we cannot make similar simplification in Eq. (8.50) since $(1-2\gamma-S_0)$ is of order α and γ, rather than of order unity (or 9/10), as are the similar terms in Eqs. (8.51) through (8.54). Introducing these simplifications in Eqs. (8.51) through (8.54) we find that Eqs. (8.50) through (8.54) are the spherical harmonic projections of the equation of transfer

$$\frac{1}{c}\frac{\partial I(\nu, \mathbf{\Omega})}{\partial t} + \mathbf{\Omega} \cdot \nabla I(\nu, \mathbf{\Omega}) = \sigma_a'(\nu)[B(\nu) - I(\nu, \mathbf{\Omega})]$$

$$-\sigma_{\mathrm{Th}} I(\nu, \mathbf{\Omega}) + \frac{3\sigma_{\mathrm{Th}}}{16\pi} \int_{4\pi} d\mathbf{\Omega}'[1 + (\mathbf{\Omega}' \cdot \mathbf{\Omega})^2]\, I(\nu, \mathbf{\Omega}')$$

$$+\frac{\sigma_{\mathrm{Th}}}{4\pi} \int_{4\pi} d\mathbf{\Omega}' \left[\alpha \nu^2 \frac{\partial^2}{\partial \nu^2} + (\gamma - 2\alpha)\nu \frac{\partial}{\partial \nu} + \gamma\right] I(\nu, \mathbf{\Omega}'). \tag{8.60}$$

To Eq. (8.60) we need add the contribution of the nonlinear induced scattering terms in Eq. (8.45). Since these terms are of order γ, they can be neglected in all but the zeroth angular moment of the equation of transfer, just as we neglected all terms of order α and γ in the linear analysis just completed except in the zeroth angular moment relationship, Eq. (8.50). This implies the replacement in the equation of transfer:

$$I(\nu, \mathbf{\Omega})\left(1 - \nu\frac{\partial}{\partial \nu}\right)\int_{4\pi} d\mathbf{\Omega}'\,[1 - (\mathbf{\Omega} \cdot \mathbf{\Omega}') + (\mathbf{\Omega} \cdot \mathbf{\Omega}')^2 - (\mathbf{\Omega} \cdot \mathbf{\Omega}')^3]\, I(\nu, \mathbf{\Omega}')$$

$$\to \frac{1}{4\pi}\int_{4\pi} d\mathbf{\Omega} I(\nu, \mathbf{\Omega})\int_{4\pi} d\mathbf{\Omega}'\,[1 - (\mathbf{\Omega} \cdot \mathbf{\Omega}') + (\mathbf{\Omega} \cdot \mathbf{\Omega}')^2 - (\mathbf{\Omega} \cdot \mathbf{\Omega}')^3]\left(1 - \nu\frac{\partial}{\partial \nu}\right) I(\nu, \mathbf{\Omega}'). \tag{8.61}$$

Thus, the full form of Eq. (8.60), including the effects of induced scattering, is

$$\frac{1}{c}\frac{\partial I(\nu, \mathbf{\Omega})}{\partial t} + \mathbf{\Omega} \cdot \nabla I(\nu, \mathbf{\Omega}) = \sigma_a'(\nu)[B(\nu) - I(\nu, \mathbf{\Omega})]$$

$$-\sigma_{\mathrm{Th}} I(\nu, \mathbf{\Omega}) + \frac{3\sigma_{\mathrm{Th}}}{16\pi} \int_{4\pi} d\mathbf{\Omega}'[1 + (\mathbf{\Omega} \cdot \mathbf{\Omega}')^2]\, I(\nu, \mathbf{\Omega}')$$

$$+\frac{\sigma_{\mathrm{Th}}}{4\pi} \int_{4\pi} d\mathbf{\Omega}' \left[\alpha \nu^2 \frac{\partial^2}{\partial \nu^2} + (\gamma - 2\alpha)\nu \frac{\partial}{\partial \nu} + \gamma\right] I(\nu, \mathbf{\Omega}')$$

$$-\frac{3\sigma_{\mathrm{Th}}}{64\pi^2}\frac{c^2}{h\nu^3}\gamma \int_{4\pi} d\mathbf{\Omega}' I(\nu, \mathbf{\Omega}') \times$$

$$\times \int_{4\pi} d\mathbf{\Omega}''[1 - (\mathbf{\Omega}' \cdot \mathbf{\Omega}'') + (\mathbf{\Omega}' \cdot \mathbf{\Omega}'')^2 - (\mathbf{\Omega}' \cdot \mathbf{\Omega}'')^3]\left(1 - \nu\frac{\partial}{\partial \nu}\right) I(\nu, \mathbf{\Omega}''),$$

$$\tag{8.62}$$

which is a simplified, but *a priori* just as accurate, form of Fraser's result, Eq. (8.45). In particular, Eq. (8.62) contains far fewer scattering terms than does Eq. (8.45), and the terms which account for energy transfer in the scattering interaction, i.e., those proportional to α and γ, are isotropic in Eq. (8.62) whereas they are angularly dependent in Eq. (8.45). Both of these facts should make Eq. (8.62) much easier to solve, either analytically or numerically, than

Eq. (8.45). Since Eqs. (8.45) and (8.62) have the same zeroth angular moment, the equilibrium solution of Eq. (8.62) is the same as that of Eq. (8.45), namely a Planck distribution at the electron temperature. However, in the limiting case of a small electron temperature ($\alpha = 0$), Eq. (8.62) is more accurate, at least qualitatively, than Eq. (8.45). Several authors (see references in Pomraning, 1968a) have shown that the description of scattering in Eq. (8.45) with $\alpha = 0$ leads to a slight increase in the frequency of some photons due to the scattering interaction. This is clearly an incorrect behavior since scattering from electrons at rest can only lead to a decrease in photon energy. The integral description of scattering in Eq. (2.170) does not, of course, exhibit this incorrect behavior. It has also been shown (Pomraning, 1969c) that Eq. (8.62) gives the proper behavior of only a decrease in photon frequency due to scattering from cold electrons.

It is clear that in the derivation of Eq. (8.62) from Eq. (8.45) a certain amount of arbitrariness is involved. That is, the terms of order α and γ in the operators $S_n + 2\gamma$, $n = 1, 2, 3$ of Eqs. (8.51) through (8.53) can be replaced with any other terms of the same order since they appear coupled with a term of order unity. Similarly, the term 2γ in Eq. (8.54) can be replaced with any other term of order α and γ. In obtaining Eq. (8.62) we made the choice which gave the simplest description of scattering, i.e., one in which the energy exchange terms are isotropic. For the purposes of obtaining an eigenfunction solution in frequency of the equation of transfer, another choice is appropriate (Pomraning and Froehlich, 1969). Namely, we replace the terms of order α and γ in all the spherical harmonic projections, Eqs. (8.50) through (8.54), with the terms of the same order found in the zeroth projection, Eq. (8.50). We then find that the resulting equations are the projections onto a spherical harmonic function space of the equation of transfer

$$\frac{1}{c}\frac{\partial I(\nu, \Omega)}{\partial t} + \Omega \cdot \nabla I(\nu, \Omega) = \sigma_a'(\nu)[B(\nu) - I(\nu, \Omega)] - \sigma_{\mathrm{Th}}\, I(\nu, \Omega)$$

$$+ \frac{3\sigma_{\mathrm{Th}}}{16\pi}\int_{4\pi} d\Omega'[1 + (\Omega \cdot \Omega')^2]\, I(\nu, \Omega')$$

$$+ \sigma_{\mathrm{Th}}\left[\alpha\nu^2\frac{\partial^2}{\partial\nu^2} + (\gamma - 2\alpha)\nu\frac{\partial}{\partial\nu} + \gamma\right] I(\nu, \Omega). \tag{8.63}$$

The nonlinear induced scattering terms should be added to this result, but for our application, to be discussed next, these terms are not required.

In summary, Eqs. (8.45), (8.62), and (8.63) are all alternate forms of the small α and γ expansion of the scattering terms in the equation of transfer. Equation (8.45) is the result of the formal expansion in these smallness parameters, and Eqs. (8.62) and (8.63) are two simplified versions of this result. Each has its own merit in applications. Equation (8.62) is simplest in form and should be most useful in obtaining a numerical solution of the equation of transfer. Equation (8.63) is useful in an analytic treatment of the frequency variable as we now show.

4. The Eigenvalue Problem—The Eigenvalues for No Absorption

If one neglects induced scattering effects, the equation of radiative transfer is linear in the specific intensity and hence more amenable to analysis. In particular, we now show that the frequency variable, within the Fokker–Planck model of scattering, can be treated by an eigenfunction expansion. We shall also obtain the appropriate eigenfunctions for the case of no absorption. The consideration of the eigenfunctions in the presence of absorption will be taken up in the next section.

Consider a physical situation in which α, and hence the electron temperature, is independent of space and time. We define the new variables

$$x = h\nu/kT = \gamma/\alpha, \tag{8.64}$$

$$\bar{I}(x, \Omega) = I(\nu, \Omega), \tag{8.65}$$

$$f(x) = \sigma_a'(\nu)/\alpha\sigma_{\mathrm{Th}}, \tag{8.66}$$

$$T(x) = \sigma_a'(\nu)\, B(\nu). \tag{8.67}$$

For the physical situation just described, Eq. (8.63) then becomes, with the bar on $\bar{I}(x, \Omega)$ dropped,

$$\frac{1}{c}\frac{\partial I(x, \Omega)}{\partial t} + \Omega \cdot \nabla I(x, \Omega) = T(x) - \sigma_{\mathrm{Th}}\, I(x, \Omega) + \frac{3\sigma_{\mathrm{Th}}}{16\pi}\int_{4\pi} d\Omega'[1 + (\Omega \cdot \Omega')^2]\, I(x, \Omega')$$

$$+ \sigma_{\mathrm{Th}}\alpha \left[x^2 \frac{\partial^2}{\partial x^2} + (x-2)x\frac{\partial}{\partial x} + x - f(x) \right] I(x, \Omega). \tag{8.68}$$

If the eigenfunctions $y_n(x)$, $n = 1, 2, 3, \ldots$, corresponding to eigenvalues λ_n and satisfying the equation

$$x^2 \frac{d^2 y_n(x)}{dx^2} + (x-2)x\frac{dy_n(x)}{dx} + [x - f(x) + \lambda_n]\, y_n(x) = 0, \quad n \geq 1 \tag{8.69}$$

form a sufficiently complete set, we can represent the specific intensity by the eigenfunction expansion

$$I(x, \Omega) = \sum_{n=1}^{\infty} I_n(\Omega)\, y_n(x). \tag{8.70}$$

We shall shortly show that the $y_n(x)$ of interest here form an orthonormal set on the interval $(0, \infty)$ with respect to the weight function e^x/x^4. Use of this orthogonality condition yields an infinite uncoupled set of grey or monochromatic radiative transfer problems for the expansion coefficients $I_n(\Omega)$, i.e.,

$$\frac{1}{c}\frac{\partial I_n(\Omega)}{\partial t} + \Omega \cdot \nabla I_n(\Omega) = T_n - \sigma_{\mathrm{Th}}(1 + \alpha\lambda_n)\, I_n(\Omega)$$

$$+ \frac{3\sigma_{\mathrm{Th}}}{16\pi}\int_{4\pi} d\Omega'[1 + (\Omega \cdot \Omega')^2]\, I_n(\Omega'), \quad n \geq 1, \tag{8.71}$$

with the T_n given by

$$T_n = \int_0^\infty dx\, x^{-4} e^x y_n(x)\, T(x), \quad n \geq 1. \tag{8.72}$$

Hence the eigenfunction expansion has effectively eliminated the frequency variable from the equation of transfer.

To complete the specification of the eigenvalue problem it is necessary to impose boundary conditions at the endpoints of the interval $(0, \infty)$. These conditions arise naturally if one requires that the allowable solutions of Eq. (8.69), i.e., the eigenfunctions, form the basis of an expansion theorem. We write Eq. (8.69) in Sturm–Liouville form

$$\frac{d}{dx}\left[\frac{e^x}{x^2}\frac{dy_n(x)}{dx}\right] + \frac{e^x}{x^4}\left[x - f(x) + \lambda_n\right] y_n(x) = 0, \tag{8.73}$$

and imagine the same equation written for another eigenvalue and eigenfunction corresponding to the index m. Application of the usual cross-multiplication, subtraction, and integration technique yields the orthogonality condition

$$\int_0^\infty dx\, x^{-4}\, e^x y_n(x) y_m(x) = 0, \quad n \neq m. \tag{8.74}$$

Equation (8.74) holds provided that the boundary terms arising from two integrations by parts vanish, i.e.,

$$x^{-2}(y_n y_m' - y_m y_n')\Big|_{x=0} = 0, \tag{8.75}$$

$$x^{-2} e^x (y_n y_m' - y_m y_n')\Big|_{x=\infty} = 0, \tag{8.76}$$

where the prime indicates differentiation. Equations (8.75) and (8.76) are the boundary conditions we shall impose upon the solution of Eq. (8.69). Equations (8.69), (8.75), and (8.76) completely specify the eigenvalue problem for an arbitrary absorption function $f(x)$.

Let us now consider the details of the eigenvalue problem in the special case of no absorption, $f(x) = 0$. We shall show that the eigenvalue spectrum consists of two discrete points and a continuum. Because of this continuum, it is inconvenient to label the eigenfunctions with an integer subscript n. Accordingly, rather than denoting the eigenfunctions and eigenvalues by $y_n(x)$ and λ_n, we shall simply let $y_\lambda(x)$ denote the eigenfunction corresponding to a particular eigenvalue λ. In the case of no absorption and with the new notation, Eq. (8.69) becomes

$$x^2 y_\lambda''(x) + (x-2)x y_\lambda'(x) + (x - \lambda) y_\lambda(x) = 0. \tag{8.77}$$

We define a new independent variable $u_\lambda(x)$ according to

$$y_\lambda(x) = x^{3+k} e^{-x} u_\lambda(x), \tag{8.78}$$

where the constant k depends upon λ and is given by the solution of

$$k^2 + 3k + \lambda = 0. \tag{8.79}$$

Use of Eq. (8.78) in Eq. (8.77) yields the equation which $u_\lambda(x)$ satisfies, namely

$$xu_\lambda''(x)+(2k+4-x)u_\lambda'(x)-ku_\lambda(x) = 0. \tag{8.80}$$

Since Eq. (8.79) defines two values of k whose sum is equal to minus three for any value of λ, we can without loss of generality restrict k such that $Re\ k \geq -3/2$. Equation (8.80) is a special case of the confluent hypergeometric equation whose general solution for $k \neq 0,\ -1$ is given by

$$u_\lambda(x) = A_\lambda\varphi(k,\ 2k+4;\ x)+B_\lambda\psi(k,\ 2k+4;\ x), \quad Re\ k \geq -3/2; \quad k \neq 0,\ -1, \tag{8.81}$$

where A_λ and B_λ are arbitrary constants. The functions $\varphi(a,\ c;\ x)$ and $\psi(a,\ c;\ x)$ are the widely studied confluent hypergeometric functions of the first and second kind, respectively. The cases $k = 0$ and $k = -1$ must be treated separately, since in these two cases the φ and ψ functions are linearly dependent. Thus, except in the two cases just noted, the general solution for $y_\lambda(x)$ is given by

$$y_\lambda(x) = x^{3+k}e^{-x}[A_\lambda\varphi(k,\ 2k+4;\ x)+B_\lambda\psi(k,\ 2k+4;\ x),$$
$$Re\ k \geq -3/2; \quad k \neq 0,\ -1. \tag{8.82}$$

We now use the boundary conditions, Eqs. (8.75) and (8.76), to determine the relationship between the constants A_λ and B_λ and the allowable values of k which, through Eq. (8.79), determine the eigenvalues λ. From the asymptotic behavior of the confluent hypergeometric functions, one can easily show (Pomraning, 1968b) that the Wronskian, $y_\lambda y_\mu' - y_\mu y_\lambda'$, goes to zero as x^{-s}, s being a positive integer, if $A_\lambda \neq 0$. Because of the exponential factor in Eq. (8.76), this means we must set $A_\lambda = 0$ for the condition at infinity to be satisfied. The resulting function

$$y_\lambda(x) = B_\lambda x^{3+k}e^{-x}\psi(k,\ 2k+4;\ x),\ Re\ k \geq -3/2; \quad k \neq 0,\ -1, \tag{8.83}$$

has a Wronskian with $y_\mu(x)$ which behaves like $x^s e^{-2x}$, s being a positive integer, as x approaches infinity for all values of the constants k and $B_\lambda \neq 0$. Hence, Eq. (8.83) is acceptable as far as the condition at infinity is concerned. From the small x behavior of the confluent hypergeometric functions, one can show (Pomraning, 1968b) that for small ε

$$x^{-2}(y_\lambda y_\mu' - y_\mu y_\lambda')_{x=\varepsilon} \sim \varepsilon^{-(k_\lambda+k_\mu+3)}. \tag{8.84}$$

Equation (8.84) implies that for the condition at $x = 0$, Eq. (8.75), to be satisfied we must reject all values of k with $Re\ k > -3/2$. Since k is restricted such that $Re\ k \geq -3/2$, this means that only those values of k with $Re\ k = -3/2$ are possibly acceptable. From Eq. (8.69) we have

$$k = -3/2 \pm i\sqrt{\lambda-9/4}, \tag{8.85}$$

and we see that $Re\ k = -3/2$ corresponds to λ real with $\lambda \geq 9/4$. It can be shown by straightforward but tedious algebra (Pomraning, 1968b) that all values of k with $Re\ k = -3/2$ are acceptable and hence we conclude that a continuum of eigenvalues, $\lambda \geq 9/4$, exists. Since either value of k given by Eq. (8.85) is acceptable, we choose, for definiteness, the positive

The Equations of Radiation Hydrodynamics

root. Thus, the continuum eigenfunctions are given by

$$y_\lambda(x) = B_\lambda x^{3/2+ia} e^{-x} \psi(-3/2+ia, 1+2ia; x), \qquad \lambda \geq 9/4, \tag{8.86}$$

where

$$a = \sqrt{\lambda - 9/4}. \tag{8.87}$$

Writing the confluent hypergeometric function of the second kind in terms of hypergeometric functions of the first kind we can alternately write the continuum eigenfunctions as

$$y_\lambda(x) = B_\lambda \frac{\Gamma(2ia)}{\Gamma(-3/2+ia)} x^{3/2-ia} e^{-x} \varphi(-3/2-ia, 1-2ia; x)$$

$$+ B_\lambda \frac{\Gamma(-2ia)}{\Gamma(-3/2-ia)} x^{3/2+ia} e^{-x} \varphi(-3/2+ia, 1+2ia; x), \qquad \lambda \geq 9/4, \tag{8.88}$$

where $\Gamma(z)$ is the usual Gamma function. If the normalization constant B_λ is real, Eq. (8.88) shows that the continuum eigenfunctions are real since the right hand side of this equation is the sum of a function of x and its complex conjugate.

As noted earlier, the cases $k = 0$ and $k = -1$, which correspond to $\lambda = 0$ and $\lambda = 2$, respectively, require special treatment. For $\lambda = 0$, the general solution of Eq. (8.77) is

$$y_0(x) = A_0 x^3 e^{-x} \int_{c_0}^{x} d\xi\, \xi^{-4} e^{\xi} + B_0 x^3 e^{-x}, \tag{8.89}$$

where A_0, B_0, and c_0 are arbitrary constants. The condition at infinity, Eq. (8.76), requires that we set $A_0 = 0$. The conditions at zero and infinity, Eqs. (8.75) and (8.76), do not restrict B_0 in any way. Hence, $\lambda = 0$ is an eigenvalue with the corresponding eigenfunction

$$y_0(x) = B_0 x^3 e^{-x}. \tag{8.90}$$

For $\lambda = 2$ the general solution of Eq. (8.77) is

$$y_2(x) = A_2 x^2 (x-2) e^{-x} \int_{c_2}^{x} d\xi\, \xi^{-2}(\xi-2) e^{\xi} + B_2 x^2 (x-2) e^{-x}, \tag{8.91}$$

where A_2, B_2, and c_2 are arbitrary constants. Again, the condition at infinity dictates $A_2 = 0$ and the conditions at zero and infinity do not restrict B_2. Hence, $\lambda = 2$ is also acceptable as an eigenvalue and we have

$$y_2(x) = B_2 x^2 (x-2) e^{-x}, \tag{8.92}$$

as a second discrete eigenfunction.

The analysis just completed shows that the spectrum of the operator of concern here is entirely real and nonnegative, consisting of two discrete points $\lambda = 0$ and $\lambda = 2$ and a continuum $9/4 \leq \lambda < \infty$. For the purposes of an eigenfunction expansion, it is convenient to normalize the eigenfunctions such that

$$\int_0^\infty dx\, x^{-4} e^x y_m(x)\, y_n(x) = \delta_{nm}; \qquad m, n = 0, 2, \tag{8.93}$$

for the discrete modes and

$$\int_0^\infty dx\, x^{-4} e^x y_\lambda(x)\, y_\mu(x) = \delta(\lambda - \mu); \quad \lambda, \mu \geq 9/4, \tag{8.94}$$

for the continuum modes. A direct computation yields

$$B_0 = B_2 = 1/\sqrt{2}. \tag{8.95}$$

Much more care is required in computing B_λ, $\lambda \geq 9/4$. Omitting the details of the calculation, the result is (Pomraning, 1968b)

$$B^2(\lambda) = \frac{\Gamma(-3/2+ia)\,\Gamma(-3/2-ia)}{4\pi a\Gamma(2ia)\,\Gamma(-2ia)}, \quad \lambda \geq 9/4. \tag{8.96}$$

Using elementary properties of the Gamma function and setting $a = \sqrt{\lambda - 9/4}$, we obtain the result

$$B(\lambda) = \left[\frac{\sinh\left(\pi \sqrt{\lambda - 9/4}\right)}{\pi \lambda(\lambda - 2)} \right]^{1/2}, \quad \lambda \geq 9/4. \tag{8.97}$$

Since a continuum of eigenvalues exists, the eigenfunction expansion indicated in Eq. (8.70) needs to be modified. That is, the proper expansion of any function $f(x)$ in the normalized eigenfunctions just computed is

$$f(x) = c_0 y_0(x) + c_2 y_2(x) + \int_{9/4}^\infty d\lambda\, c_\lambda y_\lambda(x), \tag{8.98}$$

where the expansion coefficients are given by

$$c_\lambda = \int_0^\infty dx\, x^{-4} e^x y_\lambda(x)\, f(x); \quad \lambda = 0, 2, \geq 9/4. \tag{8.99}$$

5. The Eigenvalues in the Presence of Absorption

We now consider the more difficult problem of computing the eigenfunctions and eigenvalues in the presence of a realistic absorption function $f(x)$. We characterize the class of functions $f(x)$ we shall consider by the behavior of $f(x)$ for small and large x. For small ν we assume that the absorption coefficient $\sigma_a(\nu)$ has the characteristic ν^{-3} behavior (see Chapter VII). Because of the correction for induced emission as given by Eq. (2.173), this behavior for $\sigma_a(\nu)$ for small ν implies that $f(x)$ is proportional to x^{-2} for small x. Hence we assume that in the vicinity of the origin, $f(x)$ is asymptotically represented by

$$f(x) \sim x^{-2} \sum_{k=0}^\infty r_k x^k. \tag{8.100}$$

We take $r_0 \neq 0$ unless $f(x)$ is identically zero. In general, the absorption coefficient vanishes for high frequencies, which implies that $f(x)$ vanishes for large x. However, the analysis of the eigenvalue problem is no more complex if we allow the more general behavior of possible linear growth of $f(x)$ for large x. Hence for large x we take

$$f(x) \sim x \sum_{k=0}^{\infty} s_k x^{-k} \tag{8.101}$$

as the asymptotic representation of $f(x)$. It should be emphasized that Eqs. (8.100) and (8.101) are asymptotic expressions for $f(x)$, and this apparent power series behavior does not preclude the real structure associated with an absorption coefficient, such as almost discontinuous behavior near a photoelectric edge and almost delta function behavior near certain frequencies due to bound–bound transitions.

As we shall show later, this class of functions $f(x)$ leads to an eigenvalue spectrum which is entirely discrete (for $f \neq 0$). We therefore return to the notation of Eq. (8.69), subscripting the eigenfunctions and eigenvalues with an integer n. We let $n = 1$ denote the smallest eigenvalue, $n = 2$ the next largest eigenvalue, and so forth. For an absorption function in the class just described, Eq. (8.69) has irregular singular points at both $x = 0$ and $x = \infty$. Hence, even if Eq. (8.100) or Eq. (8.101) represented a convergent expansion for $f(x)$, only asymptotic series solutions about these points would be possible. These asymptotic series are easily constructed. If we denote by $\psi_1(x)$ and $\psi_2(x)$ two linearly independent solutions of Eq. (8.69), we find as asymptotic representations about the origin

$$\psi_1(x) \sim e^{-(x+\alpha/x)}x^{2+\beta} \sum_{n=0}^{\infty} a_n x^n, \tag{8.102}$$

and

$$\psi_2(x) \sim e^{-(x-\alpha/x)} x^{2-\beta} \sum_{n=0}^{\infty} b_n x^n, \tag{8.103}$$

where α and β are defined by

$$\alpha^2 = r_0, \tag{8.104}$$

$$2\alpha\beta = r_1 \tag{8.105}$$

(α here is not to be confused with the α introduced in Section 3 of this chapter), and the coefficients a_n and b_n are given by

$$2\alpha(n+1)a_{n+1} = [\alpha+r_2-\lambda-(n+\beta+2)(n+\beta-1)]a_n$$
$$+(n+\beta+r_3-2)a_{n-1}+\sum_{k=0}^{n-2} a_k r_{n+2-k}, \; n \geq 0, \tag{8.106}$$

$$2\alpha(n+1)b_{n+1} = [\lambda+\alpha-r_2+(n-\beta+2)(n-\beta-1)]b_n$$
$$-(n-\beta+r_3-2)b_{n-1}-\sum_{k=0}^{n-2} b_k r_{n+2-k}, \; n \geq 0, \tag{8.107}$$

where $a_{-1} = b_{-1} = 0$ and the sums over k are omitted for $n < 2$. For concreteness, we assume that $\alpha \geq 0$. Similarly, if we denote by $\psi_3(x)$ and $\psi_4(x)$ two other linearly independent solutions of Eq. (8.69) we find as asymptotic representations about infinity

$$\psi_3(x) \sim e^{-(x+\alpha/x)}x^{3-\gamma} \sum_{n=0}^{\infty} c_n x^{-n}, \tag{8.108}$$

and

$$\psi_4(x) \sim e^{-\alpha/x} x^{\gamma-1} \sum_{n=0}^{\infty} d_n x^{-n}, \tag{8.109}$$

where we have set $\gamma = s_0$ (γ here is not to be confused with the γ introduced in Section 3 of this chapter) and the coefficients c_n and d_n satisfy the recurrence relationships

$$(n+1)c_{n+1} = [\alpha+s_1-\lambda-(n+\gamma-3)(n+\gamma)] \, c_n + [2\alpha(n+\gamma-2)+s_2]c_{n-1}$$

$$+(s_3-\alpha^2)c_{n-2}+ \sum_{k=0}^{n-3} c_k s_{n-k+1}, \quad n \geq 0, \tag{8.110}$$

$$(n+1)d_{n+1} = [(n-\gamma+1)(n-\gamma+4)+\lambda+\alpha-s_1]d_n - [2\alpha(n-\gamma+2)+s_2]d_{n-1}$$

$$+(\alpha^2-s_3)d_{n-2}- \sum_{k=0}^{n-3} d_k s_{n-k+1}, \quad n \geq 0, \tag{8.111}$$

where $c_{-1} = c_{-2} = d_{-1} = d_{-2} = 0$ and the sums over k are omitted for $n < 3$.

The solution acceptable at the origin, $\psi_1(x)$, is defined asymptotically by Eq. (8.102). This series is only asymptotically valid for two reasons. First, the expression used for $f(x)$, Eq. (8.100), is in general only asymptotically valid. Second, in the general case, the coefficients a_n grow as n approaches infinity and hence the series diverges. We now consider Eq. (8.100) to be a convergent expansion of $f(x)$ for all x and seek conditions under which the asymptotic series, Eq. (8.102), truncates itself. The resulting truncated series will be a solution of Eq. (8.69) with the proper behavior at both the origin and infinity to qualify as an eigenfunction. These results could also be obtained by seeking truncation conditions for the series representing $\psi_3(x)$ about infinity, Eq. (8.108). Omitting the algebraic details (Pomraning and Froehlich, 1969) we find that truncation occurs at $n = N$, i.e.,

$$a_N \neq 0; \quad a_n = 0, \quad n > N, \tag{8.112}$$

if and only if the function $f(x)$ is composed of at most four terms, i.e.,

$$f(x) = \frac{\alpha^2}{x^2} + \frac{2\alpha\beta}{x} + r_2 + r_3 x. \tag{8.113}$$

Further, β and r_3 must be related according to

$$N+\beta+r_3-1 = 0 \tag{8.114}$$

for truncation to occur. The eigenvalues then follow from the condition

$$[\alpha+r_2-\lambda-(N+\beta+2)(N+\beta-1)]a_N+(N+\beta+r_3-1)a_{N-1} = 0, \tag{8.115}$$

in conjunction with the recurrence relationship, Eq. (8.106), needed to determine a_{N-1} and a_N.

As the simplest example of these considerations, we set

$$f(x) = \frac{\alpha^2}{x^2} + \frac{2\alpha}{x}. \tag{8.116}$$

Then $r_2 = r_3 = 0$, $\beta = 1$, and Eq. (8.114) is satisfied by $N = 0$. Equation (8.115) then gives a single eigenvalue

$$\lambda_1 = \alpha, \tag{8.117}$$

and the corresponding eigenfunction is simply

$$y_1(x) = C_1 x^3 e^{-(x+\alpha/x)}, \tag{8.118}$$

where C_1 is a normalization constant. As a second example, we set

$$f(x) = \alpha^2/x^2. \tag{8.119}$$

We then find that Eqs. (8.114) and (8.115) give two eigenvalues

$$\lambda_n = 1+\alpha \pm \sqrt{1+2\alpha}, \quad n = 1, 2, \tag{8.120}$$

with corresponding eigenfunctions

$$y_n(x) = C_n[2\alpha x^2 + (\alpha - \lambda_n + 2)x^3]e^{-(x+\alpha/x)}, \quad n = 1, 2. \tag{8.121}$$

The final example we shall consider is

$$f(x) = \frac{\alpha^2}{x^2} - \frac{2\alpha}{x}. \tag{8.122}$$

This choice for $f(x)$ is somewhat nonphysical since it is negative for $x > \alpha/2$. However, if one adds a constant r_2, where $r_2 \geq 1$, to the right hand side of Eq. (8.122), the resulting $f(x)$ is nonnegative for all $x \geq 0$. Since the addition of a constant to $f(x)$ merely shifts the eigenvalue spectrum uniformly by that constant, i.e., λ and r_2 always occur together in the eigenvalue problem as $\lambda - r_2$, it is physically meaningful to consider an $f(x)$ as given by Eq. (8.122). In this case we find three eigenvalues given by

$$\lambda_n = \alpha + 2 - \eta_n, \quad n = 1, 2, 3, \tag{8.123}$$

where the η_n are the three roots of the cubic equation

$$\eta^3 - 2\eta^2 - 8\alpha\eta + 8\alpha = 0. \tag{8.124}$$

The associated eigenfunctions are

$$y_n(x) = C_n[8\alpha^2 x + 4\alpha\eta_n x^2 + (\eta_n^2 - 4\alpha)x^3]e^{-(x+\alpha/x)}, \quad n = 1, 2, 3. \tag{8.125}$$

These three eigenvalues can be shown to be no less than minus one in value for all nonnegative values of α. This is the expected behavior on physical grounds, since for a nonnegative $f(x)$ one must find all nonnegative eigenvalues (we shall prove this later), and $f(x)+1$ is nonnegative. For small values of α, Eqs. (8.123) and (8.124) give

$$\lambda_1 = -\alpha + O(\alpha^2), \tag{8.126}$$

$$\lambda_n = 2 \pm 2\sqrt{\alpha} + 2\alpha \mp \frac{3}{4}\alpha^{3/2} + O(\alpha^2), \quad n = 2, 3. \tag{8.127}$$

This case is particularly interesting because of its relationship to the pure scattering ($f = 0$) case. For pure scattering, discussed in Section 4, two discrete eigenvalues $\lambda = 0$ and $\lambda = 2$ are found. With $\alpha = 0$ in the present example, these same two eigenvalues must appear, and Eqs. (8.126) and (8.127) show that our asymptotic analysis yields them. However, we also see from Eq. (8.127) that two values of λ which are distinct for $\alpha > 0$ approach the common limit $\lambda = 2$ as α approaches zero. Hence, the introduction of absorption according to Eq. (8.122) into the pure scattering problem causes the single eigenvalue $\lambda = 2$ to split into two eigenvalues.

As an example of the eigenfunctions which the asymptotic analysis of this section has provided, we plot $y_1(x)$ as given by Eq. (8.118) for various values of α in Fig. 8.4. These curves correspond to a choice of normalization constant C_1 as

$$C_1 = [2(2\alpha)^{3/2}K_3(2\sqrt{2\alpha})]^{-1/2}, \tag{8.128}$$

where $K_3(z)$ is the modified Bessel function of the second kind of order three. With C_1 given by Eq. (8.128), we have

$$\int_0^\infty dx\, x^{-4} e^x y_1^2(x) = 1. \tag{8.129}$$

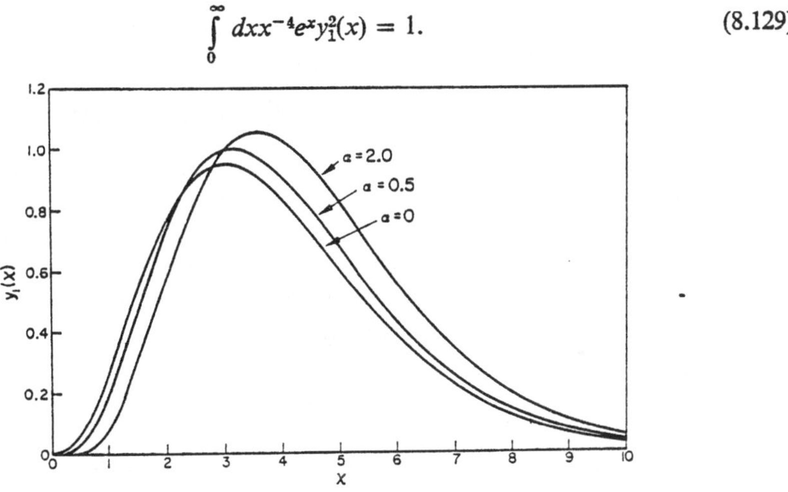

FIG. 8.4. The fundamental eigenfunction for $f(x) = \alpha^2/x^2 + 2\alpha/x$.

The value $\alpha = 0$ corresponds to no absorption, and the associated curve in Fig. 8.4 is simply the Wien equilibrium distribution of the scattering operator. As discussed earlier, one must include the induced scattering terms, quadratic in the intensity, in the description of the scattering process for the Planck distribution to be the equilibrium solution of the scattering operator. The curves for $\alpha \neq 0$ show the distortion of the Wien distribution, due to absorption, which the fundamental eigenfunction exhibits.

The asymptotic analysis provides eigenvalues and eigenfunctions only when the function $f(x)$ has a very special form. Even in these special cases one finds only the first few lowest eigenfunctions. We now present a numerical method for computing as many eigenvalues and eigenfunctions as one pleases for a general function $f(x)$ of the form discussed earlier.

Before considering this numerical method, let us briefly discuss the nature of the eigenvalue spectrum we expect. One can easily show that all the eigenvalues are real. This is proved by considering the eigenvalue problem in Sturm–Liouville form, Eq. (8.73), as well as the complex conjugate of this equation. One applies the usual cross-multiplication, integration, and subtraction procedure to these two equations. The boundary terms in the two integrations by parts vanish due to the exponential behavior, for $\alpha > 0$, of $y_n(x)$ at both zero and infinity. This leads immediately to the result that λ_n is equal to its complex conjugate; i.e., λ_n is real. Further, we can show that all the eigenvalues are nonnegative for $f(x) \geq 0$. We consider the Rayleigh quotient associated with Eq. (8.73), i.e.,

$$R[y] = \frac{\int_0^\infty dx\, x^{-4} e^x [x^2 (y')^2 + (f - x) y^2]}{\int_0^\infty dx\, x^{-4} e^x y^2}, \tag{8.130}$$

with admissible functions $y(x)$ which satisfy the eigenfunction boundary conditions, Eqs. (8.75) and (8.76). We know that for a given absorption function $f(x)$, the function $y(x)$ which minimizes $R[y]$ is the fundamental eigenfunction $y_1(x)$ and the corresponding value of the functional, $R[y_1]$, is the fundamental, or smallest, eigenvalue. We consider two different functions $f(x)$, say $f^{(a)}(x)$ and $f^{(b)}(x)$, such that $f^{(b)}(x) \geq f^{(a)}(x)$. We then have

$$\lambda_1^{(a)} = R^{(a)}[y_1^{(a)}] \leq R^{(a)}[y_1^{(b)}] \leq R^{(b)}[y_1^{(b)}] = \lambda_1^{(b)}. \tag{8.131}$$

The first inequality follows from the fact that $y_1^{(a)}(x)$ minimizes $R^{(a)}[y]$, and the second inequality follows directly by inspection of Eq. (8.130). We know from the previous section that if we set $f^{(a)}(x) = 0$, $\lambda_1^{(a)} = 0$. Hence, for any $f(x) \geq 0$, it follows from Eq. (8.131) that $\lambda_1 \geq 0$. Since λ_1 is the smallest eigenvalue, we conclude that $\lambda_n \geq 0$; i.e., all the eigenvalues are nonnegative for $f(x) \geq 0$. Finally, we can argue that all the λ_n are discrete for $\alpha > 0$. We define the normalization integral of $y_n(x)$ to be N_n, i.e.,

$$N_n = \int_0^\infty dx\, x^{-4} e^x y_n^2(x). \tag{8.132}$$

For a discrete eigenvalue λ_n, the corresponding normalization integral for $y_n(x)$ is finite, i.e., $N_n < \infty$, whereas for a continuum eigenvalue the normalization integral is divergent, $N_n = \infty$. Since all eigenfunctions are continuous functions of x which vanish as $x^{2+\beta} e^{-\alpha/x}$ at the origin and as $x^{3-\gamma} e^{-x}$ at infinity, the normalization integral N_n is finite; i.e., all the eigenvalues λ_n are discrete. Hence, for $f(x) > 0$ we expect all the eigenvalues to be real, positive, and discrete. As $f(x)$ approaches zero the spacing between the higher eigenvalues must go to zero, since for $f(x) = 0$ the eigenvalue spectrum consists of two discrete points $\lambda = 0$ and $\lambda = 2$ and a continuum $\lambda \geq 9/4$.

The numerical method we shall discuss consists of assuming that the kth eigenfunction can be represented by a sum of N terms according to

$$y(x) = \sum_{n=0}^{N-1} a_n g_n(x), \tag{8.133}$$

where the functions $g_n(x)$ are given and the expansion coefficients a_n are to be determined. For notational simplicity, we have dropped the subscript k on $y_k(x)$ and a_{kn} in Eq. (8.133).

We obtain the necessary equations for the a_n from the standard variational procedure for determining eigenvalues of a self-adjoint system, i.e., the Rayleigh quotient, Eq. (8.130). This method, for a given N, yields estimates of the first N eigenvalues and eigenfunctions, with first order errors in the eigenfunctions giving rise to second order errors in the eigenvalues. Further, this procedure yields upper bounds for the eigenvalues, and hence, as N approaches infinity, the eigenvalues of the N term variational approximation converge to the eigenvalues of the differential equation from above, provided that the $g_n(x)$ are sufficiently complete to represent the eigenfunction. The variational procedure is to substitute the trial function, Eq. (8.131), into the Sturm–Liouville form of the eigenvalue problem, Eq. (8.73), multiply this result by $g_m(x)$, $0 \leq m \leq N-1$, and integrate over all x. This leads to the matrix problem

$$A\psi = \lambda B\psi, \tag{8.134}$$

where ψ is a column vector consisting of N elements $a_0, a_1, \ldots, a_{N-2}, a_{N-1}$, and A and B are $N \times N$ square symmetric matrices with elements

$$A_{nm} = \int_0^\infty dx\, x^{-4} e^x [x^2 g_n' g_m' + (f-x)g_n g_m], \tag{8.135}$$

$$B_{nm} = \int_0^\infty dx\, x^{-4} e^x g_n g_m. \tag{8.136}$$

In writing Eq. (8.135) we have performed an integration by parts and assumed that the $g_n(x)$ are such that the surface terms vanish.

In view of our asymptotic analysis, a reasonable choice for the expansion functions $g_n(x)$ is

$$g_n(x) = x^{2+\beta} L_n(x) e^{-(x+\alpha/x)}, \tag{8.137}$$

where $L_n(x)$ is the standard Laguerre polynomial of degree n. Using elementary properties of the Laguerre polynomials, we find that the matrix elements take the form

$$A_{nm} = \int_0^\infty dx\, x^{2\beta} e^{-(x+2\alpha/x)} [h_n h_m + (f-x)L_n L_m], \tag{8.138}$$

$$B_{nm} = \int_0^\infty dx\, x^{2\beta} e^{-(x+2\alpha/x)} L_n L_m, \tag{8.139}$$

where

$$h_n(x) = (1+\beta-n+\alpha/x)L_n(x) + (n+1)L_{n+1}(x). \tag{8.140}$$

Originally, the equivalent expansion functions

$$g_n(x) = x^{2+\beta+n} e^{-(x+\alpha/x)} \tag{8.141}$$

were used (Pomraning and Froehlich, 1969), but it was found that this led to very ill conditioned matrices A and B. The Laguerre polynomials, rather than other polynomials, were substituted for the simple powers x^n because the resulting functions $g_n(x)$ lead to very simple matrix elements in the limit of $f(x) = 0$. In this limit, Eqs. (8.138) and (8.139) give B as the identity matrix and A as a tridiagonal matrix with elements

$$A_{nm} = (1-m^2)\delta_{n,\,m+1} + (2n^2-2n+1)\delta_{nm} + (1-n^2)\delta_{m,\,n+1}. \tag{8.142}$$

The Equations of Radiation Hydrodynamics

The numerical methods used to compute the matrix elements defined by Eqs. (8.138) and (8.139) and to solve the matrix problem, Eq. (8.134), are discussed by Pomraning and Froehlich (1969). Estimates of the required computer time are also given in that paper. Here we limit the discussion to typical results obtained for the eigenvalues.

As the first example of the eigenvalues obtained with this method, we discuss in some detail the numerical results corresponding to the simplest absorption function $f(x) = \alpha^2/x^2$. Calculations were performed for various values of α with $N = 5, 10, \ldots, 55, 60$. Figure 8.5 shows the first five eigenvalues as a function of α. These curves were plotted from results converged to well within the accuracy of the graph. Figure 8.5 shows, as expected, that the spacing between eigenvalues increases as α increases.

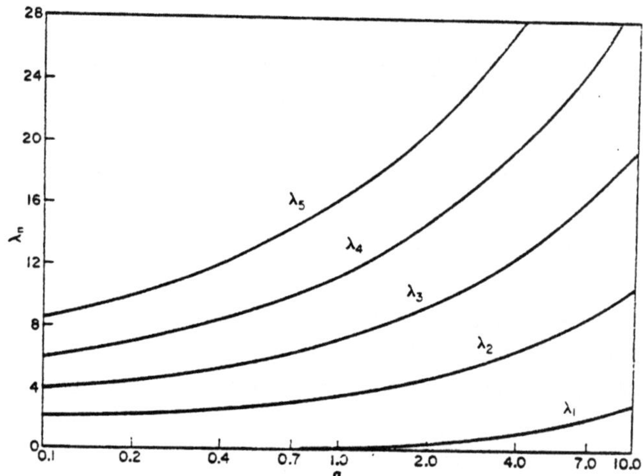

FIG. 8.5. The first five eigenvalues for $f(x) = \alpha^2/x^2$.

As an indication of the rate of convergence of the method, we give in Table 8.1 the first five eigenvalues computed with different values of N for $\alpha = 1$. It can be seen that λ_1 and λ_2 are independent of N to at least seven digits of accuracy (the number of digits printed in the output). This is a consequence of the fact that for $N \geq 2$ the assumed representation of the eigenfunction, Eq. (8.133) with the $g_n(x)$ given by Eq. (8.137), is exact for the first two eigenfunctions. This was shown by the asymptotic analysis. The eigenvalues for these first two eigenfunctions are, from Eq. (8.120) with $\alpha = 1$, $\lambda_1 = 2 - \sqrt{3}$ and $\lambda_2 = 2 + \sqrt{3}$. The numerical method has reproduced these eigenvalues for all values of N. This is particularly noteworthy for large N, since it is an indication that the method is numerically stable. As a further example of the convergence rate, Fig. 8.6 shows the higher eigenvalues as a function of N for $\alpha = 1$. Both Table 8.1 and Fig. 8.6 show the expected behavior that all eigenvalues of the matrix problem approach the eigenvalues of the differential equation from above, and that the rate of convergence is better for a low eigenvalue than for a high eigenvalue.

To show the dependence of the rate of convergence upon α, we give in Table 8.2 the calculated values of λ_3 and λ_6 for different N for three values of α. Again, we note the fact that for a given value of α the lower eigenvalue converges faster than the higher eigenvalue.

208

TABLE 8.1. *The Five Lowest Eigenvalues for $f(x) = 1/x^2$*

N	λ_1	λ_2	λ_3	λ_4	λ_5
5	0.2679491	3.732050	7.81	15.8	36.2
10			7.56	12.3	19.9
15			7.550	12.01	17.7
20			7.5496	11.94	17.14
25			7.54954	11.935	16.94
30			7.549523	11.9321	16.875
35			7.549517	11.9313	16.852
40				11.93116	16.844
45				11.93110	16.8414
50				11.93107	16.8402
55					16.83982
60	0.2679491	3.732050	7.549517	11.93107	16.83963

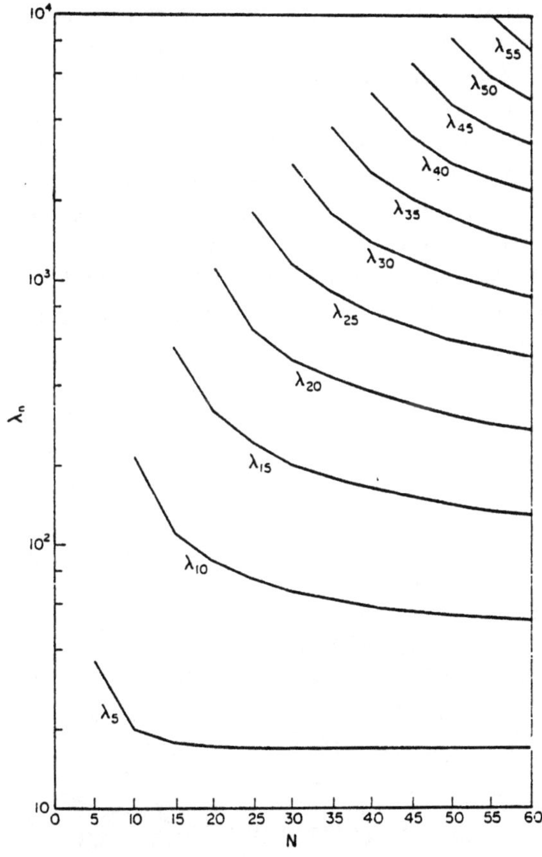

FIG. 8.6. The higher eigenvalues for $f(x) = 1/x^2$.

TABLE 8.2. *The Third and Sixth Eigenvalues for* $f(x) = \alpha^2/x^2$

N	$\alpha = 0.1$		$\alpha = 0.7$		$\alpha = 4.0$	
	λ_3	λ_6	λ_3	λ_6	λ_3	λ_6
5	4.46	—	6.88	—	14.814	—
10	3.98	26.2	6.59	30.8	14.7079	45.3
15	3.896	19.4	6.579	24.3	14.70710	40.1
20	3.868	16.5	6.5771	21.9	14.70710	39.0
25	3.858	15.0	6.57680	20.8	14.70708	38.85
30	3.8535	14.04	6.57672	20.27		38.813
35	3.8512	13.37	6.576706	19.96		38.805
40	3.8501	12.88	6.576701	19.80		38.8037
45	3.84949	12.52	6.576699	19.720		38.80344
50	3.84914	12.245	6.576698	19.673		38.80337
55	3.848940	12.028	↓	19.6487		38.80335
60	3.848817	11.856	6.576698	19.6354	14.70708	38.80335

Further, we see that convergence is much faster for large α than for small α. This is the expected behavior, since for large values of α the eigenvalues have a greater separation and hence the numerical method can more easily resolve the eigenvalues.

As a second example, we consider the somewhat more realistic absorption function $f(x) = \alpha^2(1-e^{-x})/x^3$. This corresponds to a hydrogenic model of the atom with neglect of photoelectric edges and bound–bound transitions. Figure 8.7 shows the converged values of the first five eigenvalues as a function of α. The convergence characteristics for this choice of $f(x)$ are essentially the same as those, just discussed, for $f(x) = \alpha^2/x^2$.

Generally, this numerical scheme works quite well, although for large N (> 50) together with a large α (> 8) some difficulty (numerical instability) is encountered (Pomraning and Froehlich, 1969).

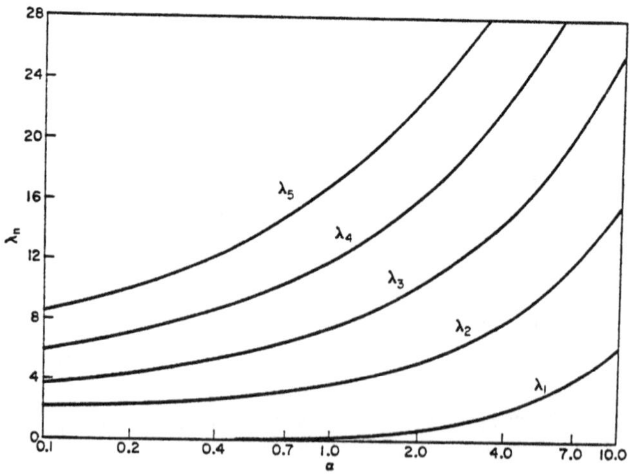

FIG. 8.7. The first five eigenvalues for $f(x) = \alpha^2(1-e^{-x})/x^3$.

IX

Relativistic Hydrodynamics in the Presence of a Radiation Field

1. Introduction

Thus far we have been concerned with the equation of transfer describing the radiation field and the interaction of this radiation field with matter. The final item we shall consider is the description of the motion of the matter. Since in radiation hydrodynamic problems the matter is generally in the gaseous state, one could envision describing the matter by a kinetic (transport) equation similar to the equation of radiative transfer. Such an equation is generally referred to as the Boltzmann equation, and the study of this equation is a field within itself.

In radiation hydrodynamic work such a detailed kinetic description of the matter is not used. Rather, one uses hydrodynamics, with a proper accounting of the effects of the radiation field, to describe the motion of the fluid. The fact that hydrodynamics, rather than kinetic theory, is used to describe the matter is almost a matter of definition in that the class of problems we are considering are those referred to as radiation *hydrodynamic* problems. As we shall see, the equations of hydrodynamics actually follow from simple kinetic theory considerations and hence constitute an approximation to the Boltzmann (kinetic) equation.

As is also implied in the term radiation hydrodynamics, it is necessary to include the effects of the radiation field in the hydrodynamic equations for this class of problems. The equations of hydrodynamics result from particle, momentum, and energy balances for a differential volume of space. If a significant radiation field is present, one must include the radiation momentum and energy in these balances. This gives rise to radiation terms in the equations of hydrodynamics. Also, in some instances the fluid velocities involved may be great enough that relativistic effects become important, and we shall concern ourselves with the relativistic equations of hydrodynamics. Further, we will allow the fluid to be compressible in all of our considerations. We shall restrict the derivations to an ideal fluid, however. This means we shall neglect viscous and heat conduction effects. The neglect of these effects generally introduces a small error in problems in the radiation hydrodynamic regime.

The relativistic hydrodynamic equations in the absence of a radiation field are given in many texts on relativity and fluid mechanics. The approach universally used to derive these equations is to employ the energy–momentum tensor. This tensor is obtained by arguing that it must have a certain form in order to undergo the proper Lorentz transformation and to reduce to the correct diagonal tensor for a fluid at rest. In our derivation of the relativistic hydrodynamic equations including radiative contributions we shall use kinetic theory arguments rather than an energy–momentum tensor containing radiation terms. This avoids the use of the transformation properties of tensors and seems to be a more basic starting point. In particular, our derivation emphasizes the assumption required to obtain a hydrodynamic description of the motion of an ideal fluid. Further, the concept of fluid pressure enters naturally.

Following some introductory kinetic theory considerations in the next section, we derive the relativistic equations of hydrodynamics for a compressible ideal fluid in the presence of a radiation field in three different forms, namely: (1) Eulerian; (2) modified Eulerian; and (3) Lagrangian. The difference between Eulerian and Lagrangian hydrodynamics is only one of viewpoint. In the Eulerian picture one asks for the state of the fluid as a function of time at a given (fixed) position in space. In the Lagrangian picture one asks for the position, as well as the state, of a particular mass of fluid as a function of time. The modified Eulerian picture of hydrodynamics contains elements of both the Eulerian and Lagrangian viewpoints. The distinction between these three equivalent sets of hydrodynamic equations will become clearer in the course of their derivations. In each case we shall also give the non-relativistic limit.

2. Kinetic Theory Considerations

We consider a fluid composed of particles of rest mass m_0 having various momenta $\mathbf{p}$ and described microscopically at time t by a distribution function per unit volume and per unit momentum $\psi(\mathbf{r}, t, \mathbf{p})$. For simplicity we drop all arguments $\mathbf{r}$ and t and simply denote this distribution function by $\psi(\mathbf{p})$. Thus the number of particles at time t in a differential volume element $d\mathbf{r}$ centered at $\mathbf{r}$ and in a differential momentum element $d\mathbf{p}$ centered at $\mathbf{p}$ is given by $\psi(\mathbf{p})\, d\mathbf{r}\, d\mathbf{p}$. The macroscopic velocity of the fluid is denoted by $\mathbf{u}$; i.e., if $\mathbf{v}$ denotes the velocity of a particle with momentum $\mathbf{p}$, then

$$\mathbf{u} = \frac{\int d\mathbf{p}\, \mathbf{v}\psi(\mathbf{p})}{\int d\mathbf{p}\,\psi(\mathbf{p})}. \tag{9.1}$$

That is, the fluid velocity $\mathbf{u}$ is just the velocity of the individual particles which make up the fluid averaged over the distribution function.

We introduce a second frame of reference, namely the frame moving with the fluid. We refer to this as the fluid rest frame and subscript all quantities in this frame of reference with a zero. In particular, we denote by $\psi_0(\mathbf{p}_0)$ the particle distribution function in the rest frame, with $\mathbf{p}_0$ denoting the momentum in this frame. It is important to note that the fluid rest frame is not in general an inertial frame of reference since the fluid can undergo acceleration at any point in space and time.

The basic assumption which leads to ideal fluid hydrodynamics is that the momentum dependence of the distribution function is isotropic in the fluid rest frame. That is, $\psi_0(\mathbf{p}_0)$ depends only upon the magnitude of the momentum and not its direction. In the fluid rest frame we define the three quantities N, E_t, and P_m by the equations

$$N \equiv \int d\mathbf{p}_0 \psi_0(\mathbf{p}_0), \tag{9.2}$$

$$E_t \equiv \int d\mathbf{p}_0 E_0 \psi_0(\mathbf{p}_0), \tag{9.3}$$

where E_0 is the total energy, including the rest energy, associated with a particle of momentum $\mathbf{p}_0$, and

$$P_m \equiv \int d\mathbf{p}_0 (\mathbf{v}_0 \cdot \mathbf{n})(\mathbf{p}_0 \cdot \mathbf{n}) \psi_0(\mathbf{p}_0), \tag{9.4}$$

where $\mathbf{v}_0$ is the velocity associated with a particle of momentum $\mathbf{p}_0$, and $\mathbf{n}$ is an arbitrary unit vector. Since $\psi_0(\mathbf{p}_0)$ is isotropic by assumption, P_m does not depend upon the choice of $\mathbf{n}$. The physical interpretation of N and E_t is immediate. N is just the particle density and E_t is the total energy density, both in the fluid rest frame. From its definition, P_m is just the rate of momentum transfer across a surface of unit area whose normal direction is $\mathbf{n}$. This quantity is conventionally called the material pressure. We emphasize that N, E_t, and P_m are all defined in the fluid rest frame, even though we have not subscripted these quantities with a zero for notational simplicity.

We now compute, in terms of the three quantities defined by Eqs. (9.2) through (9.4), the particle, momentum, and energy densities corresponding to $\psi(\mathbf{p})$; that is, we compute these quantities in a frame of reference for which the macroscopic fluid velocity is $\mathbf{u}$. The definition of the particle density (PD) in terms of the distribution function is

$$\mathrm{PD} \equiv \int d\mathbf{p} \psi(\mathbf{p}). \tag{9.5}$$

To evaluate this integral, we change variables of integration from $\mathbf{p}$ to $\mathbf{p}_0$, the rest frame momentum. Since $\mathbf{p}$ and iE/c $\left(i = \sqrt{-1}\right)$ form a four vector, they transform according to the usual Lorentz transformation as discussed in the Appendix. The transformation velocity of the Appendix, i.e., the velocity of the unadorned frame with respect to the zero frame, is in the present context given by $-\mathbf{u}$. With this identification, Eqs. (A.38) through (A.41) of the Appendix become

$$\mathbf{p} = \mathbf{p}_0 + \left[\frac{(\mathbf{u} \cdot \mathbf{p}_0)(\Lambda - 1)}{u^2} + \frac{\Lambda E_0}{c^2} \right] \mathbf{u}, \tag{9.6}$$

$$E = \Lambda(E_0 + \mathbf{u} \cdot \mathbf{p}_0), \tag{9.7}$$

where $u = |\mathbf{u}|$ and

$$\Lambda = (1 - u^2/c^2)^{-1/2}. \tag{9.8}$$

The variables $\mathbf{p}$ and E are not independent, but are related by

$$E^2 = p^2 c^2 + m_0^2 c^4, \tag{9.9}$$

with a similar relationship valid in the zero frame. Here m_0 is the rest mass of a particle comprising the fluid. From Eqs. (9.6) through (9.9) one can easily compute the Jacobian between $\mathbf{p}$ and $\mathbf{p}_0$ which relates the differentials $d\mathbf{p}$ and $d\mathbf{p}_0$. The details are outlined in the Appendix [see Eqs. (A.43) through (A.46)], and the result is

$$d\mathbf{p} = \frac{E}{E_0} d\mathbf{p}_0. \tag{9.10}$$

213

The Equations of Radiation Hydrodynamics

Further, it is shown in the Appendix [see Eqs. (A.37) through (A.54)] that the distribution function is a Lorentz invariant, i.e.,

$$\psi(\mathbf{p}) = \psi_0(\mathbf{p}_0). \tag{9.11}$$

Use of Eqs. (9.10) and (9.11) in Eq. (9.5) gives an expression for the particle density as

$$\text{PD} = \int d\mathbf{p}_0 \Lambda[1+(\mathbf{u}\cdot\mathbf{p}_0)/E_0]\,\psi_0(\mathbf{p}_0), \tag{9.12}$$

where we have used Eq. (9.7) for the ratio E/E_0. Since $\psi_0(\mathbf{p}_0)$ is isotropic, the term involving $\mathbf{p}_0$ in Eq. (9.12) has a zero integral and hence, recalling Eq. (9.2),

$$\text{PD} = \Lambda N. \tag{9.13}$$

In deriving Eq. (9.13), we have used the Lorentz transformation to transform to the fluid rest frame. As we have remarked earlier, however, the fluid rest frame is not an inertial frame of reference. Nevertheless, the Lorentz transformation can legitimately be used, based on arguments already made in Section 2 of Chapter VI.

Proceeding now to the particle momentum density (PMD), we have by definition

$$\text{PMD} = \int d\mathbf{p}\,\mathbf{p}\psi(\mathbf{p}). \tag{9.14}$$

Use of Eqs. (9.10) and (9.11) in Eq. (9.14) yields

$$\text{PMD} = \int d\mathbf{p}_0\,\frac{E}{E_0}\,\mathbf{p}\psi_0(\mathbf{p}_0), \tag{9.15}$$

and use of Eqs. (9.6) and (9.7) in this result gives

$$\text{PMD} = \int d\mathbf{p}_0\,\frac{\Lambda}{E_0}\,(E_0+\mathbf{u}\cdot\mathbf{p}_0)\left[\mathbf{p}_0+\frac{(\mathbf{u}\cdot\mathbf{p}_0)(\Lambda-1)}{u^2}\,\mathbf{u}+\frac{\Lambda E_0}{c^2}\,\mathbf{u}\right]\psi_0(\mathbf{p}_0). \tag{9.16}$$

Expanding the integrand and keeping only those terms which have a nonzero integral [again, since $\psi_0(\mathbf{p}_0)$ is isotropic several terms integrate to zero], we find

$$\text{PMD} = \int d\mathbf{p}_0\,\frac{\Lambda}{E_0}\left[\frac{\Lambda E_0^2}{c^2}\,\mathbf{u}+(\mathbf{u}\cdot\mathbf{p}_0)\,\mathbf{p}_0+\frac{(\Lambda-1)}{u^2}\,(\mathbf{u}\cdot\mathbf{p}_0)^2\,\mathbf{u}\right]\psi_0(\mathbf{p}_0). \tag{9.17}$$

Now, the momentum of a relativistic particle can be written (in the fluid rest frame or any other frame)

$$\mathbf{p}_0 = \frac{E_0}{c^2}\,\mathbf{v}_0. \tag{9.18}$$

Thus Eq. (9.17) can be written

$$\text{PMD} = \int d\mathbf{p}_0\left[\frac{\Lambda^2 E_0}{c^2}\,\mathbf{u}+\frac{\Lambda(\mathbf{u}\cdot\mathbf{v}_0)}{c^2}\,\mathbf{p}_0+\frac{\Lambda(\Lambda-1)(\mathbf{u}\cdot\mathbf{v}_0)(\mathbf{u}\cdot\mathbf{p}_0)}{u^2 c^2}\,\mathbf{u}\right]\psi_0(\mathbf{p}_0). \tag{9.19}$$

From Eqs. (9.3) and (9.4), identifying $\mathbf{n}$ in Eq. (9.4) with $\mathbf{u}/u$, we find that we can rewrite Eq. (9.19) as

$$\text{PMD} = \frac{\Lambda^2 E_t}{c^2}\,\mathbf{u}+\frac{\Lambda(\Lambda-1)\,P_m}{c^2}\,\mathbf{u}+\frac{\Lambda}{c^2}\int d\mathbf{p}_0(\mathbf{u}\cdot\mathbf{v}_0)\,\psi_0(\mathbf{p}_0)\,\mathbf{p}_0. \tag{9.20}$$

214

By going to component form, one can easily show that

$$\int d\mathbf{p}_0(\mathbf{u}\cdot\mathbf{v}_0)\,\psi_0(\mathbf{p}_0)\,\mathbf{p}_0 = \int d\mathbf{p}_0(\mathbf{u}\cdot\mathbf{v}_0)(\mathbf{u}\cdot\mathbf{p}_0)\,\psi_0(\mathbf{p}_0)\,\mathbf{u}/u^2. \tag{9.21}$$

Again identifying $\mathbf{n}$ in Eq. (9.4) with $\mathbf{u}/u$ in Eq. (9.21), we see that the right hand side of Eq. (9.21) is just the material pressure P_m times $\mathbf{u}$, and hence Eq. (9.20) can be written in final form as, combining terms,

$$\text{PMD} = \frac{\Lambda^2}{c^2}\,(E_t + P_m)\,\mathbf{u}. \tag{9.22}$$

Finally, the particle energy density (PED) is given by

$$\text{PED} = \int d\mathbf{p}\,E\psi(\mathbf{p}). \tag{9.23}$$

Use of Eqs. (9.7), (9.10), and (9.11) in Eq. (9.23) yields

$$\text{PED} = \int d\mathbf{p}_0\,\frac{\Lambda^2}{E_0}\,(E_0 + \mathbf{u}\cdot\mathbf{p}_0)^2\,\psi_0(\mathbf{p}_0). \tag{9.24}$$

Carrying out the squaring in the integrand of Eq. (9.24) and recognizing that the cross-product term has a zero integral [because $\psi_0(\mathbf{p}_0)$ is isotropic], we find

$$\text{PED} = \int d\mathbf{p}_0\,\Lambda^2[E_0 + (\mathbf{u}\cdot\mathbf{p}_0)^2/E_0]\,\psi_0(\mathbf{p}_0). \tag{9.25}$$

Use of Eq. (9.18) in Eq. (9.25) gives

$$\text{PED} = \int d\mathbf{p}_0\,\Lambda^2[E_0 + (\mathbf{u}\cdot\mathbf{p}_0)(\mathbf{u}\cdot\mathbf{v}_0)/c^2]\,\psi_0(\mathbf{p}_0), \tag{9.26}$$

and using Eqs. (9.3) and (9.4), with $\mathbf{n} = \mathbf{u}/u$ in Eq. (9.4), we find

$$\text{PED} = \Lambda^2\left(E_t + \frac{u^2}{c^2}\,P_m\right). \tag{9.27}$$

From Eq. (9.8) we obtain

$$\frac{u^2}{c^2} = 1 - \frac{1}{\Lambda^2}, \tag{9.28}$$

and hence Eq. (9.27) can be written

$$\text{PED} = \Lambda^2(E_t + P_m) - P_m, \tag{9.29}$$

which is an alternate description of the particle energy density.

We next consider the flux (or current) of particles, momentum, and energy across a unit area. The definitions of these quantities are the same as the definitions for the corresponding densities, but with an additional factor of velocity in the integrand. Considering first the particle flux (PF), we have

$$\text{PF} = \int d\mathbf{p}\,\mathbf{v}\psi(\mathbf{p}). \tag{9.30}$$

Using Eqs. (9.10) and (9.11), we can rewrite Eq. (9.30) as

$$\text{PF} = \int d\mathbf{p}_0\,\frac{E}{E_0}\,\mathbf{v}\psi_0(\mathbf{p}_0). \tag{9.31}$$

The Equations of Radiation Hydrodynamics

The velocity of a relativistic particle can be written in terms of its momentum and energy as

$$\mathbf{v} = \frac{c^2}{E}\,\mathbf{p}, \tag{9.32}$$

and thus Eq. (9.31) becomes

$$PF = \int d\mathbf{p}_0\,\frac{c^2}{E_0}\,\mathbf{p}\psi_0(\mathbf{p}_0). \tag{9.33}$$

Use of Eq. (9.6) for $\mathbf{p}$ in Eq. (9.33) and retaining only the terms which have a nonzero integral, we find

$$PF = \int d\mathbf{p}_0\,\varLambda \mathbf{u}\psi_0(\mathbf{p}_0), \tag{9.34}$$

and hence, from Eq. (9.2),

$$PF = \varLambda N\mathbf{u}. \tag{9.35}$$

Similarly, the particle momentum flux (PMF) is defined as

$$PMF = \int d\mathbf{p}\,\mathbf{v}\mathbf{p}\psi(\mathbf{p}). \tag{9.36}$$

Using the same kind of algebraic manipulation used throughout this section, we have

$$
\begin{aligned}
PMF &= \int d\mathbf{p}_0\,\frac{E}{E_0}\,\mathbf{v}\mathbf{p}\psi_0(\mathbf{p}_0) = \int d\mathbf{p}_0\,\frac{c^2}{E_0}\,\mathbf{p}\mathbf{p}\psi_0(\mathbf{p}_0)\\[2mm]
&= \int d\mathbf{p}_0\,\frac{c^2}{E_0}\left[\mathbf{p}_0 + \frac{(\mathbf{u}\bullet\mathbf{p}_0)(\varLambda-1)}{u^2}\,\mathbf{u} + \frac{\varLambda E_0}{c^2}\,\mathbf{u}\right]^2\psi_0(\mathbf{p}_0)\\[2mm]
&= \int d\mathbf{p}_0\,\frac{c^2}{E_0}\left[\mathbf{p}_0\mathbf{p}_0 + \frac{(\mathbf{u}\bullet\mathbf{p}_0)^2(\varLambda-1)^2}{u^4}\,\mathbf{u}\mathbf{u} + \frac{\varLambda^2 E_0^2}{c^4}\,\mathbf{u}\mathbf{u} + \frac{2(\mathbf{u}\bullet\mathbf{p}_0)(\varLambda-1)\,\mathbf{p}_0\mathbf{u}}{u^2}\right]\psi_0(\mathbf{p}_0)\\[2mm]
&= \int d\mathbf{p}_0\left[\mathbf{p}_0\mathbf{v}_0 + \frac{(\mathbf{u}\bullet\mathbf{p}_0)(\mathbf{u}\bullet\mathbf{v}_0)(\varLambda-1)^2}{u^4}\,\mathbf{u}\mathbf{u} + \frac{\varLambda^2 E_0}{c^2}\,\mathbf{u}\mathbf{u} + \frac{2(\mathbf{u}\bullet\mathbf{v}_0)(\varLambda-1)\,\mathbf{p}_0\mathbf{u}}{u^2}\right]\psi_0(\mathbf{p}_0).
\end{aligned}
\tag{9.37}
$$

Use of Eqs. (9.3), (9.4), and (9.21) in Eq. (9.37) gives the result

$$PMF = P_m\mathbf{I} + \frac{(\varLambda-1)^2}{u^2}\,P_m\mathbf{u}\mathbf{u} + \frac{\varLambda^2}{c^2}\,E_t\mathbf{u}\mathbf{u} + \frac{2(\varLambda-1)}{u^2}\,P_m\mathbf{u}\mathbf{u}, \tag{9.38}$$

or, grouping terms,

$$PMF = P_m\mathbf{I} + \frac{(\varLambda^2-1)}{u^2}\,P_m\mathbf{u}\mathbf{u} + \frac{\varLambda^2}{c^2}\,E_t\mathbf{u}\mathbf{u}. \tag{9.39}$$

Here $\mathbf{I}$ is the unit dyad (a three by three tensor with ones on the diagonal, zeros elsewhere). From Eq. (9.8) we find

$$\frac{\varLambda^2-1}{u^2} = \frac{\varLambda^2}{c^2}, \tag{9.40}$$

and Eq. (9.39) can be rewritten as.

$$\text{PMF} = P_m \mathbf{1} + \frac{\varLambda^2}{c^2}(E_t + P_m)\,\mathbf{uu}. \tag{9.41}$$

The final item we need consider is the particle energy flux (PEF). We have

$$\text{PEF} = \int d\mathbf{p}\,vE\psi(\mathbf{p}), \tag{9.42}$$

which leads to

$$\begin{aligned}
\text{PEF} &= \int d\mathbf{p}_0 \frac{E^2}{E_0}\,\mathbf{v}\psi_0(\mathbf{p}_0) = \int d\mathbf{p}_0 \frac{Ec^2}{E_0}\,\mathbf{p}\psi_0(\mathbf{p}_0) \\
&= \int d\mathbf{p}_0 \frac{c^2}{E_0}\varLambda(E_0 + \mathbf{u}\cdot\mathbf{p}_0)\left[\mathbf{p}_0 + \frac{(\mathbf{u}\cdot\mathbf{p}_0)(\varLambda-1)}{u^2}\,\mathbf{u} + \frac{\varLambda E_0}{c^2}\,\mathbf{u}\right]\psi_0(\mathbf{p}_0) \\
&= \int d\mathbf{p}_0 \frac{c^2\varLambda}{E_0}\left[\frac{\varLambda E_0^2}{c^2}\,\mathbf{u} + (\mathbf{u}\cdot\mathbf{p}_0)\,\mathbf{p}_0 + \frac{(\mathbf{u}\cdot\mathbf{p}_0)^2(\varLambda-1)}{u^2}\,\mathbf{u}\right]\psi_0(\mathbf{p}_0) \\
&= \int d\mathbf{p}_0 \left[\varLambda^2 E_0 \mathbf{u} + \varLambda(\mathbf{u}\cdot\mathbf{v}_0)\,\mathbf{p}_0 + \frac{(\mathbf{u}\cdot\mathbf{p}_0)(\mathbf{u}\cdot\mathbf{v}_0)\,\varLambda(\varLambda-1)}{u^2}\,\mathbf{u}\right]\psi_0(\mathbf{p}_0) \\
&= \varLambda^2 E_t \mathbf{u} + \varLambda P_m \mathbf{u} + \varLambda(\varLambda-1)P_m \mathbf{u}. \tag{9.43}
\end{aligned}$$

Combining terms, we find as the final result for the particle energy flux

$$\text{PEF} = \varLambda^2(E_t + P_m)\,\mathbf{u}. \tag{9.44}$$

We now consider the contributions of the radiation field to the energy and momentum density and flux. As in Chapter I, we let $f(v,\,\Omega)$ denote the number of photons per unit volume, frequency, and solid angle in the reference frame in which the fluid velocity is $\mathbf{u}$. The specific intensity of radiation $I(v,\,\Omega)$ is then defined as [see Eq. (1.6)]

$$I(v,\,\Omega) = chvf(v,\,\Omega). \tag{9.45}$$

In terms of the specific intensity we define three quantities, namely

$$E_r = \frac{1}{c}\int_0^\infty dv \int_{4\pi} d\Omega I(v,\,\Omega), \tag{9.46}$$

$$\mathbf{F}_r = \int_0^\infty dv \int_{4\pi} d\Omega \,\mathbf{\Omega} I(v,\,\Omega), \tag{9.47}$$

$$P_r = \frac{1}{c}\int_0^\infty dv \int_{4\pi} d\Omega \,\mathbf{\Omega}\mathbf{\Omega} I(v,\,\Omega). \tag{9.48}$$

These same three quantities were introduced in Chapter I [see Eqs. (1.9), (1.16), and (1.19)] and denoted by u, $\mathbf{F}$, and p. We have changed notation here to avoid confusion with the fluid speed (u) and the particle momentum ($\mathbf{p}$).

The Equations of Radiation Hydrodynamics

We wish to give expressions for the density and flux of energy and momentum due to the radiation field in terms of E_r, $\mathbf{F}_r$, and $\mathbf{P}_r$. The radiation energy density (RED) is given by

$$\text{RED} = \int_0^\infty d\nu \int_{4\pi} d\Omega\, h\nu f(\nu, \mathbf{\Omega}). \tag{9.49}$$

In terms of the specific intensity, Eq. (9.49) can be written

$$\text{RED} = \frac{1}{c} \int_0^\infty d\nu \int_{4\pi} d\Omega\, I(\nu, \mathbf{\Omega}), \tag{9.50}$$

or simply [see Eq. (9.46)]

$$\text{RED} = E_r. \tag{9.51}$$

Similarly, the radiation momentum density (RMD) is written

$$\text{RMD} = \int_0^\infty d\nu \int_{4\pi} d\Omega \left(\frac{h\nu}{c}\mathbf{\Omega}\right) f(\nu, \mathbf{\Omega}) = \frac{1}{c^2} \int_0^\infty d\nu \int_{4\pi} d\Omega\,\mathbf{\Omega} I(\nu, \mathbf{\Omega}), \tag{9.52}$$

which, according to Eq. (9.47), gives

$$\text{RMD} = \frac{1}{c^2} \mathbf{F}_r. \tag{9.53}$$

In a like manner, we have for the radiation energy flux (REF)

$$\text{REF} = \int_0^\infty d\nu \int_{4\pi} d\Omega\, c\mathbf{\Omega} h\nu f(\nu, \mathbf{\Omega}) = \int_0^\infty d\nu \int_{4\pi} d\Omega\,\mathbf{\Omega} I(\nu, \mathbf{\Omega}) = \mathbf{F}_r, \tag{9.54}$$

and the radiation momentum flux (RMF)

$$\text{RMF} = \int_0^\infty d\nu \int_{4\pi} d\Omega\, c\mathbf{\Omega} \left(\frac{h\nu}{c}\mathbf{\Omega}\right) f(\nu, \mathbf{\Omega}) = \frac{1}{c} \int_0^\infty d\nu \int_{4\pi} d\Omega\,\mathbf{\Omega}\mathbf{\Omega} I(\nu, \mathbf{\Omega}) = \mathbf{P}_r. \tag{9.55}$$

The sum of Eqs. (9.29) and (9.51) gives the total energy density due to the fluid particles and the radiation field. Likewise, Eqs. (9.22) and (9.53) give the total momentum density, Eqs. (9.44) and (9.54) give the total energy flux, and Eqs. (9.41) and (9.55) give the total momentum flux.

To separate the results of this section from the algebra, we give here a summary of these results. Consider a frame of reference in which a fluid moves with a macroscopic velocity $\mathbf{u}$ (which depends upon both space and time). In this frame define the radiation energy density E_r, the radiation energy flux $\mathbf{F}_r$, and the radiation pressure $\mathbf{P}_r$ by Eqs. (9.46) through (9.48). Further, in the rest frame of the fluid define the particle density N, the energy density E_t (which includes the rest energy of the fluid particles), and the fluid pressure P_m by Eqs. (9.2) through (9.4). If we define

$$\Lambda = (1 - u^2/c^2)^{-1/2}, \tag{9.56}$$

then in the reference frame for which the fluid velocity is $\mathbf{u}$, we have

$$\text{Particle density} = \varLambda N, \tag{9.57}$$

$$\text{Particle flux} = \varLambda N\mathbf{u}, \tag{9.58}$$

$$\text{Total momentum density} = \frac{\varLambda^2}{c^2}(E_t + P_m)\,\mathbf{u} + \frac{\mathbf{F}_r}{c^2}, \tag{9.59}$$

$$\text{Total momentum flux} = P_m\mathbf{I} + \frac{\varLambda^2}{c^2}(E_t + P_m)\,\mathbf{u}\mathbf{u} + \mathbf{P}_r, \tag{9.60}$$

$$\text{Total energy density} = \varLambda^2(E_t + P_m) - P_m + E_r, \tag{9.61}$$

$$\text{Total energy flux} = \varLambda^2(E_t + P_m)\,\mathbf{u} + \mathbf{F}_r. \tag{9.62}$$

3. Eulerian Hydrodynamics

The equations of relativistic hydrodynamics, in so-called Eulerian form and including effects due to the presence of a radiation field, follow immediately from Eqs. (9.57) through (9.62). In the Eulerian picture one considers the spatial and temporal coordinates $\mathbf{r}$ and t as independent variables and seeks equations describing the dependent variables, such as velocity, pressure, etc., in terms of these independent variables. These equations result from considering a differential volume element fixed in space and performing a particle, momentum, and energy balance for this volume.

We let $D \equiv D(\mathbf{r}, t)$ represent the density of the quantity under consideration (particle, momentum, or energy), and let $\mathbf{F} \equiv \mathbf{F}(\mathbf{r}, t)$, with Cartesian components F_x, F_y, and F_z, denote the corresponding flux. Further, we let $S \equiv S(\mathbf{r}, t)$ denote any external source of this quantity per unit volume. We next consider a differential volume element fixed in space at position x, y, z (i.e., at $\mathbf{r}$) of dimensions $\varDelta x$, $\varDelta y$, and $\varDelta z$. We perform a balance for this volume for the quantity under consideration. We have, omitting all but the essential arguments of the functions for notational simplicity,

$$\frac{\partial}{\partial t}(D\varDelta x\,\varDelta y\,\varDelta z) = [F_x(x) - F_x(x + \varDelta x)]\,\varDelta y\,\varDelta z + [F_y(y) - F_y(y + \varDelta y)]\,\varDelta x\,\varDelta z$$

$$+ [F_z(z) - F_z(z + \varDelta z)]\,\varDelta x\,\varDelta y + S\varDelta x\,\varDelta y\,\varDelta z. \tag{9.63}$$

The left hand side of Eq. (9.63) gives the time rate of change of the quantity contained in the volume element $\varDelta x\,\varDelta y\,\varDelta z$. This change is due to flow through the bounding surfaces of the volume element and any external source present. These two processes are described quantitatively by the right hand side of Eq. (9.63). The equality in this equation is just an expression of the conservation of the quantity under consideration. Since $\varDelta x$, $\varDelta y$, and $\varDelta z$ are differential lengths, independent of time, Eq. (9.63) can be written

$$\frac{\partial D}{\partial t} + \frac{\partial F_x}{\partial x} + \frac{\partial F_y}{\partial y} + \frac{\partial F_z}{\partial z} = S, \tag{9.64}$$

or, in vector notation,

$$\frac{\partial D}{\partial t} + \nabla \cdot \mathbf{F} = S. \tag{9.65}$$

Thus we see that the result of such a balance, for any quantity of density D, flux F, and source S, is that the time derivative of the density plus the divergence of the flux must equal the external source.

Applying this result to the three quantities of interest here, namely particles, momentum, and energy, we obtain, making use of Eqs. (9.57) through (9.62) for the relevant densities and fluxes,

$$\frac{\partial(\varLambda N)}{\partial t} + \nabla \cdot (\varLambda N \mathbf{u}) = 0, \qquad (9.66)$$

$$\frac{\partial}{\partial t}\left[\frac{\varLambda^2}{c^2}(E_t + P_m)\mathbf{u} + \frac{1}{c^2}\mathbf{F}_r\right] + \nabla P_m + \nabla \cdot \left[\frac{\varLambda^2}{c^2}(E_t + P_m)\mathbf{uu} + P_r\right] = 0, \qquad (9.67)$$

$$\frac{\partial}{\partial t}[\varLambda^2(E_t + P_m) - P_m + E_r] + \nabla \cdot [\varLambda^2(E_t + P_m)\mathbf{u} + \mathbf{F}_r] = W. \qquad (9.68)$$

In writing Eqs. (9.66) through (9.68) we have assumed no external source of particles or momentum (i.e., no external forces are present), but we have allowed an external source of energy (heat), denoted by W. This is often the situation in radiation hydrodynamic problems. However, the neglected particle and momentum sources are easily included if the need arises in any given problem. One merely adds the appropriate source terms to the right hand sides of Eqs. (9.66) and (9.67). These three equations, together with the equation of radiative transfer and two thermodynamic relations relating the fluid rest frame quantities N, E_t, T, and P_m (namely the equation of state relating N, T, and P_m, and the definition of the specific heat relating E_t and T), represent six equations for the six unknowns: I (specific intensity of radiation), T (temperature), $\mathbf{u}$ (velocity), P_m (material pressure), E_t (material energy density), and N (particle density). These six equations are the basic equations of relativistic radiation hydrodynamics.

Equations (9.66) through (9.68) can be put in a somewhat more useful form by redefining some of the variables. In particular, we eliminate N in these equations in favor of ϱ, defined as

$$\varrho = m_0 N, \qquad (9.69)$$

where m_0 is the rest mass of a fluid particle. Thus the quantity ϱ is just the mass density of the fluid if all the fluid particles were at rest. We also define E_m as

$$E_m = E_t - \varrho c^2, \qquad (9.70)$$

so that E_m is the fluid energy density (in the fluid rest frame) in excess of the rest energy. Equations (9.66) through (9.68) then become

$$\frac{\partial(\varLambda \varrho)}{\partial t} + \nabla \cdot (\varLambda \varrho \mathbf{u}) = 0, \qquad (9.71)$$

$$\frac{\partial}{\partial t}\left[\frac{\varLambda^2}{c^2}(\varrho c^2 + E_m + P_m)\mathbf{u} + \frac{1}{c^2}\mathbf{F}_r\right] + \nabla P_m + \nabla \cdot \left[\frac{\varLambda^2}{c^2}(\varrho c^2 + E_m + P_m)\mathbf{uu} + P_r\right] = 0, \qquad (9.72)$$

$$\frac{\partial}{\partial t}[\varLambda^2(\varrho c^2 + E_m + P_m) - P_m + E_r] + \nabla \cdot [\varLambda^2(\varrho c^2 + E_m + P_m)\mathbf{u} + \mathbf{F}_r] = W. \qquad (9.73)$$

An alternate form of the energy equation follows by multiplying Eq. (9.71) by c^2 and subtracting the result from Eq. (9.73). This has the effect of deleting the particle rest energy contribution from the energy equation and makes the passage to the nonrelativistic limit more straightforward. Hence an equivalent set of relativistic hydrodynamic equations, our final form for the Eulerian equations, is

$$\frac{\partial(\Lambda\varrho)}{\partial t} + \nabla\cdot(\Lambda\varrho\mathbf{u}) = 0, \qquad (9.74)$$

$$\frac{\partial}{\partial t}\left[\frac{\Lambda^2}{c^2}(\varrho c^2+E_m+P_m)\,\mathbf{u}+\frac{1}{c^2}\,\mathbf{F}_r\right]+\nabla P_m+\nabla\cdot\left[\frac{\Lambda^2}{c^2}(\varrho c^2+E_m+P_m)\,\mathbf{uu}+P_r\right]=0, \qquad (9.75)$$

$$\frac{\partial}{\partial t}[\Lambda(\Lambda-1)\varrho c^2+\Lambda^2(E_m+P_m)-P_m+E_r]+\nabla\cdot[\Lambda(\Lambda-1)\varrho c^2\mathbf{u}+\Lambda^2(E_m+P_m)\mathbf{u}+\mathbf{F}_r]=W.$$
$$(9.76)$$

Equations (9.74) through (9.76) represent the relativistic equations of hydrodynamics in so-called conservative form for a volume element fixed in space. That is, integration of these equations over an arbitrary volume fixed in space gives results which can be interpreted as particle, momentum, and energy balances for this volume. For example, consider first Eq. (9.66), recalling that ΛN is just the particle density [see Eq. (9.57)]. Integrating this equation over a closed volume V bounded by a surface S, commuting the integral operator with the time derivative, and using Gauss' theorem to rewrite the divergence term, we find

$$\frac{\partial}{\partial t}\int_V d\mathbf{r}\,\Lambda N = \int_S d\mathbf{s}\,\Lambda N(\mathbf{n}\cdot\mathbf{u}), \qquad (9.77)$$

where $\mathbf{n}$ is an *inward* normal unit vector at the surface. The interpretation of Eq. (9.77) is immediate. The left hand side is the time rate of change of the total number of particles in the volume V, whereas the right hand side is the rate at which particles enter this volume through the bounding surface S. The equality merely states conservation of particles for the volume V.

Applying this same procedure to Eqs. (9.74) through (9.76), we find the integral conservation equations of Eulerian relativistic hydrodynamics. We have

$$\frac{\partial}{\partial t}\int_V d\mathbf{r}\,\Lambda\varrho = \int_S d\mathbf{s}\,\Lambda\varrho(\mathbf{n}\cdot\mathbf{u}), \qquad (9.78)$$

$$\frac{\partial}{\partial t}\int_V d\mathbf{r}\left[\frac{\Lambda^2}{c^2}(\varrho c^2+E_m+P_m)\,\mathbf{u}+\frac{1}{c^2}\,F_r\right]$$
$$= \int_S d\mathbf{s}\left[\mathbf{n}P_m+\frac{\Lambda^2}{c^2}(\varrho c^2+E_m+P_m)(\mathbf{n}\cdot\mathbf{u})\,\mathbf{u}+\mathbf{n}\cdot P_r\right], \qquad (9.79)$$

$$\frac{\partial}{\partial t}\int_V d\mathbf{r}[\Lambda(\Lambda-1)\varrho c^2+\Lambda^2(E_m+P_m)-P_m+E_r]$$
$$= \int_S d\mathbf{s}[\Lambda(\Lambda-1)\varrho c^2(\mathbf{n}\cdot\mathbf{u})+\Lambda^2(E_m+P_m)(\mathbf{n}\cdot\mathbf{u})+\mathbf{n}\cdot\mathbf{F}_r]+\int_V d\mathbf{r}W. \qquad (9.80)$$

The Equations of Radiation Hydrodynamics

The integral conservation equations form the basis for many numerical methods of solution of the equations of hydrodynamics.

The nonrelativistic limit of the hydrodynamic equations results if one lets c, the speed of light, become infinite in the material terms of the relativistic equations. Recalling the definition of Λ as given by Eq. (9.56), we have

$$\Lambda \xrightarrow[c\to\infty]{} 1; \quad (\Lambda-1) \xrightarrow[c\to\infty]{} \frac{1}{2}\frac{u^2}{c^2}. \tag{9.81}$$

In this limit, Eqs. (9.74) through (9.76) then become

$$\frac{\partial \varrho}{\partial t} + \nabla\cdot(\varrho\mathbf{u}) = 0, \tag{9.82}$$

$$\frac{\partial}{\partial t}\left(\varrho\mathbf{u}+\frac{1}{c^2}\mathbf{F}_r\right) + \nabla P_m + \nabla\cdot(\varrho\mathbf{u}\mathbf{u}+\mathbf{P}_r) = 0, \tag{9.83}$$

$$\frac{\partial}{\partial t}\left(\frac{1}{2}\varrho u^2+E_m+E_r\right) + \nabla\cdot\left[\left(\frac{1}{2}\varrho u^2+E_m+P_m\right)\mathbf{u}+\mathbf{F}_r\right] = W. \tag{9.84}$$

Aside from the radiation terms, these are just the classical nonrelativistic equations of hydrodynamics for a compressible, ideal fluid. The corresponding integral conservation equations are

$$\frac{\partial}{\partial t}\int_V d\mathbf{r}\varrho = \int_S ds\varrho(\mathbf{n}\cdot\mathbf{u}), \tag{9.85}$$

$$\frac{\partial}{\partial t}\int_V d\mathbf{r}\left(\varrho\mathbf{u}+\frac{1}{c^2}\mathbf{F}_r\right) = \int_S ds[\mathbf{n}P_m+\varrho(\mathbf{n}\cdot\mathbf{u})\mathbf{u}+\mathbf{n}\cdot\mathbf{P}_r] = 0, \tag{9.86}$$

$$\frac{\partial}{\partial t}\int_V d\mathbf{r}\left(\frac{1}{2}\varrho u^2+E_m+E_r\right) = \int_S ds\left[\left(\frac{1}{2}\varrho u^2+E_m+P_m\right)(\mathbf{n}\cdot\mathbf{u})+(\mathbf{n}\cdot\mathbf{F}_r)\right] + \int_V d\mathbf{r}\,W. \tag{9.87}$$

As before, the vector $\mathbf{n}$ in Eqs. (9.85) through (9.87) is a unit *inward* normal vector at the surface of the volume under consideration.

4. The Eulerian Equations in Various Coordinate Systems

In this section we consider the form the Eulerian equations of hydrodynamics take in the three standard coordinate systems, namely Cartesian, cylindrical, and spherical. We consider the full three dimensional description in all cases (i.e., we assume no special symmetry).

In a Cartesian system with coordinates x, y, and z the equations can be written down without further analysis since the vector differential operators contained in the equations of hydrodynamics are basically defined in this coordinate system. In particular, for any scalar A, the x, y, and z components of the gradient of A are given by

$$(\nabla A)_x = \frac{\partial A}{\partial x}; \quad (\nabla A)_y = \frac{\partial A}{\partial y}; \quad (\nabla A)_z = \frac{\partial A}{\partial z}. \tag{9.88}$$

For any vector $\mathbf{B}$, with Cartesian components B_x, B_y, and B_z, the divergence of $\mathbf{B}$ is given by

$$\nabla \cdot \mathbf{B} = \frac{\partial B_x}{\partial x} + \frac{\partial B_y}{\partial y} + \frac{\partial B_z}{\partial z}. \tag{9.89}$$

Finally, for any dyad $\mathbf{BC}$, with Cartesian components $B_i C_j$ $(i, j = x, y, z)$, the x, y, and z components of the divergence of $\mathbf{BC}$ are defined as

$$[\nabla \cdot (\mathbf{BC})]_x = \frac{\partial}{\partial x}(B_x C_x) + \frac{\partial}{\partial y}(B_y C_x) + \frac{\partial}{\partial z}(B_z C_x), \tag{9.90}$$

$$[\nabla \cdot (\mathbf{BC})]_y = \frac{\partial}{\partial x}(B_x C_y) + \frac{\partial}{\partial y}(B_y C_y) + \frac{\partial}{\partial z}(B_z C_y), \tag{9.91}$$

$$[\nabla \cdot (\mathbf{BC})]_z = \frac{\partial}{\partial x}(B_x C_z) + \frac{\partial}{\partial y}(B_y C_z) + \frac{\partial}{\partial z}(B_z C_z). \tag{9.92}$$

We let u_x, u_y, and u_z represent the three Cartesian components of the fluid velocity $\mathbf{u}$, and similarly we let F_x, F_y, and F_z denote the three Cartesian components of the radiative flux $\mathbf{F}_r$. We denote by p_{ij} $(i, j = x, y, z)$ the nine Cartesian components of the radiation pressure tensor P_r. We note that it is the product of two vectors, namely $\mathbf{\Omega\Omega}$, which gives P_r its tensor character [see Eq. (1.20)]. Thus the x, y, and z components of the divergence of the radiation pressure tensor follow directly from Eqs. (9.90) through (9.92).

The above definitions allow us to immediately write the relativistic equations of hydrodynamics, Eqs. (9.74) through (9.76), in a Cartesian coordinate system. The continuity equation (9.74) takes the form

$$\frac{\partial(\varLambda\varrho)}{\partial t} + \frac{\partial(\varLambda\varrho u_x)}{\partial x} + \frac{\partial(\varLambda\varrho u_y)}{\partial y} + \frac{\partial(\varLambda\varrho u_z)}{\partial z} = 0. \tag{9.93}$$

The three components of the momentum equation (9.75) are

$$\frac{\partial}{\partial t}\left[\frac{\varLambda^2}{c^2}(\varrho c^2 + E_m + P_m)u_x + \frac{1}{c^2}F_x\right] + \frac{\partial P_m}{\partial x}$$
$$+ \frac{\partial}{\partial x}\left[\frac{\varLambda^2}{c^2}(\varrho c^2 + E_m + P_m)u_x u_x\right] + \frac{\partial}{\partial y}\left[\frac{\varLambda^2}{c^2}(\varrho c^2 + E_m + P_m)u_y u_x\right]$$
$$+ \frac{\partial}{\partial z}\left[\frac{\varLambda^2}{c^2}(\varrho c^2 + E_m + P_m)u_z u_x\right] + \frac{\partial p_{xx}}{\partial x} + \frac{\partial p_{yx}}{\partial y} + \frac{\partial p_{zx}}{\partial z} = 0, \tag{9.94}$$

$$\frac{\partial}{\partial t}\left[\frac{\varLambda^2}{c^2}(\varrho c^2 + E_m + P_m)u_y + \frac{1}{c^2}F_y\right] + \frac{\partial P_m}{\partial y}$$
$$+ \frac{\partial}{\partial x}\left[\frac{\varLambda^2}{c^2}(\varrho c^2 + E_m + P_m)u_x u_y\right] + \frac{\partial}{\partial y}\left[\frac{\varLambda^2}{c^2}(\varrho c^2 + E_m + P_m)u_y u_y\right]$$
$$+ \frac{\partial}{\partial z}\left[\frac{\varLambda^2}{c^2}(\varrho c^2 + E_m + P_m)u_z u_y\right] + \frac{\partial p_{xy}}{\partial x} + \frac{\partial p_{yy}}{\partial y} + \frac{\partial p_{zy}}{\partial z} = 0, \tag{9.95}$$

$$\frac{\partial}{\partial t}\left[\frac{\Lambda^2}{c^2}(\varrho c^2+E_m+P_m)u_z+\frac{1}{c^2}F_z\right]+\frac{\partial P_m}{\partial z}$$

$$+\frac{\partial}{\partial x}\left[\frac{\Lambda^2}{c^2}(\varrho c^2+E_m+P_m)u_xu_z\right]+\frac{\partial}{\partial y}\left[\frac{\Lambda^2}{c^2}(\varrho c^2+E_m+P_m)u_yu_z\right]$$

$$+\frac{\partial}{\partial z}\left[\frac{\Lambda^2}{c^2}(\varrho c^2+E_m+P_m)u_zu_z\right]+\frac{\partial p_{xz}}{\partial x}+\frac{\partial p_{yz}}{\partial y}+\frac{\partial p_{zz}}{\partial z}=0. \qquad (9.96)$$

The energy equation (9.76) is written

$$\frac{\partial}{\partial t}\left[\Lambda(\Lambda-1)\varrho c^2+\Lambda^2(E_m+P_m)-P_m+E_r\right]$$

$$+\frac{\partial}{\partial x}\left[\Lambda(\Lambda-1)\varrho c^2u_x+\Lambda^2(E_m+P_m)u_x+F_x\right]$$

$$+\frac{\partial}{\partial y}\left[\Lambda(\Lambda-1)\varrho c^2u_y+\Lambda^2(E_m+P_m)u_y+F_y\right]$$

$$+\frac{\partial}{\partial z}\left[\Lambda(\Lambda-1)\varrho c^2u_z+\Lambda^2(E_m+P_m)u_z+F_z\right]=W. \qquad (9.97)$$

By letting c approach infinity in the material terms in Eqs. (9.93) through (9.97) we obtain the nonrelativistic equations of hydrodynamics in a Cartesian coordinate system. The continuity equation (9.93) becomes

$$\frac{\partial\varrho}{\partial t}+\frac{\partial(\varrho u_x)}{\partial x}+\frac{\partial(\varrho u_y)}{\partial y}+\frac{\partial(\varrho u_z)}{\partial z}=0, \qquad (9.98)$$

the three momentum equations (9.94) through (9.96) become

$$\frac{\partial}{\partial t}\left(\varrho u_x+\frac{1}{c^2}F_x\right)+\frac{\partial P_m}{\partial x}+\frac{\partial}{\partial x}(\varrho u_xu_x)+\frac{\partial}{\partial y}(\varrho u_yu_x)$$

$$+\frac{\partial}{\partial z}(\varrho u_zu_x)+\frac{\partial p_{xx}}{\partial x}+\frac{\partial p_{yx}}{\partial y}+\frac{\partial p_{zx}}{\partial z}=0, \qquad (9.99)$$

$$\frac{\partial}{\partial t}\left(\varrho u_y+\frac{1}{c^2}F_y\right)+\frac{\partial P_m}{\partial y}+\frac{\partial}{\partial x}(\varrho u_xu_y)+\frac{\partial}{\partial y}(\varrho u_yu_y)$$

$$+\frac{\partial}{\partial z}(\varrho u_zu_y)+\frac{\partial p_{xy}}{\partial x}+\frac{\partial p_{yy}}{\partial y}+\frac{\partial p_{zy}}{\partial z}=0, \qquad (9.100)$$

$$\frac{\partial}{\partial t}\left(\varrho u_z+\frac{1}{c^2}F_z\right)+\frac{\partial P_m}{\partial z}+\frac{\partial}{\partial x}(\varrho u_xu_z)+\frac{\partial}{\partial y}(\varrho u_yu_z)$$

$$+\frac{\partial}{\partial z}(\varrho u_zu_z)+\frac{\partial p_{xz}}{\partial x}+\frac{\partial p_{yz}}{\partial y}+\frac{\partial p_{zz}}{\partial z}=0, \qquad (9.101)$$

and the energy equation (9.97) becomes

$$\frac{\partial}{\partial t}\left(\frac{1}{2}\varrho u^2 + E_m + E_r\right) + \frac{\partial}{\partial x}\left[\left(\frac{1}{2}\varrho u^2 + E_m + P_m\right)u_x + F_x\right]$$

$$+\frac{\partial}{\partial y}\left[\left(\frac{1}{2}\varrho u^2 + E_m + P_m\right)u_y + F_y\right]$$

$$+\frac{\partial}{\partial z}\left[\left(\frac{1}{2}\varrho u^2 + E_m + P_m\right)u_z + F_z\right] = W. \qquad (9.102)$$

Equations (9.98) through (9.102) are the equations in Cartesian coordinates corresponding to the general nonrelativistic Eulerian hydrodynamic equations, (9.82) through (9.84).

The analysis is somewhat more involved in curvilinear coordinate systems. Let us first consider a standard cylindrical system with coordinates r, θ, and z. These are related to the Cartesian coordinates by the relations

$$x = r\cos\theta, \qquad (9.103)$$

$$y = r\sin\theta, \qquad (9.104)$$

$$z = z. \qquad (9.105)$$

From Eqs. (9.103) through (9.105) one can easily show that for any vector $\mathbf{C}$ the Cartesian components (C_x, C_y, and C_z) are related to the cylindrical components (C_r, C_θ, C_z) of the same vector by the relations

$$C_x = C_r\cos\theta - C_\theta\sin\theta, \qquad (9.106)$$

$$C_y = C_r\sin\theta + C_\theta\cos\theta, \qquad (9.107)$$

$$C_z = C_z. \qquad (9.108)$$

The inverse relations are

$$C_r = C_x\cos\theta + C_y\sin\theta, \qquad (9.109)$$

$$C_\theta = -C_x\sin\theta + C_y\cos\theta, \qquad (9.110)$$

$$C_z = C_z. \qquad (9.111)$$

The expressions for the gradient of any scalar and the divergence of any vector in a cylindrical coordinate system are well known and we give the results here without derivation. For any scalar A, the r, θ, and z components of the gradient of A are given by

$$(\nabla A)_r = \frac{\partial A}{\partial r}; \quad (\nabla A)_\theta = \frac{1}{r}\frac{\partial A}{\partial\theta}; \quad (\nabla A)_z = \frac{\partial A}{\partial z}. \qquad (9.112)$$

For any vector $\mathbf{B}$, with components B_r, B_θ, and B_z in a cylindrical coordinate system, the divergence of $\mathbf{B}$ is given by

$$\nabla\cdot\mathbf{B} = \frac{1}{r}\frac{\partial(rB_r)}{\partial r} + \frac{1}{r}\frac{\partial B_\theta}{\partial\theta} + \frac{\partial B_z}{\partial z}. \qquad (9.113)$$

We now derive expressions for the components of the divergence of the dyad $\mathbf{BC}$ in a cylindrical coordinate system. The x component of this divergence in a Cartesian coordinate

The Equations of Radiation Hydrodynamics

system is given by

$$[\nabla \cdot (\mathbf{BC})]_x = \frac{\partial}{\partial x}(B_x C_x) + \frac{\partial}{\partial y}(B_y C_x) + \frac{\partial}{\partial z}(B_z C_x), \tag{9.114}$$

or, in vector notation,

$$[\nabla \cdot (\mathbf{BC})]_x = \nabla \cdot (\mathbf{B} C_x). \tag{9.115}$$

Considering the vector $\mathbf{B}$ in Eq. (9.115) to be represented by cylindrical components, and using Eq. (9.113) as the expression for the divergence, we find Eq. (9.115) becomes

$$[\nabla \cdot (\mathbf{BC})]_x = \frac{1}{r}\frac{\partial(rB_r C_x)}{\partial r} + \frac{1}{r}\frac{\partial(B_\theta C_x)}{\partial \theta} + \frac{\partial(B_z C_x)}{\partial z}. \tag{9.116}$$

Use of Eq. (9.106), which expresses C_x in terms of the cylindrical components of $\mathbf{C}$, in Eq. (9.116) gives

$$[\nabla \cdot (\mathbf{BC})]_x = \frac{\cos\theta}{r}\frac{\partial(rB_r C_r)}{\partial r} - \frac{\sin\theta}{r}\frac{\partial(rB_r C_\theta)}{\partial r}$$
$$+ \frac{\cos\theta}{r}\frac{\partial(B_\theta C_r)}{\partial \theta} - \frac{\sin\theta}{r}\frac{\partial(B_\theta C_\theta)}{\partial \theta} - \frac{\sin\theta}{r}B_\theta C_r$$
$$- \frac{\cos\theta}{r}B_\theta C_\theta + \cos\theta\frac{\partial(B_z C_r)}{\partial z} - \sin\theta\frac{\partial(B_z C_\theta)}{\partial z}. \tag{9.117}$$

Similarly, we can find expressions for the y and z components of $\nabla \cdot (\mathbf{BC})$. These are

$$[\nabla \cdot (\mathbf{BC})]_y = \frac{\sin\theta}{r}\frac{\partial(rB_r C_r)}{\partial r} + \frac{\cos\theta}{r}\frac{\partial(rB_r C_\theta)}{\partial r}$$
$$+ \frac{\sin\theta}{r}\frac{\partial(B_\theta C_r)}{\partial \theta} + \frac{\cos\theta}{r}\frac{\partial(B_\theta C_\theta)}{\partial \theta} + \frac{\cos\theta}{r}B_\theta C_r$$
$$- \frac{\sin\theta}{r}B_\theta C_\theta + \sin\theta\frac{\partial(B_z C_r)}{\partial z} + \cos\theta\frac{\partial(B_z C_\theta)}{\partial z}, \tag{9.118}$$

$$[\nabla \cdot (\mathbf{BC})]_z = \frac{1}{r}\frac{\partial(rB_r C_z)}{\partial r} + \frac{1}{r}\frac{\partial(B_\theta C_z)}{\partial \theta} + \frac{\partial(B_z C_z)}{\partial z}. \tag{9.119}$$

Now, since the divergence of the dyad $\mathbf{BC}$ is a vector, Eqs. (9.109) through (9.111) can be used to obtain the cylindrical components in terms of the Cartesian components, i.e.,

$$[\nabla \cdot (\mathbf{BC})]_r = [\nabla \cdot (\mathbf{BC})]_x \cos\theta + [\nabla \cdot (\mathbf{BC})]_y \sin\theta, \tag{9.120}$$
$$[\nabla \cdot (\mathbf{BC})]_\theta = -[\nabla \cdot (\mathbf{BC})]_x \sin\theta + [\nabla \cdot (\mathbf{BC})]_y \cos\theta, \tag{9.121}$$
$$[\nabla \cdot (\mathbf{BC})]_z = [\nabla \cdot (\mathbf{BC})]_z. \tag{9.122}$$

Use of Eqs. (9.117) through (9.119) in Eqs. (9.120) through (9.122) gives the final results for the cylindrical components of the divergence of the dyad $\mathbf{BC}$. These are

$$[\nabla \cdot (\mathbf{BC})]_r = \frac{1}{r}\frac{\partial(rB_r C_r)}{\partial r} + \frac{1}{r}\frac{\partial(B_\theta C_r)}{\partial \theta} + \frac{\partial(B_z C_r)}{\partial z} - \frac{1}{r}B_\theta C_\theta, \tag{9.123}$$

$$[\nabla \cdot (\mathbf{BC})]_\theta = \frac{1}{r}\frac{\partial(rB_r C_\theta)}{\partial r} + \frac{1}{r}\frac{\partial(B_\theta C_\theta)}{\partial \theta} + \frac{\partial(B_z C_\theta)}{\partial z} + \frac{1}{r}B_\theta C_r, \tag{9.124}$$

$$[\nabla \cdot (\mathbf{BC})]_z = \frac{1}{r}\frac{\partial(rB_r C_z)}{\partial r} + \frac{1}{r}\frac{\partial(B_\theta C_z)}{\partial \theta} + \frac{\partial(B_z C_z)}{\partial z}. \tag{9.125}$$

To write the equations of hydrodynamics in a cylindrical coordinate system, we let u_r, u_θ, and u_z represent the three components of the fluid velocity $\mathbf{u}$, and similarly we let F_r, F_θ, and F_z denote the three cylindrical components of the radiative flux $\mathbf{F}_r$. We denote by p_{ij} $(i, j = r, \theta, z)$ the nine cylindrical components of the radiation pressure tensor $\mathbf{P}_r$. Explicitly we have

$$F_i \equiv F_i(r, \theta, z, t) = \int_0^\infty d\nu \int_{4\pi} d\Omega \, \Omega_i I(r, \theta, z, \nu, \Omega, t); \quad i = r, \theta, z, \tag{9.126}$$

and

$$p_{ij} \equiv p_{ij}(r, \theta, z, t) = \frac{1}{c} \int_0^\infty d\nu \int_{4\pi} d\Omega \, \Omega_i \Omega_j I(r, \theta, z, \nu, \Omega, t); \quad i, j = r, \theta, z, \tag{9.127}$$

where Ω_i is the ith component of the unit vector $\mathbf{\Omega}$. Equation (9.127) shows that the pressure tensor is a dyad and thus Eqs. (9.123) through (9.125) can be used to obtain the components of the divergence of the radiation pressure tensor in cylindrical coordinates.

With these results, we can immediately write the relativistic hydrodynamic equations, (9.74) through (9.76), in a cylindrical coordinate system. The continuity equation (9.74) is written

$$\frac{\partial(\Lambda\varrho)}{\partial t} + \frac{1}{r} \frac{\partial(r\Lambda\varrho u_r)}{\partial r} + \frac{1}{r} \frac{\partial(\Lambda\varrho u_\theta)}{\partial \theta} + \frac{\partial(\Lambda\varrho u_z)}{\partial z} = 0. \tag{9.128}$$

The three components of the momentum equation (9.75) become

$$\frac{\partial}{\partial t}\left[\frac{\Lambda^2}{c^2}(\varrho c^2 + E_m + P_m)u_r + \frac{1}{c^2} F_r\right] + \frac{\partial P_m}{\partial r}$$

$$+ \frac{1}{r}\frac{\partial}{\partial r}\left[r\frac{\Lambda^2}{c^2}(\varrho c^2 + E_m + P_m)u_r u_r\right] + \frac{1}{r}\frac{\partial}{\partial \theta}\left[\frac{\Lambda^2}{c^2}(\varrho c^2 + E_m + P_m)u_\theta u_r\right]$$

$$+ \frac{\partial}{\partial z}\left[\frac{\Lambda^2}{c^2}(\varrho c^2 + E_m + P_m)u_z u_r\right] - \frac{1}{r}\frac{\Lambda^2}{c^2}(\varrho c^2 + E_m + P_m)u_\theta u_\theta$$

$$+ \frac{1}{r}\frac{\partial(rp_{rr})}{\partial r} + \frac{1}{r}\frac{\partial p_{\theta r}}{\partial \theta} + \frac{\partial p_{zr}}{\partial z} - \frac{1}{r} p_{\theta\theta} = 0, \tag{9.129}$$

$$\frac{\partial}{\partial t}\left[\frac{\Lambda^2}{c^2}(\varrho c^2 + E_m + P_m)u_\theta + \frac{1}{c^2} F_\theta\right] + \frac{1}{r}\frac{\partial P_m}{\partial \theta}$$

$$+ \frac{1}{r}\frac{\partial}{\partial r}\left[r\frac{\Lambda^2}{c^2}(\varrho c^2 + E_m + P_m)u_r u_\theta\right] + \frac{1}{r}\frac{\partial}{\partial \theta}\left[\frac{\Lambda^2}{c^2}(\varrho c^2 + E_m + P_m)u_\theta u_\theta\right]$$

$$+ \frac{\partial}{\partial z}\left[\frac{\Lambda^2}{c^2}(\varrho c^2 + E_m + P_m)u_z u_\theta\right] + \frac{1}{r}\frac{\Lambda^2}{c^2}(\varrho c^2 + E_m + P_m)u_\theta u_r$$

$$+ \frac{1}{r}\frac{\partial(rp_{r\theta})}{\partial r} + \frac{1}{r}\frac{\partial p_{\theta\theta}}{\partial \theta} + \frac{\partial p_{z\theta}}{\partial z} + \frac{1}{r} p_{\theta r} = 0, \tag{9.130}$$

$$\frac{\partial}{\partial t}\left[\frac{\Lambda^2}{c^2}(\varrho c^2 + E_m + P_m)u_z + \frac{1}{c^2} F_z\right] + \frac{\partial P_m}{\partial z}$$

$$+ \frac{1}{r}\frac{\partial}{\partial r}\left[r\frac{\Lambda^2}{c^2}(\varrho c^2 + E_m + P_m)u_r u_z\right] + \frac{1}{r}\frac{\partial}{\partial \theta}\left[\frac{\Lambda^2}{c^2}(\varrho c^2 + E_m + P_m)u_\theta u_z\right]$$

$$+ \frac{\partial}{\partial z}\left[\frac{\Lambda^2}{c^2}(\varrho c^2 + E_m + P_m)u_z u_z\right] + \frac{1}{r}\frac{\partial(rp_{rz})}{\partial r} + \frac{1}{r}\frac{\partial p_{\theta z}}{\partial \theta} + \frac{\partial p_{zz}}{\partial z} = 0. \tag{9.131}$$

The Equations of Radiation Hydrodynamics

The energy equation (9.76) is written

$$\frac{\partial}{\partial t}\left[\Lambda(\Lambda-1)\varrho c^2+\Lambda^2(E_m+P_m)-P_m+E_r\right]$$

$$+\frac{1}{r}\frac{\partial}{\partial r}\left[r\Lambda(\Lambda-1)\varrho c^2 u_r+r\Lambda^2(E_m+P_m)u_r+rF_r\right]$$

$$+\frac{1}{r}\frac{\partial}{\partial\theta}\left[\Lambda(\Lambda-1)\varrho c^2 u_\theta+\Lambda^2(E_m+P_m)u_\theta+F_\theta\right]$$

$$+\frac{\partial}{\partial z}\left[\Lambda(\Lambda-1)\varrho c^2 u_z+\Lambda^2(E_m+P_m)u_z+F_z\right]=W. \qquad (9.132)$$

The nonrelativistic limits ($u^2/c^2\to 0$) of Eqs. (9.128) through (9.132) are given by the five equations

$$\frac{\partial\varrho}{\partial t}+\frac{1}{r}\frac{\partial(r\varrho u_r)}{\partial r}+\frac{1}{r}\frac{\partial(\varrho u_\theta)}{\partial\theta}+\frac{\partial(\varrho u_z)}{\partial z}=0, \qquad (9.133)$$

$$\frac{\partial}{\partial t}\left(\varrho u_r+\frac{1}{c^2}F_r\right)+\frac{\partial P_m}{\partial r}+\frac{1}{r}\frac{\partial}{\partial r}(r\varrho u_r u_r)+\frac{1}{r}\frac{\partial}{\partial\theta}(\varrho u_\theta u_r)$$

$$+\frac{\partial}{\partial z}(\varrho u_z u_r)-\frac{1}{r}\varrho u_\theta u_\theta+\frac{1}{r}\frac{\partial(rp_{rr})}{\partial r}+\frac{1}{r}\frac{\partial p_{\theta r}}{\partial\theta}+\frac{\partial p_{zr}}{\partial z}-\frac{1}{r}p_{\theta\theta}=0, \quad (9.134)$$

$$\frac{\partial}{\partial t}\left(\varrho u_\theta+\frac{1}{c^2}F_\theta\right)+\frac{1}{r}\frac{\partial P_m}{\partial\theta}+\frac{1}{r}\frac{\partial}{\partial r}(r\varrho u_r u_\theta)+\frac{1}{r}\frac{\partial}{\partial\theta}(\varrho u_\theta u_\theta)+\frac{\partial}{\partial z}(\varrho u_z u_\theta)$$

$$+\frac{1}{r}\varrho u_\theta u_r+\frac{1}{r}\frac{\partial(rp_{r\theta})}{\partial r}+\frac{1}{r}\frac{\partial p_{\theta\theta}}{\partial\theta}+\frac{\partial p_{z\theta}}{\partial z}+\frac{1}{r}p_{\theta r}=0, \qquad (9.135)$$

$$\frac{\partial}{\partial t}\left(\varrho u_z+\frac{1}{c^2}F_z\right)+\frac{\partial P_m}{\partial z}+\frac{1}{r}\frac{\partial}{\partial r}(r\varrho u_r u_z)+\frac{1}{r}\frac{\partial}{\partial\theta}(\varrho u_\theta u_z)$$

$$+\frac{\partial}{\partial z}(\varrho u_z u_z)+\frac{1}{r}\frac{\partial(rp_{rz})}{\partial r}+\frac{1}{r}\frac{\partial p_{\theta z}}{\partial\theta}+\frac{\partial p_{zz}}{\partial z}=0, \qquad (9.136)$$

$$\frac{\partial}{\partial t}\left(\frac{1}{2}\varrho u^2+E_m+E_r\right)+\frac{1}{r}\frac{\partial}{\partial r}\left[r\left(\frac{1}{2}\varrho u^2+E_m+P_m\right)u_r+rF_r\right]$$

$$+\frac{1}{r}\frac{\partial}{\partial\theta}\left[\left(\frac{1}{2}\varrho u^2+E_m+P_m\right)u_\theta+F_\theta\right]+\frac{\partial}{\partial z}\left[\left(\frac{1}{2}\varrho u^2+E_m+P_m\right)u_z+F_z\right]=W.$$

$$(9.137)$$

Equations (9.133) through (9.137) are the continuity, momentum, and energy equations in a cylindrical coordinate system corresponding to the general nonrelativistic Eulerian hydrodynamic equations, (9.82) through (9.84).

The final coordinate system we shall consider is the standard spherical system with coordinates r, θ, and φ, related to the Cartesian coordinates by the relations

$$x = r\sin\theta\cos\varphi, \qquad (9.138)$$

$$y = r\sin\theta\sin\varphi, \qquad (9.139)$$

$$z = r\cos\theta. \qquad (9.140)$$

The Cartesian components C_x, C_y, and C_z of any vector $\mathbf{C}$ are related to the spherical components C_r, C_θ, and C_φ of the same vector by the relations

$$C_x = C_r \sin\theta\cos\varphi + C_\theta\cos\theta\cos\varphi - C_\varphi\sin\varphi, \tag{9.141}$$

$$C_y = C_r \sin\theta\sin\varphi + C_\theta\cos\theta\sin\varphi + C_\varphi\cos\varphi, \tag{9.142}$$

$$C_z = C_r \cos\theta - C_\theta\sin\theta. \tag{9.143}$$

The inverse relations, giving C_r, C_θ, and C_φ in terms of C_x, C_y, and C_z, are

$$C_r = C_x \sin\theta\cos\varphi + C_y\sin\theta\sin\varphi + C_z\cos\theta, \tag{9.144}$$

$$C_\theta = C_x \cos\theta\cos\varphi + C_y\cos\theta\sin\varphi - C_z\sin\theta, \tag{9.145}$$

$$C_\varphi = -C_x \sin\varphi + C_y\cos\varphi. \tag{9.146}$$

For any scalar A, the r, θ, and φ components of the gradient of A are given by

$$(\nabla A)_r = \frac{\partial A}{\partial r}\;; \quad (\nabla A)_\theta = \frac{1}{r}\frac{\partial A}{\partial\theta}\;; \quad (\nabla A)_\varphi = \frac{1}{r\sin\theta}\frac{\partial A}{\partial\varphi}\,. \tag{9.147}$$

For any vector $\mathbf{B}$, with components B_r, B_θ, and B_φ in a spherical coordinate system, the divergence of $\mathbf{B}$ is given by

$$\nabla\cdot\mathbf{B} = \frac{1}{r^2}\frac{\partial}{\partial r}(r^2 B_r) + \frac{1}{r\sin\theta}\frac{\partial}{\partial\theta}(\sin\theta\, B_\theta) + \frac{1}{r\sin\theta}\frac{\partial B_\varphi}{\partial\varphi}\,. \tag{9.148}$$

To obtain the expression for the divergence of any dyad $\mathbf{BC}$ in spherical coordinates, one can follow the procedure previously used in cylindrical coordinates. Since in the case of spherical coordinates the algebra involved is rather significant, we shall omit it and give only the results. These results are, for the r, θ, and φ components of the divergence of the dyad $\mathbf{BC}$,

$$[\nabla\cdot(\mathbf{BC})]_r = \frac{1}{r^2}\frac{\partial}{\partial r}(r^2 B_r C_r) + \frac{1}{r\sin\theta}\frac{\partial}{\partial\theta}(\sin\theta\, B_\theta C_r)$$

$$+\frac{1}{r\sin\theta}\frac{\partial}{\partial\varphi}(B_\varphi C_r) - \frac{1}{r}(B_\theta C_\theta + B_\varphi C_\varphi), \tag{9.149}$$

$$[\nabla\cdot(\mathbf{BC})]_\theta = \frac{1}{r^2}\frac{\partial}{\partial r}(r^2 B_r C_\theta) + \frac{1}{r\sin\theta}\frac{\partial}{\partial\theta}(\sin\theta\, B_\theta C_\theta)$$

$$+\frac{1}{r\sin\theta}\frac{\partial}{\partial\varphi}(B_\varphi C_\theta) + \frac{1}{r}B_\theta C_r - \frac{1}{r}\cot\theta\, B_\varphi C_\varphi, \tag{9.150}$$

$$[\nabla\cdot(\mathbf{BC})]_\varphi = \frac{1}{r^2}\frac{\partial}{\partial r}(r^2 B_r C_\varphi) + \frac{1}{r\sin\theta}\frac{\partial}{\partial\theta}(\sin\theta\, B_\theta C_\varphi)$$

$$+\frac{1}{r\sin\theta}\frac{\partial}{\partial\varphi}(B_\varphi C_\varphi) + \frac{1}{r}B_\varphi C_r + \frac{1}{r}\cot\theta\, B_\varphi C_\theta. \tag{9.151}$$

In spherical coordinates we denote by u_r, u_θ, and u_φ the three components of the fluid velocity $\mathbf{u}$, and similarly we let F_r, F_θ, and F_φ denote the three spherical components of the radiative flux vector $\mathbf{F}_r$. We have

$$F_i \equiv F_i(r,\theta,\varphi,t) = \int_0^\infty d\nu \int_{4\pi} d\Omega\,\Omega_i I(r,\theta,\varphi,\nu,\mathbf{\Omega},t); \quad i = r,\theta,\varphi. \tag{9.152}$$

229

The Equations of Radiation Hydrodynamics

We let p_{ij} represent the nine spherical components of the radiation pressure tensor, P_r. These components are defined as

$$p_{ij} \equiv p_{ij}(r, \theta, \varphi, t) = \frac{1}{c} \int_0^\infty dv \int_{4\pi} d\Omega\, \Omega_i \Omega_j I(r, \theta, \varphi, v, \mathbf{\Omega}, t); \quad i, j = r, \theta, \varphi. \quad (9.153)$$

In both Eqs. (9.152) and (9.153) Ω_i represents the projection of the vector $\mathbf{\Omega}$ along the ith direction. The vector $\mathbf{\Omega}$ is the unit vector in the direction of photon propagation. Clearly the pressure tensor is a dyad (its tensor character arises from the product of two vectors, namely $\mathbf{\Omega}\mathbf{\Omega}$) and hence Eqs. (9.149) through (9.151) give without further analysis the divergence of the radiation pressure tensor (we identify p_{ij} with $B_i C_j$).

Having given all of the pertinent differential operators in spherical coordinates [namely Eqs. (9.147) through (9.151)], we can easily write the relativistic equations of hydrodynamics, Eqs. (9.74) through (9.76), in a spherical coordinate system. The continuity equation (9.74) takes the form

$$\frac{\partial(\Lambda\varrho)}{\partial t} + \frac{1}{r^2}\frac{\partial}{\partial r}(r^2\Lambda\varrho u_r) + \frac{1}{r\sin\theta}\frac{\partial}{\partial\theta}(\sin\theta\Lambda\varrho u_\theta) + \frac{1}{r\sin\theta}\frac{\partial}{\partial\varphi}(\Lambda\varrho u_\varphi) = 0. \quad (9.154)$$

The three components of the momentum equation (9.75) are written

$$\frac{\partial}{\partial t}\left[\frac{\Lambda^2}{c^2}(\varrho c^2 + E_m + P_m)u_r + \frac{1}{c^2}F_r\right] + \frac{\partial P_m}{\partial r}$$

$$+ \frac{1}{r^2}\frac{\partial}{\partial r}\left[r^2\frac{\Lambda^2}{c^2}(\varrho c^2 + E_m + P_m)u_r u_r\right] + \frac{1}{r\sin\theta}\frac{\partial}{\partial\theta}\left[\sin\theta\frac{\Lambda^2}{c^2}(\varrho c^2 + E_m + P_m)u_\theta u_r\right]$$

$$+ \frac{1}{r\sin\theta}\frac{\partial}{\partial\varphi}\left[\frac{\Lambda^2}{c^2}(\varrho c^2 + E_m + P_m)u_\varphi u_r\right] - \frac{1}{r}\frac{\Lambda^2}{c^2}(\varrho c^2 + E_m + P_m)(u_\theta u_\theta + u_\varphi u_\varphi)$$

$$+ \frac{1}{r^2}\frac{\partial}{\partial r}(r^2 p_{rr}) + \frac{1}{r\sin\theta}\frac{\partial}{\partial\theta}(\sin\theta p_{\theta r}) + \frac{1}{r\sin\theta}\frac{\partial p_{\varphi r}}{\partial\varphi} - \frac{1}{r}(p_{\theta\theta} + p_{\varphi\varphi}) = 0, \quad (9.155)$$

$$\frac{\partial}{\partial t}\left[\frac{\Lambda^2}{c^2}(\varrho c^2 + E_m + P_m)u_\theta + \frac{1}{c^2}F_\theta\right] + \frac{1}{r}\frac{\partial P_m}{\partial\theta}$$

$$+ \frac{1}{r^2}\frac{\partial}{\partial r}\left[r^2\frac{\Lambda^2}{c^2}(\varrho c^2 + E_m + P_m)u_r u_\theta\right] + \frac{1}{r\sin\theta}\frac{\partial}{\partial\theta}\left[\sin\theta\frac{\Lambda^2}{c^2}(\varrho c^2 + E_m + P_m)u_\theta u_\theta\right]$$

$$+ \frac{1}{r\sin\theta}\frac{\partial}{\partial\varphi}\left[\frac{\Lambda^2}{c^2}(\varrho c^2 + E_m + P_m)u_\varphi u_\theta\right] + \frac{1}{r}\frac{\Lambda^2}{c^2}(\varrho c^2 + E_m + P_m)(u_\theta u_r - \cot\theta\, u_\varphi u_\varphi)$$

$$+ \frac{1}{r^2}\frac{\partial}{\partial r}(r^2 p_{r\theta}) + \frac{1}{r\sin\theta}\frac{\partial}{\partial\theta}(\sin\theta p_{\theta\theta}) + \frac{1}{r\sin\theta}\frac{\partial p_{\varphi\theta}}{\partial\varphi} + \frac{1}{r}(p_{\theta r} - \cot\theta\, p_{\varphi\varphi}) = 0, \quad (9.156)$$

$$\frac{\partial}{\partial t}\left[\frac{\Lambda^2}{c^2}(\varrho c^2 + E_m + P_m)u_\varphi + \frac{1}{c^2}F_\varphi\right] + \frac{1}{r\sin\theta}\frac{\partial P_m}{\partial\varphi}$$

$$+ \frac{1}{r^2}\frac{\partial}{\partial r}\left[r^2\frac{\Lambda^2}{c^2}(\varrho c^2 + E_m + P_m)u_r u_\varphi\right] + \frac{1}{r\sin\theta}\frac{\partial}{\partial\theta}\left[\sin\theta\frac{\Lambda^2}{c^2}(\varrho c^2 + E_m + P_m)u_\theta u_\varphi\right]$$

$$+\frac{1}{r\sin\theta}\frac{\partial}{\partial\varphi}\left[\frac{\varLambda^2}{c^2}\left(\varrho c^2+E_m+P_m\right)u_\varphi u_\varphi\right]+\frac{1}{r}\frac{\varLambda^2}{c^2}\left(\varrho c^2+E_m+P_m\right)\left(u_\varphi u_r+\cot\theta u_\varphi u_\theta\right)$$

$$+\frac{1}{r^2}\frac{\partial}{\partial r}\left(r^2 p_{r\varphi}\right)+\frac{1}{r\sin\theta}\frac{\partial}{\partial\theta}\left(\sin\theta p_{\theta\varphi}\right)+\frac{1}{r\sin\theta}\frac{\partial p_{\varphi\varphi}}{\partial\varphi}+\frac{1}{r}\left(p_{\varphi r}+\cot\theta p_{\varphi\theta}\right)=0.\qquad(9.157)$$

The energy equation (9.76) in spherical coordinates is

$$\frac{\partial}{\partial t}\left[\varLambda(\varLambda-1)\varrho c^2+\varLambda^2(E_m+P_m)-P_m+E_r\right]$$

$$+\frac{1}{r^2}\frac{\partial}{\partial r}\left[r^2\varLambda(\varLambda-1)\varrho c^2 u_r+r^2\varLambda^2(E_m+P_m)u_r+r^2 F_r\right]$$

$$+\frac{1}{r\sin\theta}\frac{\partial}{\partial\theta}\left[\sin\theta\varLambda(\varLambda-1)\varrho c^2 u_\theta+\sin\theta\varLambda^2(E_m+P_m)u_\theta+\sin\theta F_\theta\right]$$

$$+\frac{1}{r\sin\theta}\frac{\partial}{\partial\varphi}\left[\varLambda(\varLambda-1)\varrho c^2 u_\varphi+\varLambda^2(E_m+P_m)u_\varphi+F_\varphi\right]=W.\qquad(9.158)$$

By letting c approach infinity in the material terms in Eqs. (9.154) through (9.158) we obtain the nonrelativistic equations of hydrodynamics in a spherical coordinate system. The continuity equation (9.154) becomes

$$\frac{\partial\varrho}{\partial t}+\frac{1}{r^2}\frac{\partial}{\partial r}\left(r^2\varrho u_r\right)+\frac{1}{r\sin\theta}\frac{\partial}{\partial\theta}\left(\sin\theta\varrho u_\theta\right)+\frac{1}{r\sin\theta}\frac{\partial}{\partial\varphi}\left(\varrho u_\varphi\right)=0,\qquad(9.159)$$

the three momentum equations (9.155) through (9.157) become

$$\frac{\partial}{\partial t}\left(\varrho u_r+\frac{1}{c^2}F_r\right)+\frac{\partial P_m}{\partial r}+\frac{1}{r^2}\frac{\partial}{\partial r}\left(r^2\varrho u_r u_r\right)+\frac{1}{r\sin\theta}\frac{\partial}{\partial\theta}\left(\sin\theta\varrho u_\theta u_r\right)$$

$$+\frac{1}{r\sin\theta}\frac{\partial}{\partial\varphi}\left(\varrho u_\varphi u_r\right)-\frac{1}{r}\varrho\left(u_\theta u_\theta+u_\varphi u_\varphi\right)+\frac{1}{r^2}\frac{\partial}{\partial r}\left(r^2 p_{rr}\right)$$

$$+\frac{1}{r\sin\theta}\frac{\partial}{\partial\theta}\left(\sin\theta p_{\theta r}\right)+\frac{1}{r\sin\theta}\frac{\partial p_{\varphi r}}{\partial\varphi}-\frac{1}{r}\left(p_{\theta\theta}+p_{\varphi\varphi}\right)=0,\quad(9.160)$$

$$\frac{\partial}{\partial t}\left(\varrho u_\theta+\frac{1}{c^2}F_\theta\right)+\frac{1}{r}\frac{\partial P_m}{\partial\theta}+\frac{1}{r^2}\frac{\partial}{\partial r}\left(r^2\varrho u_r u_\theta\right)+\frac{1}{r\sin\theta}\frac{\partial}{\partial\theta}\left(\sin\theta\varrho u_\theta u_\theta\right)$$

$$+\frac{1}{r\sin\theta}\frac{\partial}{\partial\varphi}\left(\varrho u_\varphi u_\theta\right)+\frac{1}{r}\varrho\left(u_\theta u_r-\cot\theta u_\varphi u_\varphi\right)+\frac{1}{r^2}\frac{\partial}{\partial r}\left(r^2 p_{r\theta}\right)$$

$$+\frac{1}{r\sin\theta}\frac{\partial}{\partial\theta}\left(\sin\theta p_{\theta\theta}\right)+\frac{1}{r\sin\theta}\frac{\partial p_{\varphi\theta}}{\partial\varphi}+\frac{1}{r}\left(p_{\theta r}-\cot\theta p_{\varphi\varphi}\right)=0,$$

$$(9.161)$$

$$\frac{\partial}{\partial t}\left(\varrho u_\varphi+\frac{1}{c^2}F_\varphi\right)+\frac{1}{r\sin\theta}\frac{\partial P_m}{\partial\varphi}+\frac{1}{r^2}\frac{\partial}{\partial r}\left(r^2\varrho u_r u_\varphi\right)+\frac{1}{r\sin\theta}\frac{\partial}{\partial\theta}\left(\sin\theta\varrho u_\theta u_\varphi\right)$$

$$+\frac{1}{r\sin\theta}\frac{\partial}{\partial\varphi}\left(\varrho u_\varphi u_\varphi\right)+\frac{1}{r}\varrho\left(u_\varphi u_r+\cot\theta u_\varphi u_\theta\right)+\frac{1}{r^2}\frac{\partial}{\partial r}\left(r^2 p_{r\varphi}\right)$$

$$+\frac{1}{r\sin\theta}\frac{\partial}{\partial\theta}\left(\sin\theta p_{\theta\varphi}\right)+\frac{1}{r\sin\theta}\frac{\partial p_{\varphi\varphi}}{\partial\varphi}+\frac{1}{r}\left(p_{\varphi r}+\cot\theta p_{\varphi\theta}\right)=0,\qquad(9.162)$$

The Equations of Radiation Hydrodynamics

and the energy equation (9.158) becomes

$$\frac{\partial}{\partial t}\left(\frac{1}{2}\varrho u^2+E_m+E_r\right)+\frac{1}{r^2}\frac{\partial}{\partial r}\left[r^2\left(\frac{1}{2}\varrho u^2+E_m+E_r\right)u_r+r^2F_r\right]$$
$$+\frac{1}{r\sin\theta}\frac{\partial}{\partial\theta}\left[\sin\theta\left(\frac{1}{2}\varrho u^2+E_m+E_r\right)u_\theta+\sin\theta F_\theta\right]$$
$$+\frac{1}{r\sin\theta}\frac{\partial}{\partial\varphi}\left[\left(\frac{1}{2}\varrho u^2+E_m+E_r\right)u_\varphi+F_\varphi\right]=W. \qquad (9.163)$$

Equations (9.159) through (9.163) are the continuity, momentum, and energy equations in a spherical coordinate system corresponding to the general nonrelativistic Eulerian hydrodynamic equations, (9.82) through (9.84).

5. Modified Eulerian Hydrodynamics

In Section 3 of this chapter the relativistic equations of hydrodynamics, including radiative terms, were written in so-called conservative form for a volume element fixed in space. That is, integration of these equations over an arbitrary volume fixed in space gives results which can be interpreted as particle, momentum, and energy balances for this volume. In this section we show how the equations of hydrodynamics can be written in a form which is similarly conservative for a volume element moving with the fluid.

To obtain the modified Eulerian equations of hydrodynamics, we define the operator $D(\ )/Dt$ as

$$\frac{D(\ )}{Dt}=\frac{\partial(\ )}{\partial t}+\mathbf{u}\cdot\nabla(\ ). \qquad (9.164)$$

We wish to eliminate $\partial(\ )/\partial t$ in the Eulerian equations (9.74) through (9.76) in favor of $D(\ )/Dt$. As we shall show, the resulting equations have the property of being conservative for a volume element moving with the fluid.

To facilitate this elimination of $\partial(\ )/\partial t$ in the Eulerian equations, we first derive an intermediate result. For any quantity Q, the rule for differentiating a product gives the identity

$$\frac{\partial Q}{\partial t}=\Lambda\varrho\,\frac{\partial}{\partial t}(Q/\Lambda\varrho)+\frac{Q}{\Lambda\varrho}\,\frac{\partial}{\partial t}(\Lambda\varrho), \qquad (9.165)$$

or, making use of Eq. (9.164) to rewrite the $\partial(Q/\Lambda\varrho)/\partial t$ term, we find

$$\frac{\partial Q}{\partial t}=\Lambda\varrho\,\frac{D}{Dt}(Q/\Lambda\varrho)-\Lambda\varrho\mathbf{u}\cdot\nabla(Q/\Lambda\varrho)+\frac{Q}{\Lambda\varrho}\,\frac{\partial}{\partial t}(\Lambda\varrho). \qquad (9.166)$$

Use of the continuity equation (9.74) to eliminate $\partial(\Lambda\varrho)/\partial t$ in Eq. (9.166) yields

$$\frac{\partial Q}{\partial t}=\Lambda\varrho\,\frac{D}{Dt}(Q/\Lambda\varrho)-\Lambda\varrho\,\mathbf{u}\cdot\nabla(Q/\Lambda\varrho)-\frac{Q}{\Lambda\varrho}\,\nabla\cdot(\Lambda\varrho\mathbf{u}). \qquad (9.167)$$

232

Finally, we have the identity

$$\nabla \cdot Q\mathbf{u} = \nabla \cdot \left[\left(\frac{Q}{\Delta\varrho}\right)(\Delta\varrho\mathbf{u})\right] = \frac{Q}{\Delta\varrho}\nabla \cdot (\Delta\varrho\mathbf{u}) + \Delta\varrho\mathbf{u}\cdot\nabla\left(\frac{Q}{\Delta\varrho}\right), \qquad (9.168)$$

and hence Eq. (9.167) can be written

$$\frac{\partial Q}{\partial t} = \Delta\varrho\frac{D}{Dt}(Q/\Delta\varrho) - \nabla \cdot (Q\mathbf{u}). \qquad (9.169)$$

Equation (9.169) is an identity for any quantity Q, but only within the framework of hydrodynamics since the continuity equation (9.74) was used in its derivation.

The modified Eulerian equations follow very easily from Eq. (9.169). In particular, setting $Q = 1$ in Eq. (9.169) we find the modified continuity equation

$$\Delta\varrho\frac{D}{Dt}\left(\frac{1}{\Delta\varrho}\right) - \nabla \cdot \mathbf{u} = 0, \qquad (9.170)$$

and use of Eq. (9.169) to eliminate the $\partial(\ )/\partial t$ terms in Eqs. (9.75) and (9.76) gives the modified momentum and energy equations

$$\Delta\varrho\frac{D}{Dt}\left[\frac{\Delta^2(\varrho c^2 + E_m + P_m)\mathbf{u} + \mathbf{F}_r}{\Delta\varrho c^2}\right] + \nabla P_m + \nabla \cdot \left(P_r - \frac{1}{c^2}\mathbf{u}\mathbf{F}_r\right) = 0, \qquad (9.171)$$

$$\Delta\varrho\frac{D}{Dt}\left[\frac{\Delta(\Delta-1)\varrho c^2 + \Delta^2(E_m + P_m) - P_m + E_r}{\Delta\varrho}\right] + \nabla \cdot [\mathbf{F}_r + (P_m - E_r)\mathbf{u}] = W. \qquad (9.172)$$

In Eq. (9.171) the meaning of $\nabla \cdot \mathbf{u}\mathbf{F}_r$ is that, in a Cartesian coordinate system with coordinates x, y, and z, the x component of this quantity is given by

$$(\nabla \cdot \mathbf{u}\mathbf{F}_r)_x = \frac{\partial}{\partial x}(u_x F_x) + \frac{\partial}{\partial y}(u_y F_x) + \frac{\partial}{\partial z}(u_z F_x), \qquad (9.173)$$

where u_x, u_y, and u_z are the Cartesian components of the fluid velocity $\mathbf{u}$, and F_x is the x component of the radiative flux $\mathbf{F}_r$. Similar relations hold for the y and z components of $\nabla \cdot \mathbf{u}\mathbf{F}_r$.

The advantage of introducing $D(\ )/Dt$ in the equations of hydrodynamics is that the resulting equations, (9.170) through (9.172), are in conservative form for a volume element moving with the fluid velocity. To show this, we first derive a commutation rule for the derivative $D(\ )/Dt$. We consider a closed volume V bounded by a surface S whose boundary points $\mathbf{r}_s$ move with the local fluid velocity $\mathbf{u}$, i.e.,

$$\frac{d\mathbf{r}_s}{dt} = \mathbf{u}. \qquad (9.174)$$

Hence the volume V is a function of time. Consider then the integral

$$\int_V d\mathbf{r}\Delta\varrho\frac{DQ}{Dt} = \int_V d\mathbf{r}\Delta\varrho\left(\frac{\partial Q}{\partial t} + \mathbf{u}\cdot\nabla Q\right), \qquad (9.175)$$

The Equations of Radiation Hydrodynamics

where Q is an arbitrary quantity and V is the moving volume, which is time dependent, just defined. Use of the vector identity

$$\Lambda\varrho\mathbf{u}\cdot\nabla Q = \nabla\cdot(\Lambda\varrho\mathbf{u}Q)-Q\nabla\cdot(\Lambda\varrho\mathbf{u}) = \nabla\cdot(\Lambda\varrho\mathbf{u}Q)+Q\frac{\partial}{\partial t}(\Lambda\varrho), \qquad (9.176)$$

where the last equality follows from the use of the continuity equation (9.74), in Eq. (9.175) gives the result

$$\int_V d\mathbf{r}\Lambda\varrho\frac{DQ}{Dt} = \int_V d\mathbf{r}\left[\Lambda\varrho\frac{\partial Q}{\partial t}+Q\frac{\partial}{\partial t}(\Lambda\varrho)+\nabla\cdot(\Lambda\varrho\mathbf{u}Q)\right]. \qquad (9.177)$$

Combining the first two terms on the right hand side of Eq. (9.177), and using Gauss' theorem on the divergence term, we find that Eq. (9.177) can be rewritten

$$\int_V d\mathbf{r}\Lambda\varrho\frac{DQ}{Dt} = \int_V d\mathbf{r}\frac{\partial}{\partial t}(\Lambda\varrho Q)-\int_S d\mathbf{s}(\mathbf{n}\cdot\mathbf{u})\varrho\Lambda Q, \qquad (9.178)$$

where $\mathbf{n}$ is an *inward* normal unit vector at the surface. Since the bounding surface S moves with time, bringing the $\partial(\quad)/\partial t$ operator outside the integral in Eq. (9.178) gives an additional surface contribution, and we find

$$\int_V d\mathbf{r}\Lambda\varrho\frac{DQ}{Dt} = \frac{\partial}{\partial t}\int_V d\mathbf{r}\Lambda\varrho Q+\int_S d\mathbf{s}\left(\mathbf{n}\cdot\frac{d\mathbf{r}_s}{dt}\right)\Lambda\varrho Q-\int_S d\mathbf{s}(\mathbf{n}\cdot\mathbf{u})\Lambda\varrho Q, \qquad (9.179)$$

where $d\mathbf{r}_s/dt$ is the velocity of the surface points of the volume V. Equation (9.174) shows that the two surface integrals in Eq. (9.179) cancel, and we obtain our final result

$$\int_V dt\Lambda\varrho\frac{DQ}{Dt} = \frac{\partial}{\partial t}\int_V d\mathbf{r}\Lambda\varrho Q. \qquad (9.180)$$

Equation (9.180) shows that for a volume element moving with the fluid, the time derivative and the integral over this volume commute in this special sense.

We apply this result to the equations of hydrodynamics, Eqs. (9.170) through (9.172). We integrate these equations over a moving volume V and commute the integral operator and time derivative as in Eq. (9.180). We also use Gauss' theorem to rewrite the divergence term. From Eq. (9.170) we then find

$$\frac{\partial V}{\partial t}+\int_S d\mathbf{s}(\mathbf{n}\cdot\mathbf{u}) = 0, \qquad (9.181)$$

where $\mathbf{n}$ is again an *inward* normal unit vector at the surface S. This result merely states that the change, with time, in the volume encompassing a fixed number of particles is due to the change in the position of the boundary surface points, since these points move with

234

velocity $\mathbf{u}$. Similarly, Eqs. (9.171) and (9.172) yield

$$\frac{\partial}{\partial t}\int_V d\mathbf{r}\,\frac{1}{c^2}\left[\Lambda^2(\varrho c^2+E_m+P_m)\mathbf{u}+\mathbf{F}_r\right] = \int_S ds\left[\mathbf{n}P_m+\mathbf{n}\cdot P_r-\frac{1}{c^2}\,(\mathbf{n}\cdot\mathbf{u})\mathbf{F}_r\right], \qquad (9.182)$$

$$\frac{\partial}{\partial t}\int_V d\mathbf{r}[\Lambda(\Lambda-1)\,\varrho c^2+\Lambda^2(E_m+P_m)-P_m+E_r]$$

$$= \int_S ds[\mathbf{n}\cdot\mathbf{F}_r+(\mathbf{n}\cdot\mathbf{u})\,(P_m-E_r)]+\int_V d\mathbf{r}W. \qquad (9.183)$$

Equations (9.181) through (9.183) represent the integral conservation equations for a volume V moving with the fluid.

The nonrelativistic limit of the hydrodynamic equations discussed in this section is found by letting c, the speed of light, increase without bound in the material terms of the relativistic equations. We have

$$(\Lambda-1)\xrightarrow[c\to\infty]{}\frac{1}{2}\frac{u^2}{c^2}, \qquad (9.184)$$

and thus the nonrelativistic limits of Eqs. (9.170) through (9.172) are given by

$$\varrho\frac{D}{Dt}\left(\frac{1}{\varrho}\right)-\nabla\cdot\mathbf{u} = 0, \qquad (9.185)$$

$$\varrho\frac{D}{Dt}\left(\mathbf{u}+\frac{\mathbf{F}_r}{\varrho c^2}\right)+\nabla P_m+\nabla\cdot\left(P_r-\frac{1}{c^2}\,\mathbf{u}\mathbf{F}_r\right) = 0, \qquad (9.186)$$

$$\varrho\frac{D}{Dt}\left[\frac{u^2}{2}+\frac{1}{\varrho}(E_m+E_r)\right]+\nabla\cdot[\mathbf{F}_r+(P_m-E_r)\mathbf{u}] = W. \qquad (9.187)$$

Aside from the radiation terms, this form of the nonrelativistic hydrodynamic equations is the form found in any number of standard textbooks. The corresponding integral conservation equations are, letting u/c approach infinity in Eqs. (9.181) through (9.183),

$$\frac{\partial V}{\partial t}+\int_S ds(\mathbf{n}\cdot\mathbf{u}) = 0, \qquad (9.188)$$

$$\frac{\partial}{\partial t}\int_V d\mathbf{r}\left(\varrho\mathbf{u}+\frac{1}{c^2}\mathbf{F}_r\right) = \int_S ds\left[\mathbf{n}P_m+\mathbf{n}\cdot P_r-\frac{1}{c^2}\,(\mathbf{n}\cdot\mathbf{u})\mathbf{F}_r\right], \qquad (9.189)$$

$$\frac{\partial}{\partial t}\int_V d\mathbf{r}\left[\frac{1}{2}\,\varrho u^2+E_m+E_r\right] = \int_S ds[\mathbf{n}\cdot\mathbf{F}_r+(\mathbf{n}\cdot\mathbf{u})(P_m-E_r)]+\int_V d\mathbf{r}W. \qquad (9.190)$$

As before, the vector $\mathbf{n}$ in Eqs. (9.188) through (9.190) is a unit *inward* normal vector at the surface of the volume under consideration.

Before leaving this section, we derive another form of the energy equation, in the nonrelativistic limit, which is often seen in hydrodynamic textbooks. Referring to Eq. (9.187), we see that the operator $D(\ )/Dt$ operates on the total energy, which is the sum of the kinetic energy $(u^2/2)$, the material or internal energy (E_m/ϱ), and the radiation energy (E_r/ϱ), all on a per unit mass basis. One can eliminate the kinetic energy in this term as follows. We take the dot product of the momentum equation (9.186) with the velocity $\mathbf{u}$. This gives

$$\varrho \frac{D}{Dt}\left(\frac{1}{2}u^2\right)+\varrho\mathbf{u}\cdot\frac{D}{Dt}\left(\frac{\mathbf{F}_r}{\varrho c^2}\right)+\mathbf{u}\cdot\nabla P_m+\mathbf{u}\cdot\left[\nabla\cdot\left(P_r-\frac{1}{c^2}\mathbf{u}\mathbf{F}_r\right)\right]=0. \qquad (9.191)$$

Subtracting Eq. (9.191) from the energy equation (9.187) has the effect of deleting the kinetic energy from the energy equation. We find

$$\varrho\frac{D}{Dt}\left[\frac{1}{\varrho}(E_m+E_r)\right]-\varrho\mathbf{u}\cdot\frac{D}{Dt}\left(\frac{\mathbf{F}_r}{\varrho c^2}\right)+\nabla\cdot(\mathbf{F}_r-E_r\mathbf{u})+P_m\nabla\cdot\mathbf{u}$$

$$-\mathbf{u}\cdot\left[\nabla\cdot\left(P_r-\frac{1}{c^2}\mathbf{u}\mathbf{F}_r\right)\right]=W. \qquad (9.192)$$

Aside from the radiation terms, Eq. (9.192) is probably the most common form of the energy equation for an ideal fluid found in textbooks.

6. The Modified Eulerian Equations in Various Coordinate Systems

To write the modified Eulerian equations, (9.170) through (9.172), in the three standard coordinate systems (Cartesian, cylindrical, and spherical), we require expressions for the gradient of a scalar, the divergence of a vector, the components of the divergence of a dyad, and the time derivative $D(\ )/Dt$ of a scalar and a vector in these coordinate systems. Aside from the $D(\ )/Dt$ terms, all of these expressions have been given in Section 4 of this chapter. Recalling that $D(\ )/Dt$ is defined as [see Eq. (9.164)]

$$D(\ )/Dt = \frac{\partial(\ )}{\partial t}+\mathbf{u}\cdot\nabla(\ ), \qquad (9.193)$$

we see that what is additionally needed are the expressions for $\mathbf{u}\cdot\nabla(\ )$ in the three standard coordinate systems. For any scalar A, $\mathbf{u}\cdot\nabla A$ is just the vector $\mathbf{u}$ dotted with the gradient of A. For a vector $\mathbf{B}$, the components of $(\mathbf{u}\cdot\nabla)\mathbf{B}$ can be computed from the identity

$$(\mathbf{u}\cdot\nabla)\mathbf{B} = \nabla\cdot(\mathbf{u}\mathbf{B})-(\nabla\cdot\mathbf{u})\mathbf{B}, \qquad (9.194)$$

making use of the expressions for the divergence of a dyad and the divergence of a vector for the coordinate systems of interest given in Section 4.

Thus in a Cartesian system with coordinates x, y, and z, we have for any scalar A

$$\frac{DA}{Dt} = \frac{\partial A}{\partial t}+\mathbf{u}\cdot\nabla A = \frac{\partial A}{\partial t}+u_x\frac{\partial A}{\partial x}+u_y\frac{\partial A}{\partial y}+u_z\frac{\partial A}{\partial z}, \qquad (9.195)$$

where u_x, u_y, and u_z are the Cartesian components of the fluid velocity $\mathbf{u}$. For any vector $\mathbf{B}$

with Cartesian components B_x, B_y, and B_z, the x, y, and z components of DB/Dt are given by

$$\left(\frac{DB}{Dt}\right)_x = \left[\frac{\partial \mathbf{B}}{\partial t}+(\mathbf{u}\cdot\nabla)\mathbf{B}\right]_x = \frac{\partial B_x}{\partial t}+u_x\frac{\partial B_x}{\partial x}+u_y\frac{\partial B_x}{\partial y}+u_z\frac{\partial B_x}{\partial z}, \qquad (9.196)$$

$$\left(\frac{DB}{Dt}\right)_y = \left[\frac{\partial \mathbf{B}}{\partial t}+(\mathbf{u}\cdot\nabla)\mathbf{B}\right]_y = \frac{\partial B_y}{\partial t}+u_x\frac{\partial B_y}{\partial x}+u_y\frac{\partial B_y}{\partial y}+u_z\frac{\partial B_y}{\partial z}, \qquad (9.197)$$

$$\left(\frac{DB}{Dt}\right)_z = \left[\frac{\partial \mathbf{B}}{\partial t}+(\mathbf{u}\cdot\nabla)\mathbf{B}\right]_z = \frac{\partial B_z}{\partial t}+u_x\frac{\partial B_z}{\partial x}+u_y\frac{\partial B_z}{\partial y}+u_z\frac{\partial B_z}{\partial z}. \qquad (9.198)$$

In a cylindrical system with coordinates r, θ, and z, we have for any scalar A

$$\frac{DA}{Dt} = \frac{\partial A}{\partial t}+\mathbf{u}\cdot\nabla A = \frac{\partial A}{\partial t}+u_r\frac{\partial A}{\partial r}+\frac{1}{r}u_\theta\frac{\partial A}{\partial \theta}+u_z\frac{\partial A}{\partial z}, \qquad (9.199)$$

where u_r, u_θ, and u_z are the cylindrical components of the fluid velocity $\mathbf{u}$. For any vector $\mathbf{B}$ with cylindrical components B_r, B_θ, and B_z, the r, θ, and z components of DB/Dt are given by

$$\left(\frac{DB}{Dt}\right)_r = \left[\frac{\partial \mathbf{B}}{\partial t}+(\mathbf{u}\cdot\nabla)\mathbf{B}\right]_r = \frac{\partial B_r}{\partial t}+u_r\frac{\partial B_r}{\partial r}+\frac{1}{r}u_\theta\frac{\partial B_r}{\partial \theta}+u_z\frac{\partial B_r}{\partial z}-\frac{1}{r}u_\theta B_\theta, \qquad (9.200)$$

$$\left(\frac{DB}{Dt}\right)_\theta = \left[\frac{\partial \mathbf{B}}{\partial t}+(\mathbf{u}\cdot\nabla)\mathbf{B}\right]_\theta = \frac{\partial B_\theta}{\partial t}+u_r\frac{\partial B_\theta}{\partial r}+\frac{1}{r}u_\theta\frac{\partial B_\theta}{\partial \theta}+u_z\frac{\partial B_\theta}{\partial z}+\frac{1}{r}u_\theta B_r, \qquad (9.201)$$

$$\left(\frac{DB}{Dt}\right)_z = \left[\frac{\partial \mathbf{B}}{\partial t}+(\mathbf{u}\cdot\nabla)\mathbf{B}\right]_z = \frac{\partial B_z}{\partial t}+u_r\frac{\partial B_z}{\partial r}+\frac{1}{r}u_\theta\frac{\partial B_z}{\partial \theta}+u_z\frac{\partial B_z}{\partial z}. \qquad (9.202)$$

In a spherical system with coordinates r, θ, and φ, we have for any scalar A

$$\frac{DA}{Dt} = \frac{\partial A}{\partial t}+\mathbf{u}\cdot\nabla A = \frac{\partial A}{\partial t}+u_r\frac{\partial A}{\partial r}+\frac{1}{r}u_\theta\frac{\partial A}{\partial \theta}+\frac{1}{r\sin\theta}u_\varphi\frac{\partial A}{\partial \varphi}, \qquad (9.203)$$

where u_r, u_θ, and u_φ are the spherical components of the fluid velocity $\mathbf{u}$. For any vector $\mathbf{B}$ with spherical components B_r, B_θ, and B_φ, the r, θ, and φ components of DB/Dt are given by

$$\left(\frac{DB}{Dt}\right)_r = \left[\frac{\partial \mathbf{B}}{\partial t}+(\mathbf{u}\cdot\nabla)\mathbf{B}\right]_r = \frac{\partial B_r}{\partial t}+u_r\frac{\partial B_r}{\partial r}+\frac{1}{r}u_\theta\frac{\partial B_r}{\partial \theta}$$
$$+\frac{1}{r\sin\theta}u_\varphi\frac{\partial B_r}{\partial \varphi}-\frac{1}{r}(u_\theta B_\theta+u_\varphi B_\varphi), \qquad (9.204)$$

$$\left(\frac{DB}{Dt}\right)_\theta = \left[\frac{\partial \mathbf{B}}{\partial t}+(\mathbf{u}\cdot\nabla)\mathbf{B}\right]_\theta = \frac{\partial B_\theta}{\partial t}+u_r\frac{\partial B_\theta}{\partial r}+\frac{1}{r}u_\theta\frac{\partial B_\theta}{\partial \theta}$$
$$+\frac{1}{r\sin\theta}u_\varphi\frac{\partial B_\theta}{\partial \varphi}+\frac{1}{r}(u_\theta B_r-\cot\theta\, u_\varphi B_\varphi), \qquad (9.205)$$

$$\left(\frac{DB}{Dt}\right)_\varphi = \left[\frac{\partial \mathbf{B}}{\partial t}+(\mathbf{u}\cdot\nabla)\mathbf{B}\right]_\varphi = \frac{\partial B_\varphi}{\partial t}+u_r\frac{\partial B_\varphi}{\partial r}+\frac{1}{r}u_\theta\frac{\partial B_\varphi}{\partial \theta}$$
$$+\frac{1}{r\sin\theta}u_\varphi\frac{\partial B_\varphi}{\partial \varphi}+\frac{1}{r}(u_\varphi B_r+\cot\theta\, u_\varphi B_\theta). \qquad (9.206)$$

The Equations of Radiation Hydrodynamics

With the results of this section and those of Section 4 for the differential operations in Cartesian, cylindrical, and spherical coordinate systems, one can immediately write the modified Eulerian equations of hydrodynamics, either in relativistic form, Eqs. (9.170) through (9.172), or in the nonrelativistic limit, Eqs. (9.185) through (9.187), in these coordinate systems.

In Cartesian coordinates we denote the three components of the radiative flux $\mathbf{F}_r$ by F_i ($i = x, y, z$) and the nine components of the radiative pressure tensor $\mathbf{P}_r$ by p_{ij} ($i, j = x, y, z$) [see Eqs. (1.17) and (1.20)]. The relativistic continuity equation (9.170) in this coordinate system is written

$$\Lambda\varrho\left[\frac{\partial}{\partial t}\left(\frac{1}{\Lambda\varrho}\right) + u_x\frac{\partial}{\partial x}\left(\frac{1}{\Lambda\varrho}\right) + u_y\frac{\partial}{\partial y}\left(\frac{1}{\Lambda\varrho}\right) + u_z\frac{\partial}{\partial z}\left(\frac{1}{\Lambda\varrho}\right)\right]$$

$$-\left[\frac{\partial u_x}{\partial x} + \frac{\partial u_y}{\partial y} + \frac{\partial u_z}{\partial z}\right] = 0. \tag{9.207}$$

The three components of the relativistic momentum equation (9.171) are written

$$\Lambda\varrho\frac{\partial}{\partial t}\left[\frac{\Lambda^2(\varrho c^2+E_m+P_m)u_x+F_x}{\Lambda\varrho c^2}\right] + \Lambda\varrho u_x\frac{\partial}{\partial x}\left[\frac{\Lambda^2(\varrho c^2+E_m+P_m)u_x+F_x}{\Lambda\varrho c^2}\right]$$

$$+ \Lambda\varrho u_y\frac{\partial}{\partial y}\left[\frac{\Lambda^2(\varrho c^2+E_m+P_m)u_x+F_x}{\Lambda\varrho c^2}\right] + \Lambda\varrho u_z\frac{\partial}{\partial z}\left[\frac{\Lambda^2(\varrho c^2+E_m+P_m)u_x+F_x}{\Lambda\varrho c^2}\right]$$

$$+ \frac{\partial P_m}{\partial x} + \frac{\partial}{\partial x}\left(p_{xx} - \frac{1}{c^2}u_xF_x\right) + \frac{\partial}{\partial y}\left(p_{yx} - \frac{1}{c^2}u_yF_x\right)$$

$$+ \frac{\partial}{\partial z}\left(p_{zx} - \frac{1}{c^2}u_zF_x\right) = 0, \tag{9.208}$$

$$\Lambda\varrho\frac{\partial}{\partial t}\left[\frac{\Lambda^2(\varrho c^2+E_m+P_m)u_y+F_y}{\Lambda\varrho c^2}\right] + \Lambda\varrho u_x\frac{\partial}{\partial x}\left[\frac{\Lambda^2(\varrho c^2+E_m+P_m)u_y+F_y}{\Lambda\varrho c^2}\right]$$

$$+ \Lambda\varrho u_y\frac{\partial}{\partial y}\left[\frac{\Lambda^2(\varrho c^2+E_m+P_m)u_y+F_y}{\Lambda\varrho c^2}\right] + \Lambda\varrho u_z\frac{\partial}{\partial z}\left[\frac{\Lambda^2(\varrho c^2+E_m+P_m)u_y+F_y}{\Lambda\varrho c^2}\right]$$

$$+ \frac{\partial P_m}{\partial y} + \frac{\partial}{\partial x}\left(p_{xy} - \frac{1}{c^2}u_xF_y\right) + \frac{\partial}{\partial y}\left(p_{yy} - \frac{1}{c^2}u_yF_y\right) + \frac{\partial}{\partial z}\left(p_{zy} - \frac{1}{c^2}u_zF_y\right) = 0, \tag{9.209}$$

$$\Lambda\varrho\frac{\partial}{\partial t}\left[\frac{\Lambda^2(\varrho c^2+E_m+P_m)u_z+F_z}{\Lambda\varrho c^2}\right] + \Lambda\varrho u_x\frac{\partial}{\partial x}\left[\frac{\Lambda^2(\varrho c^2+E_m+P_m)u_z+F_z}{\Lambda\varrho c^2}\right]$$

$$+ \Lambda\varrho u_y\frac{\partial}{\partial y}\left[\frac{\Lambda^2(\varrho c^2+E_m+P_m)u_z+F_z}{\Lambda\varrho c^2}\right] + \Lambda\varrho u_z\frac{\partial}{\partial z}\left[\frac{\Lambda^2(\varrho c^2+E_m+P_m)u_z+F_z}{\Lambda\varrho c^2}\right]$$

$$+ \frac{\partial P_m}{\partial z} + \frac{\partial}{\partial x}\left(p_{xz} - \frac{1}{c^2}u_xF_z\right) + \frac{\partial}{\partial y}\left(p_{yz} - \frac{1}{c^2}u_yF_z\right) + \frac{\partial}{\partial z}\left(p_{zz} - \frac{1}{c^2}u_zF_z\right) = 0. \tag{9.210}$$

The relativistic energy equation (9.172) is written

$$
\Delta\varrho\,\frac{\partial}{\partial t}\left[\frac{\Delta(\Delta-1)\varrho c^2+\Delta^2(E_m+P_m)-P_m+E_r}{\Delta\varrho}\right]
$$

$$
+\Delta\varrho u_x\,\frac{\partial}{\partial x}\left[\frac{\Delta(\Delta-1)\varrho c^2+\Delta^2(E_m+P_m)-P_m+E_r}{\Delta\varrho}\right]
$$

$$
+\Delta\varrho u_y\,\frac{\partial}{\partial y}\left[\frac{\Delta(\Delta-1)\varrho c^2+\Delta^2(E_m+P_m)-P_m+E_r}{\Delta\varrho}\right]
$$

$$
+\Delta\varrho u_z\,\frac{\partial}{\partial z}\left[\frac{\Delta(\Delta-1)\varrho c^2+\Delta^2(E_m+P_m)-P_m+E_r}{\Delta\varrho}\right]
$$

$$
+\frac{\partial}{\partial x}[F_x+(P_m-E_r)\,u_x]+\frac{\partial}{\partial y}[F_y+(P_m-E_r)\,u_y]
$$

$$
+\frac{\partial}{\partial z}[F_z+(P_m-E_r)\,u_z] = W. \tag{9.211}
$$

The nonrelativistic continuity equation (9.185) in a Cartesian coordinate system is written

$$
\varrho\left[\frac{\partial}{\partial t}\left(\frac{1}{\varrho}\right)+u_x\frac{\partial}{\partial x}\left(\frac{1}{\varrho}\right)+u_y\frac{\partial}{\partial y}\left(\frac{1}{\varrho}\right)+u_z\frac{\partial}{\partial z}\left(\frac{1}{\varrho}\right)\right]-\left[\frac{\partial u_x}{\partial x}+\frac{\partial u_y}{\partial y}+\frac{\partial u_z}{\partial z}\right]=0. \tag{9.212}
$$

The three components of the nonrelativistic momentum equation (9.186) are written

$$
\varrho\frac{\partial}{\partial t}\left(u_x+\frac{F_x}{\varrho c^2}\right)+\varrho u_x\frac{\partial}{\partial x}\left(u_x+\frac{F_x}{\varrho c^2}\right)+\varrho u_y\frac{\partial}{\partial y}\left(u_x+\frac{F_x}{\varrho c^2}\right)
$$

$$
+\varrho u_z\frac{\partial}{\partial z}\left(u_x+\frac{F_x}{\varrho c^2}\right)+\frac{\partial P_m}{\partial x}+\frac{\partial}{\partial x}\left(p_{xx}-\frac{1}{c^2}u_xF_x\right)
$$

$$
+\frac{\partial}{\partial y}\left(p_{yx}-\frac{1}{c^2}u_yF_x\right)+\frac{\partial}{\partial z}\left(p_{zx}-\frac{1}{c^2}u_zF_x\right)=0, \tag{9.213}
$$

$$
\varrho\frac{\partial}{\partial t}\left(u_y+\frac{F_y}{\varrho c^2}\right)+\varrho u_x\frac{\partial}{\partial x}\left(u_y+\frac{F_y}{\varrho c^2}\right)+\varrho u_y\frac{\partial}{\partial y}\left(u_y+\frac{F_y}{\varrho c^2}\right)
$$

$$
+\varrho u_z\frac{\partial}{\partial z}\left(u_y+\frac{F_y}{\varrho c^2}\right)+\frac{\partial P_m}{\partial y}+\frac{\partial}{\partial x}\left(p_{xy}-\frac{1}{c^2}u_xF_y\right)
$$

$$
+\frac{\partial}{\partial y}\left(p_{yy}-\frac{1}{c^2}u_yF_y\right)+\frac{\partial}{\partial z}\left(p_{zy}-\frac{1}{c^2}u_zF_y\right)=0, \tag{9.214}
$$

$$
\varrho\frac{\partial}{\partial t}\left(u_z+\frac{F_z}{\varrho c^2}\right)+\varrho u_x\frac{\partial}{\partial x}\left(u_z+\frac{F_z}{\varrho c^2}\right)+\varrho u_y\frac{\partial}{\partial y}\left(u_z+\frac{F_z}{\varrho c^2}\right)
$$

$$
+\varrho u_z\frac{\partial}{\partial z}\left(u_z+\frac{F_z}{\varrho c^2}\right)+\frac{\partial P_m}{\partial z}+\frac{\partial}{\partial x}\left(p_{xz}-\frac{1}{c^2}u_xF_z\right)
$$

$$
+\frac{\partial}{\partial y}\left(p_{yz}-\frac{1}{c^2}u_yF_z\right)+\frac{\partial}{\partial z}\left(p_{zz}-\frac{1}{c^2}u_zF_z\right)=0. \tag{9.215}
$$

The Equations of Radiation Hydrodynamics

The nonrelativistic energy equation (9.187) is written

$$\varrho\frac{\partial}{\partial t}\left[\frac{u^2}{2}+\frac{1}{\varrho}(E_m+E_r)\right]+\varrho u_x\frac{\partial}{\partial x}\left[\frac{u^2}{2}+\frac{1}{\varrho}(E_m+E_r)\right]$$

$$+\varrho u_y\frac{\partial}{\partial y}\left[\frac{u^2}{2}+\frac{1}{\varrho}(E_m+E_r)\right]+\varrho u_z\frac{\partial}{\partial z}\left[\frac{u^2}{2}+\frac{1}{\varrho}(E_m+E_r)\right]$$

$$+\frac{\partial}{\partial x}[F_x+(P_m-E_r)u_x]+\frac{\partial}{\partial y}[F_y+(P_m-E_r)u_y]$$

$$+\frac{\partial}{\partial z}[F_z+(P_m-E_r)u_z]=W. \tag{9.216}$$

In cylindrical coordinates, we denote the three components of the radiative flux $\mathbf{F}$, by F_i ($i=r,\theta,z$) and the nine components of the radiative pressure tensor $\mathbf{P}$, by p_{ij} ($i,j=r,\theta,z$) [see Eqs. (9.126) and (9.127)]. The relativistic continuity equation (9.170) in this coordinate system is written

$$\Lambda\varrho\left[\frac{\partial}{\partial t}\left(\frac{1}{\Lambda\varrho}\right)+u_r\frac{\partial}{\partial r}\left(\frac{1}{\Lambda\varrho}\right)+\frac{1}{r}u_\theta\frac{\partial}{\partial\theta}\left(\frac{1}{\Lambda\varrho}\right)+u_z\frac{\partial}{\partial z}\left(\frac{1}{\Lambda\varrho}\right)\right]$$

$$-\left[\frac{1}{r}\frac{\partial}{\partial r}(ru_r)+\frac{1}{r}\frac{\partial u_\theta}{\partial\theta}+\frac{\partial u_z}{\partial z}\right]=0. \tag{9.217}$$

The three components of the relativistic momentum equation (9.171) are written

$$\Lambda\varrho\frac{\partial}{\partial t}\left[\frac{\Lambda^2(\varrho c^2+E_m+P_m)u_r+F_r}{\Lambda\varrho c^2}\right]+\Lambda\varrho u_r\frac{\partial}{\partial r}\left[\frac{\Lambda^2(\varrho c^2+E_m+P_m)u_r+F_r}{\Lambda\varrho c^2}\right]$$

$$+\Lambda\varrho\frac{u_\theta}{r}\frac{\partial}{\partial\theta}\left[\frac{\Lambda^2(\varrho c^2+E_m+P_m)u_r+F_r}{\Lambda\varrho c^2}\right]+\Lambda\varrho u_z\frac{\partial}{\partial z}\left[\frac{\Lambda^2(\varrho c^2+E_m+P_m)u_r+F_r}{\Lambda\varrho c^2}\right]$$

$$-\frac{u_\theta}{r}\left[\frac{\Lambda^2(\varrho c^2+E_m+P_m)u_\theta+F_\theta}{c^2}\right]+\frac{\partial P_m}{\partial r}+\frac{1}{r}\frac{\partial}{\partial r}\left[r\left(p_{rr}-\frac{1}{c^2}u_rF_r\right)\right]$$

$$+\frac{1}{r}\frac{\partial}{\partial\theta}\left(p_{\theta r}-\frac{1}{c^2}u_\theta F_r\right)+\frac{\partial}{\partial z}\left(p_{zr}-\frac{1}{c^2}u_zF_r\right)-\frac{1}{r}\left(p_{\theta\theta}-\frac{1}{c^2}u_\theta F_\theta\right)=0, \tag{9.218}$$

$$\Lambda\varrho\frac{\partial}{\partial t}\left[\frac{\Lambda^2(\varrho c^2+E_m+P_m)u_\theta+F_\theta}{\Lambda\varrho c^2}\right]+\Lambda\varrho u_r\frac{\partial}{\partial r}\left[\frac{\Lambda^2(\varrho c^2+E_m+P_m)u_\theta+F_\theta}{\Lambda\varrho c^2}\right]$$

$$+\Lambda\varrho\frac{u_\theta}{r}\frac{\partial}{\partial\theta}\left[\frac{\Lambda^2(\varrho c^2+E_m+P_m)u_\theta+F_\theta}{\Lambda\varrho c^2}\right]+\Lambda\varrho u_z\frac{\partial}{\partial z}\left[\frac{\Lambda^2(\varrho c^2+E_m+P_m)u_\theta+F_\theta}{\Lambda\varrho c^2}\right]$$

$$+\frac{u_\theta}{r}\left[\frac{\Lambda^2(\varrho c^2+E_m+P_m)u_r+F_r}{c^2}\right]+\frac{1}{r}\frac{\partial P_m}{\partial\theta}+\frac{1}{r}\frac{\partial}{\partial r}\left[r\left(p_{r\theta}-\frac{1}{c^2}u_rF_\theta\right)\right]$$

$$+\frac{1}{r}\frac{\partial}{\partial\theta}\left(p_{\theta\theta}-\frac{1}{c^2}u_\theta F_\theta\right)+\frac{\partial}{\partial z}\left(p_{z\theta}-\frac{1}{c^2}u_zF_\theta\right)+\frac{1}{r}\left(p_{\theta r}-\frac{1}{c^2}u_\theta F_r\right)=0, \tag{9.219}$$

$$\Lambda\varrho\,\frac{\partial}{\partial t}\left[\frac{\Lambda^2(\varrho c^2+E_m+P_m)\,u_z+F_z}{\Lambda\varrho c^2}\right]+\Lambda\varrho u_r\,\frac{\partial}{\partial r}\left[\frac{\Lambda^2(\varrho c^2+E_m+P_m)\,u_z+F_z}{\Lambda\varrho c^2}\right]$$

$$+\Lambda\varrho\,\frac{u_\theta}{r}\,\frac{\partial}{\partial\theta}\left[\frac{\Lambda^2(\varrho c^2+E_m+P_m)\,u_z+F_z}{\Lambda\varrho c^2}\right]+\Lambda\varrho u_z\,\frac{\partial}{\partial z}\left[\frac{\Lambda^2(\varrho c^2+E_m+P_m)\,u_z+F_z}{\Lambda\varrho c^2}\right]$$

$$+\frac{\partial P_m}{\partial z}+\frac{1}{r}\,\frac{\partial}{\partial r}\left[r\Big(p_{rz}-\frac{1}{c^2}\,u_r F_z\Big)\right]+\frac{1}{r}\,\frac{\partial}{\partial\theta}\left(p_{\theta z}-\frac{1}{c^2}\,u_\theta F_z\right)+\frac{\partial}{\partial z}\left(p_{zz}-\frac{1}{c^2}\,u_z F_z\right)=0.$$

$$(9.220)$$

The relativistic energy equation (9.172) is written

$$\Lambda\varrho\,\frac{\partial}{\partial t}\left[\frac{\Lambda(\Lambda-1)\,\varrho c^2+\Lambda^2(E_m+P_m)-P_m+E_r}{\Lambda\varrho}\right]$$

$$+\Lambda\varrho u_r\,\frac{\partial}{\partial r}\left[\frac{\Lambda(\Lambda-1)\,\varrho c^2+\Lambda^2(E_m+P_m)-P_m+E_r}{\Lambda\varrho}\right]$$

$$+\Lambda\varrho\,\frac{u_\theta}{r}\,\frac{\partial}{\partial\theta}\left[\frac{\Lambda(\Lambda-1)\,\varrho c^2+\Lambda^2(E_m+P_m)-P_m+E_r}{\Lambda\varrho}\right]$$

$$+\Lambda\varrho u_z\,\frac{\partial}{\partial z}\left[\frac{\Lambda(\Lambda-1)\,\varrho c^2+\Lambda^2(E_m+P_m)-P_m+E_r}{\Lambda\varrho}\right]$$

$$+\frac{1}{r}\,\frac{\partial}{\partial r}\left[rF_r+r(P_m-E_r)\,u_r\right]+\frac{1}{r}\,\frac{\partial}{\partial\theta}\left[F_\theta+(P_m-E_r)\,u_\theta\right]$$

$$+\frac{\partial}{\partial z}\left[F_z+(P_m-E_r)\,u_z\right]=W.\qquad(9.221)$$

The nonrelativistic continuity equation (9.185) in a cylindrical coordinate system is written

$$\varrho\left[\frac{\partial}{\partial t}\left(\frac{1}{\varrho}\right)+u_r\,\frac{\partial}{\partial r}\left(\frac{1}{\varrho}\right)+\frac{u_\theta}{r}\,\frac{\partial}{\partial\theta}\left(\frac{1}{\varrho}\right)+u_z\,\frac{\partial}{\partial z}\left(\frac{1}{\varrho}\right)\right]$$

$$-\left[\frac{1}{r}\,\frac{\partial}{\partial r}\,(ru_r)+\frac{1}{r}\,\frac{\partial u_\theta}{\partial\theta}+\frac{\partial u_z}{\partial z}\right]=0.\qquad(9.222)$$

The three components of the nonrelativistic momentum equation (9.186) are written

$$\varrho\,\frac{\partial}{\partial t}\left(u_r+\frac{F_r}{\varrho c^2}\right)+\varrho u_r\,\frac{\partial}{\partial r}\left(u_r+\frac{F_r}{\varrho c^2}\right)+\varrho\,\frac{u_\theta}{r}\,\frac{\partial}{\partial\theta}\left(u_r+\frac{F_r}{\varrho c^2}\right)$$

$$+\varrho u_z\,\frac{\partial}{\partial z}\left(u_r+\frac{F_r}{\varrho c^2}\right)-\varrho\,\frac{u_\theta}{r}\left(u_\theta+\frac{F_\theta}{\varrho c^2}\right)+\frac{\partial P_m}{\partial r}$$

$$+\frac{1}{r}\,\frac{\partial}{\partial r}\left[r\left(p_{rr}-\frac{1}{c^2}\,u_r F_r\right)\right]+\frac{1}{r}\,\frac{\partial}{\partial\theta}\left(p_{\theta r}-\frac{1}{c^2}\,u_\theta F_r\right)$$

$$+\frac{\partial}{\partial z}\left(p_{zr}-\frac{1}{c^2}\,u_z F_r\right)-\frac{1}{r}\left(p_{\theta\theta}-\frac{1}{c^2}\,u_\theta F_\theta\right)=0,\qquad(9.223)$$

The Equations of Radiation Hydrodynamics

$$\varrho\,\frac{\partial}{\partial t}\left(u_\theta+\frac{F_\theta}{\varrho c^2}\right)+\varrho u_r\,\frac{\partial}{\partial r}\left(u_\theta+\frac{F_\theta}{\varrho c^2}\right)+\varrho\,\frac{u_\theta}{r}\,\frac{\partial}{\partial\theta}\left(u_\theta+\frac{F_\theta}{\varrho c^2}\right)$$

$$+\varrho u_z\,\frac{\partial}{\partial z}\left(u_\theta+\frac{F_\theta}{\varrho c^2}\right)+\varrho\,\frac{u_\theta}{r}\left(u_r+\frac{F_r}{\varrho c^2}\right)+\frac{1}{r}\,\frac{\partial P_m}{\partial\theta}$$

$$+\frac{1}{r}\,\frac{\partial}{\partial r}\left[r\left(p_{r\theta}-\frac{1}{c^2}\,u_rF_\theta\right)\right]+\frac{1}{r}\,\frac{\partial}{\partial\theta}\left(p_{\theta\theta}-\frac{1}{c^2}\,u_\theta F_\theta\right)$$

$$+\frac{\partial}{\partial z}\left(p_{z\theta}-\frac{1}{c^2}\,u_z F_\theta\right)+\frac{1}{r}\left(p_{\theta r}-\frac{1}{c^2}\,u_\theta F_r\right)=0, \tag{9.224}$$

$$\varrho\,\frac{\partial}{\partial t}\left(u_z+\frac{F_z}{\varrho c^2}\right)+\varrho u_r\,\frac{\partial}{\partial r}\left(u_z+\frac{F_z}{\varrho c^2}\right)+\varrho\,\frac{u_\theta}{r}\,\frac{\partial}{\partial\theta}\left(u_z+\frac{F_z}{\varrho c^2}\right)$$

$$+\varrho u_z\,\frac{\partial}{\partial z}\left(u_z+\frac{F_z}{\varrho c^2}\right)+\frac{\partial P_m}{\partial z}+\frac{1}{r}\,\frac{\partial}{\partial r}\left[r\left(p_{rz}-\frac{1}{c^2}\,u_rF_z\right)\right]$$

$$+\frac{1}{r}\,\frac{\partial}{\partial\theta}\left(p_{\theta z}-\frac{1}{c^2}\,u_\theta F_z\right)+\frac{\partial}{\partial z}\left(p_{zz}-\frac{1}{c^2}\,u_z F_z\right)=0. \tag{9.225}$$

The nonrelativistic energy equation (9.187) is written

$$\varrho\,\frac{\partial}{\partial t}\left[\frac{u^2}{2}+\frac{1}{\varrho}\,(E_m+E_r)\right]+\varrho u_r\,\frac{\partial}{\partial r}\left[\frac{u^2}{2}+\frac{1}{\varrho}\,(E_m+E_r)\right]$$

$$+\varrho\,\frac{u_\theta}{r}\,\frac{\partial}{\partial\theta}\left[\frac{u^2}{2}+\frac{1}{\varrho}\,(E_m+E_r)\right]+\varrho u_z\,\frac{\partial}{\partial z}\left[\frac{u^2}{2}+\frac{1}{\varrho}\,(E_m+E_r)\right]$$

$$+\frac{1}{r}\,\frac{\partial}{\partial r}\,[rF_r+r(P_m-E_r)u_r]+\frac{1}{r}\,\frac{\partial}{\partial\theta}\,[F_\theta+(P_m-E_r)u_\theta]$$

$$+\frac{\partial}{\partial z}\,[F_z+(P_m-E_r)u_z]=W. \tag{9.226}$$

In spherical coordinates, we denote the three components of the radiative flux $\mathbf{F}$, by F ($i=r,\theta,\varphi$) and the nine components of the radiative pressure tensor $\mathbf{P}$, by p_{ij} ($i,j=r,\theta,\varphi$) [see Eqs. (9.152) and (9.153)]. The relativistic continuity equation (9.170) in this coordinate system is written

$$\Delta\varrho\left[\frac{\partial}{\partial t}\left(\frac{1}{\Delta\varrho}\right)+u_r\,\frac{\partial}{\partial r}\left(\frac{1}{\Delta\varrho}\right)+\frac{u_\theta}{r}\,\frac{\partial}{\partial\theta}\left(\frac{1}{\Delta\varrho}\right)+\frac{u_\varphi}{r\sin\theta}\,\frac{\partial}{\partial\varphi}\left(\frac{1}{\Delta\varrho}\right)\right]$$

$$-\left[\frac{1}{r^2}\,\frac{\partial}{\partial r}\,(r^2 u_r)+\frac{1}{r\sin\theta}\,\frac{\partial}{\partial\theta}\,(\sin\theta u_\theta)+\frac{1}{r\sin\theta}\,\frac{\partial u_\varphi}{\partial\varphi}\right]=0. \tag{9.227}$$

The three components of the relativistic momentum equation (9.171) are written

$$\Delta\varrho\,\frac{\partial}{\partial t}\left[\frac{\Delta^2(\varrho c^2+E_m+P_m)u_r+F_r}{\Delta\varrho c^2}\right]+\Delta\varrho u_r\,\frac{\partial}{\partial r}\left[\frac{\Delta^2(\varrho c^2+E_m+P_m)u_r+F_r}{\Delta\varrho c^2}\right]$$

$$+\Delta\varrho\,\frac{u_\theta}{r}\,\frac{\partial}{\partial\theta}\left[\frac{\Delta^2(\varrho c^2+E_m+P_m)u_r+F_r}{\Delta\varrho c^2}\right]+\Delta\varrho\,\frac{u_\varphi}{r\sin\theta}\,\frac{\partial}{\partial\varphi}\left[\frac{\Delta^2(\varrho c^2+E_m+P_m)u_r+F_r}{\Delta\varrho c^2}\right]$$

$$-\frac{u_\theta}{r}\left[\frac{\varLambda^2(\varrho c^2+E_m+P_m)u_\theta+F_\theta}{c^2}\right]-\frac{u_\varphi}{r}\left[\frac{\varLambda^2(\varrho c^2+E_m+P_m)u_\varphi+F_\varphi}{c^2}\right]$$

$$+\frac{\partial P_m}{\partial r}+\frac{1}{r^2}\frac{\partial}{\partial r}\left[r^2\left(p_{rr}-\frac{1}{c^2}u_rF_r\right)\right]+\frac{1}{r\sin\theta}\frac{\partial}{\partial\theta}\left[\sin\theta\left(p_{\theta r}-\frac{1}{c^2}u_\theta F_r\right)\right]$$

$$+\frac{1}{r\sin\theta}\frac{\partial}{\partial\varphi}\left(p_{\varphi r}-\frac{1}{c^2}u_\varphi F_r\right)-\frac{1}{r}\left[p_{\theta\theta}+p_{\varphi\varphi}-\frac{1}{c^2}(u_\theta F_\theta+u_\varphi F_\varphi)\right]=0,\tag{9.228}$$

$$\varLambda\varrho\frac{\partial}{\partial t}\left[\frac{\varLambda^2(\varrho c^2+E_m+P_m)u_\theta+F_\theta}{\varLambda\varrho c^2}\right]+\varLambda\varrho u_r\frac{\partial}{\partial r}\left[\frac{\varLambda^2(\varrho c^2+E_m+P_m)u_\theta+F_\theta}{\varLambda\varrho c^2}\right]$$

$$+\varLambda\varrho\frac{u_\theta}{r}\frac{\partial}{\partial\theta}\left[\frac{\varLambda^2(\varrho c^2+E_m+P_m)u_\theta+F_\theta}{\varLambda\varrho c^2}\right]+\varLambda\varrho\frac{u_\varphi}{r\sin\theta}\frac{\partial}{\partial\varphi}\left[\frac{\varLambda^2(\varrho c^2+E_m+P_m)u_\theta+F_\theta}{\varLambda\varrho c^2}\right]$$

$$+\frac{u_\theta}{r}\left[\frac{\varLambda^2(\varrho c^2+E_m+P_m)u_r+F_r}{c^2}\right]-\frac{\cot\theta u_\varphi}{r}\left[\frac{\varLambda^2(\varrho c^2+E_m+P_m)u_\varphi+F_\varphi}{c^2}\right]$$

$$+\frac{1}{r}\frac{\partial P_m}{\partial\theta}+\frac{1}{r^2}\frac{\partial}{\partial r}\left[r^2\left(p_{r\theta}-\frac{1}{c^2}u_rF_\theta\right)\right]+\frac{1}{r\sin\theta}\frac{\partial}{\partial\theta}\left[\sin\theta\left(p_{\theta\theta}-\frac{1}{c^2}u_\theta F_\theta\right)\right]$$

$$+\frac{1}{r\sin\theta}\frac{\partial}{\partial\varphi}\left(p_{\varphi\theta}-\frac{1}{c^2}u_\varphi F_\theta\right)+\frac{1}{r}\left(p_{\theta r}-\frac{1}{c^2}u_\theta F_r\right)-\frac{\cot\theta}{r}\left(p_{\varphi\varphi}-\frac{1}{c^2}u_\varphi F_\varphi\right)=0,\tag{9.229}$$

$$\varLambda\varrho\frac{\partial}{\partial t}\left[\frac{\varLambda^2(\varrho c^2+E_m+P_m)u_\varphi+F_\varphi}{\varLambda\varrho c^2}\right]+\varLambda\varrho u_r\frac{\partial}{\partial r}\left[\frac{\varLambda^2(\varrho c^2+E_m+P_m)u_\varphi+F_\varphi}{\varLambda\varrho c^2}\right]$$

$$+\varLambda\varrho\frac{u_\varphi}{r}\frac{\partial}{\partial\theta}\left[\frac{\varLambda^2(\varrho c^2+E_m+P_m)u_\varphi+F_\varphi}{\varLambda\varrho c^2}\right]+\varLambda\varrho\frac{u_\varphi}{r\sin\theta}\frac{\partial}{\partial\varphi}\left[\frac{\varLambda^2(\varrho c^2+E_m+P_m)u_\varphi+F_\varphi}{\varLambda\varrho c^2}\right]$$

$$+\frac{u_\varphi}{r}\left[\frac{\varLambda^2(\varrho c^2+E_m+P_m)u_r+F_r}{c^2}\right]+\frac{\cot\theta\,u_\varphi}{r}\left[\frac{\varLambda^2(\varrho c^2+E_m+P_m)u_\theta+F_\theta}{c^2}\right]$$

$$+\frac{1}{r\sin\theta}\frac{\partial P_m}{\partial\varphi}+\frac{1}{r^2}\frac{\partial}{\partial r}\left[r^2\left(p_{r\varphi}-\frac{1}{c^2}u_rF_\varphi\right)\right]+\frac{1}{r\sin\theta}\frac{\partial}{\partial\theta}\left[\sin\theta\left(p_{\theta\varphi}-\frac{1}{c^2}u_\theta F_\varphi\right)\right]$$

$$+\frac{1}{r\sin\theta}\frac{\partial}{\partial\varphi}\left(p_{\varphi\varphi}-\frac{1}{c^2}u_\varphi F_\varphi\right)+\frac{1}{r}\left(p_{\varphi r}-\frac{1}{c^2}u_\varphi F_r\right)+\frac{\cot\theta}{r}\left(p_{\varphi\theta}-\frac{1}{c^2}u_\varphi F_\theta\right)=0.\tag{9.230}$$

The relativistic energy equation (9.172) is written

$$\varLambda\varrho\frac{\partial}{\partial t}\left[\frac{\varLambda(\varLambda-1)\varrho c^2+\varLambda^2(E_m+P_m)-P_m+E_r}{\varLambda\varrho}\right]$$

$$+\varLambda\varrho u_r\frac{\partial}{\partial r}\left[\frac{\varLambda(\varLambda-1)\varrho c^2+\varLambda^2(E_m+P_m)-P_m+E_r}{\varLambda\varrho}\right]$$

$$+\varLambda\varrho\frac{u_\theta}{r}\frac{\partial}{\partial\theta}\left[\frac{\varLambda(\varLambda-1)\varrho c^2+\varLambda^2(E_m+P_m)-P_m+E_r}{\varLambda\varrho}\right]$$

$$+\varLambda\varrho\frac{u_\varphi}{r\sin\theta}\frac{\partial}{\partial\varphi}\left[\frac{\varLambda(\varLambda-1)\varrho c^2+\varLambda^2(E_m+P_m)-P_m+E_r}{\varLambda\varrho}\right]$$

$$+\frac{1}{r^2}\frac{\partial}{\partial r}[r^2F_r+r^2(P_m-E_r)u_r]+\frac{1}{r\sin\theta}\frac{\partial}{\partial\theta}[\sin\theta\,F_\theta+\sin\theta(P_m-E_r)u_\theta]$$

$$+\frac{1}{r\sin\theta}\frac{\partial}{\partial\varphi}[F_\varphi+(P_m-E_r)u_\varphi]=\dot{W}.\tag{9.231}$$

The Equations of Radiation Hydrodynamics

The nonrelativistic continuity equation (9.185) in a spherical coordinate system is written

$$\varrho\left[\frac{\partial}{\partial t}\left(\frac{1}{\varrho}\right)+u_r\frac{\partial}{\partial r}\left(\frac{1}{\varrho}\right)+\frac{u_\theta}{r}\frac{\partial}{\partial\theta}\left(\frac{1}{\varrho}\right)+\frac{u_\varphi}{r\sin\theta}\frac{\partial}{\partial\varphi}\left(\frac{1}{\varrho}\right)\right]$$

$$-\left[\frac{1}{r^2}\frac{\partial}{\partial r}(r^2u_r)+\frac{1}{r\sin\theta}\frac{\partial}{\partial\theta}(\sin\theta u_\theta)+\frac{1}{r\sin\theta}\frac{\partial u_\varphi}{\partial\varphi}\right]=0. \qquad (9.232)$$

The three components of the nonrelativistic momentum equation (9.186) are written

$$\varrho\frac{\partial}{\partial t}\left(u_r+\frac{F_r}{\varrho c^2}\right)+\varrho u_r\frac{\partial}{\partial r}\left(u_r+\frac{F_r}{\varrho c^2}\right)+\varrho\frac{u_\theta}{r}\frac{\partial}{\partial\theta}\left(u_r+\frac{F_r}{\varrho c^2}\right)$$

$$+\varrho\frac{u_\varphi}{r\sin\theta}\frac{\partial}{\partial\varphi}\left(u_r+\frac{F_r}{\varrho c^2}\right)-\varrho\frac{u_\theta}{r}\left(u_\theta+\frac{F_\theta}{\varrho c^2}\right)-\varrho\frac{u_\varphi}{r}\left(u_\varphi+\frac{F_\varphi}{\varrho c^2}\right)$$

$$+\frac{\partial P_m}{\partial r}+\frac{1}{r^2}\frac{\partial}{\partial r}\left[r^2\left(P_{rr}-\frac{1}{c^2}u_rF_r\right)\right]+\frac{1}{r\sin\theta}\frac{\partial}{\partial\theta}\left[\sin\theta\left(P_{\theta r}-\frac{1}{c^2}u_\theta F_r\right)\right]$$

$$+\frac{1}{r\sin\theta}\frac{\partial}{\partial\varphi}\left(P_{\varphi r}-\frac{1}{c^2}u_\varphi F_r\right)-\frac{1}{r}\left[P_{\theta\theta}+P_{\varphi\varphi}-\frac{1}{c^2}(u_\theta F_\theta+u_\varphi F_\varphi)\right]=0, \qquad (9.233)$$

$$\varrho\frac{\partial}{\partial t}\left(u_\theta+\frac{F_\theta}{\varrho c^2}\right)+\varrho u_r\frac{\partial}{\partial r}\left(u_\theta+\frac{F_\theta}{\varrho c^2}\right)+\varrho\frac{u_\theta}{r}\frac{\partial}{\partial\theta}\left(u_\theta+\frac{F_\theta}{\varrho c^2}\right)$$

$$+\varrho\frac{u_\varphi}{r\sin\theta}\frac{\partial}{\partial\varphi}\left(u_\theta+\frac{F_\theta}{\varrho c^2}\right)+\varrho\frac{u_\theta}{r}\left(u_r+\frac{F_r}{\varrho c^2}\right)-\varrho\frac{\cot\theta u_\varphi}{r}\left(u_\varphi+\frac{F_\varphi}{\varrho c^2}\right)$$

$$+\frac{1}{r}\frac{\partial P_m}{\partial\theta}+\frac{1}{r^2}\frac{\partial}{\partial r}\left[r^2\left(P_{r\theta}-\frac{1}{c^2}u_rF_\theta\right)\right]+\frac{1}{r\sin\theta}\frac{\partial}{\partial\theta}\left[\sin\theta\left(P_{\theta\theta}-\frac{1}{c^2}u_\theta F_\theta\right)\right]$$

$$+\frac{1}{r\sin\theta}\frac{\partial}{\partial\varphi}\left(P_{\varphi\theta}-\frac{1}{c^2}u_\varphi F_\theta\right)+\frac{1}{r}\left(P_{\theta r}-\frac{1}{c^2}u_\theta F_r\right)-\frac{\cot\theta}{r}\left(P_{\varphi\varphi}-\frac{1}{c^2}u_\varphi F_\varphi\right)=0,$$

$$(9.234)$$

$$\varrho\frac{\partial}{\partial t}\left(u_\varphi+\frac{F_\varphi}{\varrho c^2}\right)+\varrho u_r\frac{\partial}{\partial r}\left(u_\varphi+\frac{F_\varphi}{\varrho c^2}\right)+\varrho\frac{u_\theta}{r}\frac{\partial}{\partial\theta}\left(u_\varphi+\frac{F_\varphi}{\varrho c^2}\right)$$

$$+\varrho\frac{u_\varphi}{r\sin\theta}\frac{\partial}{\partial\varphi}\left(u_\varphi+\frac{F_\varphi}{\varrho c^2}\right)+\varrho\frac{u_\varphi}{r}\left(u_r+\frac{F_r}{\varrho c^2}\right)+\varrho\frac{\cot\theta u_\varphi}{r}\left(u_\theta+\frac{F_\theta}{\varrho c^2}\right)$$

$$+\frac{1}{r\sin\theta}\frac{\partial P_m}{\partial\varphi}+\frac{1}{r^2}\frac{\partial}{\partial r}\left[r^2\left(P_{r\varphi}-\frac{1}{c^2}u_rF_\varphi\right)\right]+\frac{1}{r\sin\theta}\frac{\partial}{\partial\theta}\left[\sin\theta\left(P_{\theta\varphi}-\frac{1}{c^2}u_\theta F_\varphi\right)\right]$$

$$+\frac{1}{r\sin\theta}\frac{\partial}{\partial\varphi}\left(P_{\varphi\varphi}-\frac{1}{c^2}u_\varphi F_\varphi\right)+\frac{1}{r}\left(P_{\varphi r}-\frac{1}{c^2}u_\varphi F_r\right)+\frac{\cot\theta}{r}\left(P_{\varphi\theta}-\frac{1}{c^2}u_\varphi F_\theta\right)=0.$$

$$(9.235)$$

The nonrelativistic energy equation (9.187) is written

$$\varrho\frac{\partial}{\partial t}\left[\frac{u^2}{2}+\frac{1}{\varrho}(E_m+E_r)\right]+\varrho u_r\frac{\partial}{\partial r}\left[\frac{u^2}{2}+\frac{1}{\varrho}(E_m+E_r)\right]$$

$$+\varrho\,\frac{u_\theta}{r}\,\frac{\partial}{\partial\theta}\left[\frac{u^2}{2}+\frac{1}{\varrho}\,(E_m+E_r)\right]+\varrho\,\frac{u_\varphi}{r\sin\theta}\,\frac{\partial}{\partial\varphi}\left[\frac{u^2}{2}+\frac{1}{\varrho}\,(E_m+E_r)\right]$$

$$+\frac{1}{r^2}\,\frac{\partial}{\partial r}\,[r^2F_r+r^2(P_m-E_r)\,u_r]+\frac{1}{r\sin\theta}\,\frac{\partial}{\partial\theta}\,[\sin\theta\,F_\theta+\sin\theta(P_m-E_r)\,u_\theta]$$

$$+\frac{1}{r\sin\theta}\,\frac{\partial}{\partial\varphi}\,[F_\varphi+(P_m-E_r)\,u_\varphi]=W. \tag{9.236}$$

Equations (9.207) through (9.236) are the explicit expressions for the continuity, momentum, and energy equations, in both the relativistic and nonrelativistic forms, in Cartesian, cylindrical, and spherical coordinates, written in modified Eulerian form (conservative for a volume element moving with the fluid).

7. Lagrangian Hydrodynamics

In Eulerian (or modified Eulerian) hydrodynamics, the dependent variables such as velocity, pressure, etc., are considered to be functions of the spatial and temporal coordinates $\mathbf{r}$ and t. In so-called Lagrangian hydrodynamics an entirely different point of view is employed. One labels a given differential mass of fluid by its position, say $\mathbf{r}_0$, at an initial time, say $t = 0$. One then asks for the position $\mathbf{r}$ as well as the velocity, pressure, etc., of this same differential mass as a function of time. (For convenience, we use the term mass, but actually we mean here a differential number of particles, since relativistically mass is not conserved.)

Mathematically, this viewpoint is introduced by changing independent variables from the set $(\mathbf{r}, t)$ to the set $(\mathbf{r}_0, \tau)$, i.e.,

$$t \equiv t(\mathbf{r}_0,\, \tau), \tag{9.237}$$

$$\mathbf{r} \equiv \mathbf{r}(\mathbf{r}_0,\, \tau). \tag{9.238}$$

Specifically, the appropriate change of variables is

$$t = \tau, \tag{9.239}$$

$$\left(\frac{\partial \mathbf{r}}{\partial \tau}\right)_{\mathbf{r}_0} = \mathbf{u}, \tag{9.240}$$

$$\mathbf{r}(\mathbf{r}_0,\, 0) = \mathbf{r}_0, \tag{9.241}$$

where $\mathbf{u}$ is the fluid velocity. That is, the time transformation is simply the identity transformation, while the spatial transformation is defined by a differential equation (9.240) with Eq. (9.241) acting as the boundary condition on this differential equation. We emphasize that physically these transformations mean that rather than specifying all dependent variables as functions of space and time as in the Eulerian picture, we focus on a particular differential mass of fluid and ask for its position and properties as a function of time. [The fact that these transformations imply a fixed mass of fluid will become clear shortly—see Eq. (9.276).] Specifically, Eq. (9.240) just states that the position $\mathbf{r}$ of a differential mass of fluid changes with time τ, and this rate of change is just the velocity $\mathbf{u}$. The content of Eq. (9.241) is that at the initial time $\tau = 0$ the position $\mathbf{r}$ of the fluid mass is the initial position $\mathbf{r}_0$. It is worth noting that in a certain sense the variable $\mathbf{r}_0$ plays a dual role. It is first of all a

The Equations of Radiation Hydrodynamics

label specifying the particular mass of fluid under consideration. Secondly, it is the initial position of this same fluid mass. This dual role is evident in Eq. (9.241).

By the chain rule of differentiation, we have

$$\frac{\partial(\ )}{\partial\tau} = \left(\frac{\partial t}{\partial\tau}\right)_{\mathbf{r}_0}\frac{\partial(\ )}{\partial t} + \left(\frac{\partial\mathbf{r}}{\partial\tau}\right)_{\mathbf{r}_0}\cdot\nabla(\ ). \tag{9.242}$$

Use of Eqs. (9.239) and (9.240) in Eq. (9.242) yields

$$\frac{\partial(\ )}{\partial\tau} = \frac{\partial(\ )}{\partial t} + \mathbf{u}\cdot\nabla(\ ). \tag{9.243}$$

Equation (9.243) is the essence of Lagrangian hydrodynamics and gives $\partial(\ )/\partial\tau$, a time derivative following the motion of the fluid, in terms of the usual (Eulerian) time and space derivatives. This time derivative is often called the Lagrangian, or substantive, derivative.

It should be noted that $\partial(\ )/\partial\tau$ given by Eq. (9.243) is identical to the derivative $D(\ )/Dt$ introduced in Section 5 of this chapter [see Eq. (9.164)] in connection with the modified Eulerian equations. This is not surprising since both the Lagrangian and modified Eulerian pictures concentrate on a moving mass of fluid. However, in the Lagrangian picture one proceeds further than in the modified Eulerian method. In addition to introducing $\partial(\ )/\partial\tau$ into the equations of hydrodynamics, one completes the change of variables given by Eqs. (9.239) through (9.241) and eliminates all spatial derivatives with respect to the components of $\mathbf{r}$ in favor of derivatives with respect to the components of $\mathbf{r}_0$.

To effect this change of variables, we resort to the basic coordinate system in which the differential vector operations occurring in the equations of hydrodynamics are defined, namely a Cartesian system with coordinates x, y, and z. In this coordinate system, the change of variables given by Eqs. (9.239) through (9.241) is written explicitly as

$$t = \tau, \tag{9.244}$$

$$\frac{\partial x}{\partial\tau} = u_x; \qquad \frac{\partial y}{\partial\tau} = u_y; \qquad \frac{\partial z}{\partial\tau} = u_z, \tag{9.245}$$

$$x(x_0, y_0, z_0, 0) = x_0; \quad y(x_0, y_0, z_0, 0) = y_0; \quad z(x_0, y_0, z_0, 0) = z_0. \tag{9.246}$$

Here u_x, u_y, and u_z are the Cartesian components of the fluid velocity $\mathbf{u}$. These component velocities, as well as all other dependent variables, are considered to be functions of x_0, y_0, z_0, and τ, i.e.,

$$u_i \equiv u_i(x_0, y_0, z_0, \tau), \qquad (i = x, y, z). \tag{9.247}$$

We have previously pointed out that the expression giving $\partial(\ )/\partial\tau$ in terms of Eulerian derivatives, Eq. (9.243), is a fundamental relationship between the Lagrangian and Eulerian descriptions. Another fundamental relationship between these two descriptions is the Lagrangian expression for $\nabla\cdot\mathbf{u}$, the divergence of the velocity field. To this end, we first obtain differential relationships between the coordinates x_0, y_0, z_0 and the coordinates x, y, z. The chain rule of differentiation gives

$$\frac{\partial(\ )}{\partial x} = \left(\frac{\partial x_0}{\partial x}\right)\frac{\partial(\ )}{\partial x_0} + \left(\frac{\partial y_0}{\partial x}\right)\frac{\partial(\ )}{\partial y_0} + \left(\frac{\partial z_0}{\partial x}\right)\frac{\partial(\ )}{\partial z_0}. \tag{9.248}$$

246

Applying Eq. (9.248) to the functions x, y, and z respectively, we obtain the three equations

$$\left(\frac{\partial x}{\partial x_0}\right)\left(\frac{\partial x_0}{\partial x}\right) + \left(\frac{\partial x}{\partial y_0}\right)\left(\frac{\partial y_0}{\partial x}\right) + \left(\frac{\partial x}{\partial z_0}\right)\left(\frac{\partial z_0}{\partial x}\right) = 1, \qquad (9.249)$$

$$\left(\frac{\partial y}{\partial x_0}\right)\left(\frac{\partial x_0}{\partial x}\right) + \left(\frac{\partial y}{\partial y_0}\right)\left(\frac{\partial y_0}{\partial x}\right) + \left(\frac{\partial y}{\partial z_0}\right)\left(\frac{\partial z_0}{\partial x}\right) = 0, \qquad (9.250)$$

$$\left(\frac{\partial z}{\partial x_0}\right)\left(\frac{\partial x_0}{\partial x}\right) + \left(\frac{\partial z}{\partial y_0}\right)\left(\frac{\partial y_0}{\partial x}\right) + \left(\frac{\partial z}{\partial z_0}\right)\left(\frac{\partial z_0}{\partial x}\right) = 0. \qquad (9.251)$$

Equations (9.249) through (9.251) can be considered as three linear algebraic equations for the three unknowns $\partial x_0/\partial x$, $\partial y_0/\partial x$, and $\partial z_0/\partial x$. The solution is

$$\frac{\partial x_0}{\partial x} = \frac{1}{J}\left[\left(\frac{\partial y}{\partial y_0}\right)\left(\frac{\partial z}{\partial z_0}\right) - \left(\frac{\partial y}{\partial z_0}\right)\left(\frac{\partial z}{\partial y_0}\right)\right], \qquad (9.252)$$

$$\frac{\partial y_0}{\partial x} = -\frac{1}{J}\left[\left(\frac{\partial y}{\partial x_0}\right)\left(\frac{\partial z}{\partial z_0}\right) - \left(\frac{\partial y}{\partial z_0}\right)\left(\frac{\partial z}{\partial x_0}\right)\right], \qquad (9.253)$$

$$\frac{\partial z_0}{\partial x} = \frac{1}{J}\left[\left(\frac{\partial y}{\partial x_0}\right)\left(\frac{\partial z}{\partial y_0}\right) - \left(\frac{\partial y}{\partial y_0}\right)\left(\frac{\partial z}{\partial x_0}\right)\right], \qquad (9.254)$$

where J is the Jacobian of the transformation from (x, y, z) to (x_0, y_0, z_0), i.e.,

$$J = \begin{vmatrix} \dfrac{\partial x}{\partial x_0} & \dfrac{\partial x}{\partial y_0} & \dfrac{\partial x}{\partial z_0} \\[2mm] \dfrac{\partial y}{\partial x_0} & \dfrac{\partial y}{\partial y_0} & \dfrac{\partial y}{\partial z_0} \\[2mm] \dfrac{\partial z}{\partial x_0} & \dfrac{\partial z}{\partial y_0} & \dfrac{\partial z}{\partial z_0} \end{vmatrix}. \qquad (9.255)$$

Similarly, one finds for the derivatives with respect to y and z

$$\frac{\partial x_0}{\partial y} = -\frac{1}{J}\left[\left(\frac{\partial x}{\partial y_0}\right)\left(\frac{\partial z}{\partial z_0}\right) - \left(\frac{\partial x}{\partial z_0}\right)\left(\frac{\partial z}{\partial y_0}\right)\right], \qquad (9.256)$$

$$\frac{\partial y_0}{\partial y} = \frac{1}{J}\left[\left(\frac{\partial x}{\partial x_0}\right)\left(\frac{\partial z}{\partial z_0}\right) - \left(\frac{\partial x}{\partial z_0}\right)\left(\frac{\partial z}{\partial x_0}\right)\right], \qquad (9.257)$$

$$\frac{\partial z_0}{\partial y} = \frac{1}{J}\left[\left(\frac{\partial x}{\partial y_0}\right)\left(\frac{\partial z}{\partial x_0}\right) - \left(\frac{\partial x}{\partial x_0}\right)\left(\frac{\partial z}{\partial y_0}\right)\right], \qquad (9.258)$$

and

$$\frac{\partial x_0}{\partial z} = \frac{1}{J}\left[\left(\frac{\partial y}{\partial z_0}\right)\left(\frac{\partial x}{\partial y_0}\right) - \left(\frac{\partial y}{\partial y_0}\right)\left(\frac{\partial x}{\partial z_0}\right)\right], \qquad (9.259)$$

$$\frac{\partial y_0}{\partial z} = -\frac{1}{J}\left[\left(\frac{\partial y}{\partial z_0}\right)\left(\frac{\partial x}{\partial x_0}\right) - \left(\frac{\partial y}{\partial x_0}\right)\left(\frac{\partial x}{\partial z_0}\right)\right], \qquad (9.260)$$

$$\frac{\partial z_0}{\partial z} = \frac{1}{J}\left[\left(\frac{\partial y}{\partial y_0}\right)\left(\frac{\partial x}{\partial x_0}\right) - \left(\frac{\partial y}{\partial x_0}\right)\left(\frac{\partial x}{\partial y_0}\right)\right]. \qquad (9.261)$$

The Equations of Radiation Hydrodynamics

To obtain the divergence of the velocity field, we first consider $\partial u_x/\partial x$. By the chain rule of differentiation we have

$$\frac{\partial u_x}{\partial x} = \left(\frac{\partial x_0}{\partial x}\right)\left(\frac{\partial u_x}{\partial x_0}\right) + \left(\frac{\partial y_0}{\partial x}\right)\left(\frac{\partial u_x}{\partial y_0}\right) + \left(\frac{\partial z_0}{\partial x}\right)\left(\frac{\partial u_x}{\partial z_0}\right). \tag{9.262}$$

Making use of Eq. (9.245) for u_x, we see that Eq. (9.262) can be written

$$\frac{\partial u_x}{\partial x} = \left(\frac{\partial x_0}{\partial x}\right)\left(\frac{\partial^2 x}{\partial x_0\,\partial\tau}\right) + \left(\frac{\partial y_0}{\partial x}\right)\left(\frac{\partial^2 x}{\partial y_0\,\partial\tau}\right) + \left(\frac{\partial z_0}{\partial x}\right)\left(\frac{\partial^2 x}{\partial z_0\,\partial\tau}\right). \tag{9.263}$$

Use of Eqs. (9.252) through (9.254) in Eq. (9.263) yields

$$\begin{aligned}
\frac{\partial u_x}{\partial x} = {}& \frac{1}{J}\frac{\partial^2 x}{\partial x_0\,\partial\tau}\left[\left(\frac{\partial y}{\partial y_0}\right)\left(\frac{\partial z}{\partial z_0}\right) - \left(\frac{\partial y}{\partial z_0}\right)\left(\frac{\partial z}{\partial y_0}\right)\right] \\
& - \frac{1}{J}\frac{\partial^2 x}{\partial y_0\,\partial\tau}\left[\left(\frac{\partial y}{\partial x_0}\right)\left(\frac{\partial z}{\partial z_0}\right) - \left(\frac{\partial y}{\partial z_0}\right)\left(\frac{\partial z}{\partial x_0}\right)\right] \\
& + \frac{1}{J}\frac{\partial^2 x}{\partial z_0\,\partial\tau}\left[\left(\frac{\partial y}{\partial x_0}\right)\left(\frac{\partial z}{\partial y_0}\right) - \left(\frac{\partial y}{\partial y_0}\right)\left(\frac{\partial z}{\partial x_0}\right)\right],
\end{aligned} \tag{9.264}$$

or, in determinant form,

$$\frac{\partial u_x}{\partial x} = \frac{1}{J}\begin{vmatrix} \dfrac{\partial^2 x}{\partial x_0\,\partial\tau} & \dfrac{\partial^2 x}{\partial y_0\,\partial\tau} & \dfrac{\partial^2 x}{\partial z_0\,\partial\tau} \\[2ex] \dfrac{\partial y}{\partial x_0} & \dfrac{\partial y}{\partial y_0} & \dfrac{\partial y}{\partial z_0} \\[2ex] \dfrac{\partial z}{\partial x_0} & \dfrac{\partial z}{\partial y_0} & \dfrac{\partial z}{\partial z_0} \end{vmatrix}. \tag{9.265}$$

Similar expressions can be found for $\partial u_y/\partial y$ and $\partial u_z/\partial z$. These results give

$$\begin{aligned}
\nabla\cdot\mathbf{u} = \frac{\partial u_x}{\partial x} + \frac{\partial u_y}{\partial y} + \frac{\partial u_z}{\partial z} = {}& \frac{1}{J}\begin{vmatrix} \dfrac{\partial^2 x}{\partial x_0\,\partial\tau} & \dfrac{\partial^2 x}{\partial y_0\,\partial\tau} & \dfrac{\partial^2 x}{\partial z_0\,\partial\tau} \\[2ex] \dfrac{\partial y}{\partial x_0} & \dfrac{\partial y}{\partial y_0} & \dfrac{\partial y}{\partial z_0} \\[2ex] \dfrac{\partial z}{\partial x_0} & \dfrac{\partial z}{\partial y_0} & \dfrac{\partial z}{\partial z_0} \end{vmatrix} \\[2ex]
& + \frac{1}{J}\begin{vmatrix} \dfrac{\partial x}{\partial x_0} & \dfrac{\partial x}{\partial y_0} & \dfrac{\partial x}{\partial z_0} \\[2ex] \dfrac{\partial^2 y}{\partial x_0\,\partial\tau} & \dfrac{\partial^2 y}{\partial y_0\,\partial\tau} & \dfrac{\partial^2 y}{\partial z_0\,\partial\tau} \\[2ex] \dfrac{\partial z}{\partial x_0} & \dfrac{\partial z}{\partial y_0} & \dfrac{\partial z}{\partial z_0} \end{vmatrix} + \frac{1}{J}\begin{vmatrix} \dfrac{\partial x}{\partial x_0} & \dfrac{\partial x}{\partial y_0} & \dfrac{\partial x}{\partial z_0} \\[2ex] \dfrac{\partial y}{\partial x_0} & \dfrac{\partial y}{\partial y_0} & \dfrac{\partial y}{\partial z_0} \\[2ex] \dfrac{\partial^2 z}{\partial x_0\,\partial\tau} & \dfrac{\partial^2 z}{\partial y_0\,\partial\tau} & \dfrac{\partial^2 z}{\partial z_0\,\partial\tau} \end{vmatrix}.
\end{aligned} \tag{9.266}$$

Recalling the rule for differentiating a determinant, we find that Eq. (9.266) can be written

$$\nabla \cdot \mathbf{u} = \frac{1}{J} \frac{\partial}{\partial \tau} \begin{vmatrix} \dfrac{\partial x}{\partial x_0} & \dfrac{\partial x}{\partial y_0} & \dfrac{\partial x}{\partial z_0} \\[2mm] \dfrac{\partial y}{\partial x_0} & \dfrac{\partial y}{\partial y_0} & \dfrac{\partial y}{\partial z_0} \\[2mm] \dfrac{\partial z}{\partial x_0} & \dfrac{\partial z}{\partial y_0} & \dfrac{\partial z}{\partial z_0} \end{vmatrix}, \tag{9.267}$$

or, since the determinant in Eq. (9.267) is just the Jacobian J [see Eq. (9.255)],

$$\nabla \cdot \mathbf{u} = \frac{1}{J} \frac{\partial J}{\partial \tau}. \tag{9.268}$$

Equation (9.267) is the desired expression for the divergence of the velocity field and this result, together with Eq. (9.243), constitute the two fundamental connections between the Lagrangian and Eulerian descriptions of hydrodynamics.

We now turn to the task of expressing the relativistic continuity, momentum, and energy equations (9.170) through (9.172) in Lagrangian form. Considering first the continuity equation (9.170), we have

$$\varLambda\varrho \frac{\partial}{\partial \tau}\left(\frac{1}{\varLambda\varrho}\right) - \frac{1}{J}\frac{\partial J}{\partial \tau} = 0, \tag{9.269}$$

where we have used Eq. (9.268) for $\nabla \cdot \mathbf{u}$, together with the fact that $D(\ )/Dt = \partial(\ )/\partial \tau$ [see Eqs. (9.164) and (9.243)]. Rewriting Eq. (9.269), we have

$$\frac{1}{\varLambda\varrho}\frac{\partial(\varLambda\varrho)}{\partial \tau} + \frac{1}{J}\frac{\partial J}{\partial \tau} = 0, \tag{9.270}$$

or

$$\frac{\partial}{\partial \tau}\ln(\varLambda\varrho) + \frac{\partial}{\partial \tau}\ln J = 0. \tag{9.271}$$

Integration of Eq. (9.271) gives

$$\ln(\varLambda\varrho) + \ln J = \ln(\varLambda\varrho J) = \ln C, \tag{9.272}$$

or

$$\varLambda\varrho J = C. \tag{9.273}$$

In both Eqs. (9.272) and (9.273) the "constant" of integration C is a function of x_0, y_0, and z_0, but not τ. We can evaluate C by considering $\tau = 0$. At this time the Jacobian J is unity, as is clear from Eq. (9.246). This gives $C = (\varLambda\varrho)_0$, where the subscript zero implies $\tau = 0$. Thus Eq. (9.273) becomes

$$\varLambda\varrho J = (\varLambda\varrho)_0. \tag{9.274}$$

Thus we see that in Lagrangian hydrodynamics the relativistic continuity equation (9.270) can be solved in simple closed form in the general case. The solution is just Eq. (9.274). The solution has a simple physical interpretation. If we multiply Eq. (9.274) by $d\tau_0$ (a differential volume element at $\tau = 0$) and use the fact that $d\mathbf{r}$ (a differential volume element at the time under consideration) and $d\mathbf{r}_0$ are related by the Jacobian of the transformation from $\mathbf{r}$ to $\mathbf{r}_0$ according to

$$d\mathbf{r} = J\, d\mathbf{r}_0, \tag{9.275}$$

we have

$$\Lambda\varrho\, d\mathbf{r} = (\Lambda\varrho)_0\, d\mathbf{r}_0. \tag{9.276}$$

Since $\Lambda\varrho$ is, to within a constant, the particle density [see Eqs. (9.57) and (9.69)], the equality in Eq. (9.276) indicates that Lagrangian hydrodynamics corresponds to following a fixed differential number of particles (or mass in the nonrelativistic limit), and asking for the properties of these particles, such as position, velocity, etc., as a function of time.

To write the relativistic momentum and energy equations (9.171) and (9.172) in Lagrangian form, we require Lagrangian expressions for the gradient of a scalar, the divergence of a vector, and the components of the divergence of a dyad. The $D(\ )/Dt$ operator appearing in Eqs. (9.171) and (9.172) gives no trouble since it can immediately be replaced by the Lagrangian derivative $\partial(\ )/\partial\tau$ [see Eqs. (9.164) and (9.243)]. To obtain the above mentioned differential operators in Lagrangian form, we first note that the chain rule of differentiation gives

$$\frac{\partial(\ )}{\partial x} = \left(\frac{\partial x_0}{\partial x}\right)\frac{\partial(\ )}{\partial x_0} + \left(\frac{\partial y_0}{\partial x}\right)\frac{\partial(\ )}{\partial y_0} + \left(\frac{\partial z_0}{\partial x}\right)\frac{\partial(\ )}{\partial z_0}. \tag{9.277}$$

Use of Eqs. (9.252) through (9.254) in Eq. (9.277) gives

$$\begin{aligned}
\frac{\partial(\ )}{\partial x} &= \frac{1}{J}\left[\left(\frac{\partial y}{\partial y_0}\right)\left(\frac{\partial z}{\partial z_0}\right) - \left(\frac{\partial y}{\partial z_0}\right)\left(\frac{\partial z}{\partial y_0}\right)\right]\frac{\partial(\ )}{\partial x_0} \\
&\quad - \frac{1}{J}\left[\left(\frac{\partial y}{\partial x_0}\right)\left(\frac{\partial z}{\partial z_0}\right) - \left(\frac{\partial y}{\partial z_0}\right)\left(\frac{\partial z}{\partial x_0}\right)\right]\frac{\partial(\ )}{\partial y_0} \\
&\quad + \frac{1}{J}\left[\left(\frac{\partial y}{\partial x_0}\right)\left(\frac{\partial z}{\partial y_0}\right) - \left(\frac{\partial y}{\partial y_0}\right)\left(\frac{\partial z}{\partial x_0}\right)\right]\frac{\partial(\ )}{\partial z_0}.
\end{aligned} \tag{9.278}$$

Equation (9.278) can be written in determinant form as

$$\frac{\partial(\ )}{\partial x} = \frac{1}{J}\begin{vmatrix} \dfrac{\partial(\ )}{\partial x_0} & \dfrac{\partial(\ )}{\partial y_0} & \dfrac{\partial(\ )}{\partial z_0} \\[2ex] \dfrac{\partial y}{\partial x_0} & \dfrac{\partial y}{\partial y_0} & \dfrac{\partial y}{\partial z_0} \\[2ex] \dfrac{\partial z}{\partial x_0} & \dfrac{\partial z}{\partial y_0} & \dfrac{\partial z}{\partial z_0} \end{vmatrix} \equiv D_x(\ ). \tag{9.279}$$

We have defined the right hand side of the first equality in Eq. (9.279) as the Lagrangian operator $D_x(\ )$. Similarly, we find for $\partial(\ )/\partial y$ and $\partial(\ )/\partial z$

$$\frac{\partial(\ )}{\partial y} = \frac{1}{J}\begin{vmatrix} \dfrac{\partial x}{\partial x_0} & \dfrac{\partial x}{\partial y_0} & \dfrac{\partial x}{\partial z_0} \\[2mm] \dfrac{\partial(\ )}{\partial x_0} & \dfrac{\partial(\ )}{\partial y_0} & \dfrac{\partial(\ )}{\partial z_0} \\[2mm] \dfrac{\partial z}{\partial x_0} & \dfrac{\partial z}{\partial y_0} & \dfrac{\partial z}{\partial z_0} \end{vmatrix} \equiv D_y(\ ), \tag{9.280}$$

and

$$\frac{\partial(\ )}{\partial z} = \frac{1}{J}\begin{vmatrix} \dfrac{\partial x}{\partial x_0} & \dfrac{\partial x}{\partial y_0} & \dfrac{\partial x}{\partial z_0} \\[2mm] \dfrac{\partial y}{\partial x_0} & \dfrac{\partial y}{\partial y_0} & \dfrac{\partial y}{\partial z_0} \\[2mm] \dfrac{\partial(\ )}{\partial x_0} & \dfrac{\partial(\ )}{\partial y_0} & \dfrac{\partial(\ )}{\partial z_0} \end{vmatrix} \equiv D_z(\ ). \tag{9.281}$$

We now define the vector Lagrangian operator $\mathbf{D}(\ )$ as

$$\mathbf{D}(\ ) = \mathbf{i}D_x(\ ) + \mathbf{j}D_y(\ ) + \mathbf{k}D_z(\ ), \tag{9.282}$$

where $\mathbf{i}$, $\mathbf{j}$, and $\mathbf{k}$ are unit vectors along the x, y, and z axes, respectively. Then, for any scalar A we clearly have

$$\nabla A = \mathbf{D}A, \tag{9.283}$$

for any vector $\mathbf{B}$ we have

$$\nabla \cdot \mathbf{B} = \mathbf{D} \cdot \mathbf{B}, \tag{9.284}$$

and the x, y, and z components of the divergence of the dyad $\mathbf{BC}$ are given by

$$[\nabla \cdot (\mathbf{BC})]_i = [\mathbf{D} \cdot (\mathbf{BC})]_i = \mathbf{D} \cdot (\mathbf{B}C_i), \qquad i = x, y, z, \tag{9.285}$$

where the C_i $(i = x, y, z)$ are the Cartesian components of the vector $\mathbf{C}$. The right hand sides of Eqs. (9.283) through (9.285) are the Lagrangian equivalents of the Eulerian operators on the corresponding left hand sides of these equations.

With these operator identities, one can immediately write the relativistic momentum and energy equations in Lagrangian form. The momentum equation (9.171) becomes

$$\varLambda\varrho\,\frac{\partial}{\partial\tau}\left[\frac{\varLambda^2(\varrho c^2 + E_m + P_m)\mathbf{u} + \mathbf{F}_r}{\varLambda\varrho c^2}\right] + \mathbf{D}P_m + \mathbf{D}\cdot\left(P_r - \frac{1}{c^2}\mathbf{u}\mathbf{F}_r\right) = 0. \tag{9.286}$$

The energy equation (9.172) in Lagrangian form is written

$$\varLambda\varrho\,\frac{\partial}{\partial\tau}\left[\frac{\varLambda(\varLambda-1)\,\varrho c^2 + \varLambda^2(E_m + P_m) - P_m + E_r}{\varLambda\varrho}\right] + \mathbf{D}\cdot[\mathbf{F}_r + (P_m - E_r)\mathbf{u}] = W. \tag{9.287}$$

In both Eqs. (9.286) and (9.287) the vector operator $\mathbf{D}(\ )$ is defined by Eqs. (9.279) through (9.282).

The Equations of Radiation Hydrodynamics

In Lagrangian hydrodynamics in the presence of a radiation field, one is concerned with seven dependent variables. These are: (1) the specific intensity of radiation I (and its angular moments E_r, F_r, and P_r); (2) the fluid density in the fluid rest frame ϱ; (3) the fluid pressure in the fluid rest frame P_m; (4) the internal energy of the fluid E_m; (5) the fluid temperature in the fluid rest frame T; (6) the fluid velocity $\mathbf{u}$; and (7) the fluid position $\mathbf{r}$ (in Eulerian hydrodynamics one does not have $\mathbf{r}$ as a dependent variable). The seven equations for these seven unknowns are: (1) the equation of radiative transfer, such as Eq. (2.170); (2) the continuity equation or its solution, Eq. (9.269) or Eq. (9.274); (3) the equation of state relating P_m, T, and ϱ; (4) the energy equation (9.287); (5) the equation relating E_m and T through the specific heat; (6) the momentum equation (9.286); and (7) the position equation (9.240) (in Eulerian hydrodynamics one does not have the position equation). Hence whereas Eulerian hydrodynamics deals with six equations for six unknowns [see the discussion below Eq. (9.68)], Lagrangian hydrodynamics deals with seven equations for seven unknowns.

The nonrelativistic limits of the position, continuity, momentum, and energy equations are obtained if one lets c, the vacuum speed of light, approach infinity in the fluid terms of these equations. In this limit we have

$$\Lambda \xrightarrow[c\to\infty]{} 1; \quad (\Lambda-1) \xrightarrow[c\to\infty]{} \frac{1}{2}\frac{u^2}{c^2}. \tag{9.288}$$

In this limit the position equation (9.240) remains unchanged, i.e.,

$$\frac{\partial \mathbf{r}}{\partial \tau} = \mathbf{u}, \tag{9.289}$$

the continuity equation (9.269) and its solution, Eq. (9.274), become

$$\varrho\frac{\partial}{\partial \tau}\left(\frac{1}{\varrho}\right) - \frac{1}{J}\frac{\partial J}{\partial \tau} = 0, \tag{9.290}$$

$$\varrho J = \varrho_0, \tag{9.291}$$

the momentum equation (9.286) is written

$$\varrho\frac{\partial}{\partial \tau}\left(\mathbf{u}+\frac{\mathbf{F}_r}{\varrho c^2}\right)+\mathbf{D}P_m+\mathbf{D}\cdot\left(P_r-\frac{1}{c^2}\,\mathbf{u}\mathbf{F}_r\right) = 0, \tag{9.292}$$

and the energy equation (9.287) limits to

$$\varrho\frac{\partial}{\partial \tau}\left[\frac{u^2}{2}+\frac{1}{\varrho}(E_m+E_r)\right]+\mathbf{D}\cdot[\mathbf{F}_r+(P_m-E_r)\mathbf{u}] = W. \tag{9.293}$$

Equations (9.290) through (9.293) are the Lagrangian forms of the nonrelativistic (modified) Eulerian equations given by Eqs. (9.185) through (9.187).

8. The Lagrangian Equations in Various Coordinate Systems

In this section we give the Lagrangian equations of hydrodynamics in the standard Cartesian, cylindrical, and spherical coordinate systems. What we basically require in these coordinate systems are explicit expressions for the Lagrangian forms of the gradient of a scalar, the divergence of a vector, the divergence of a dyad, and the time derivative of a vector. In addition, we require the expression for the Jacobian J [see Eq. (9.255)] in these coordinate systems.

The Lagrangian continuity equation and its solution [Eqs. (9.269) and (9.274) in the relativistic case; Eqs. (9.290) and (9.291) in the nonrelativistic limit] clearly read the same in all coordinate systems. One must, of course, use the appropriate expression for the Jacobian for the coordinate system under consideration. Accordingly, insofar as the Lagrangian continuity equation is concerned, no further discussion or analysis is required except to give the expressions for the Jacobians in the three coordinate systems we will consider. The position, momentum, and energy equations will require separate discussion in each coordinate system.

We first consider a Cartesian system with coordinates x, y, and z. This case is particularly simple since the differential spatial operators contained in the Lagrangian equations of hydrodynamics are basically defined in this coordinate system, and have been given explicitly in the last section of this chapter. In particular, the Jacobian J is given by Eq. (9.255), the x, y, and z components of the Lagrangian gradient of any scalar A are given by [see Eq. (9.283)]

$$(\mathbf{D}A)_x = D_xA; \quad (\mathbf{D}A)_y = D_yA; \quad (\mathbf{D}A)_z = D_zA, \tag{9.294}$$

the Lagrangian divergence of any vector $\mathbf{B}$ with Cartesian components B_x, B_y, and B_z is [see Eq. (9.284)]

$$\mathbf{D}\cdot\mathbf{B} = D_xB_x + D_yB_y + D_zB_z, \tag{9.295}$$

and the three components of the Lagrangian divergence of the dyad $\mathbf{BC}$, with Cartesian components B_iC_j $(i, j = x, y, z)$, are given by [see Eq. (9.285)]

$$[\mathbf{D}\cdot(\mathbf{BC})]_x = D_x(B_xC_x) + D_y(B_yC_x) + D_z(B_zC_x), \tag{9.296}$$
$$[\mathbf{D}\cdot(\mathbf{BC})]_y = D_x(B_xC_y) + D_y(B_yC_y) + D_z(B_zC_y), \tag{9.297}$$
$$[\mathbf{D}\cdot(\mathbf{BC})]_z = D_x(B_xC_z) + D_y(B_yC_z) + D_z(B_zC_z). \tag{9.298}$$

The operators D_x, D_y, and D_z in Eqs. (9.294) through (9.298) are defined by Eqs. (9.279) through (9.281). Since the unit vectors $\mathbf{i}$, $\mathbf{j}$, and $\mathbf{k}$, lying along the x, y, and z axes respectively, are fixed in space, they are independent of τ. Thus we have for the Lagrangian time derivative of any vector $\mathbf{B}$, with Cartesian coordinates B_x, B_y, and B_z,

$$\frac{\partial\mathbf{B}}{\partial\tau} = \frac{\partial}{\partial\tau}(B_x\mathbf{i} + B_y\mathbf{j} + B_z\mathbf{k}) = \mathbf{i}\frac{\partial B_x}{\partial\tau} + \mathbf{j}\frac{\partial B_y}{\partial\tau} + \mathbf{k}\frac{\partial B_z}{\partial\tau}, \tag{9.299}$$

and the x, y, and z components of $\partial\mathbf{B}/\partial\tau$ are then clearly given by

$$\left(\frac{\partial\mathbf{B}}{\partial\tau}\right)_x = \frac{\partial B_x}{\partial\tau}; \quad \left(\frac{\partial\mathbf{B}}{\partial\tau}\right)_y = \frac{\partial B_y}{\partial\tau}; \quad \left(\frac{\partial\mathbf{B}}{\partial\tau}\right)_z = \frac{\partial B_z}{\partial\tau}. \tag{9.300}$$

The coordinates x, y, and z represent the three Cartesian components of the position vector $\mathbf{r}$, and we let u_x, u_y, and u_z represent the three Cartesian components of the fluid velocity $\mathbf{u}$. Similarly, we let F_x, F_y, and F_z denote the three Cartesian components of the radiative flux

The Equations of Radiation Hydrodynamics

F_r. We denote by p_{ij} $(i, j = x, y, z)$ the nine Cartesian components of the radiation pressure tensor P_r. We note that it is the product of two vectors, namely $\Omega\Omega$, which gives P_r its tensor character [see Eq. (1.20)]. Thus the x, y, and z components of the Lagrangian divergence of the radiation pressure tensor follow directly from Eqs. (9.296) through (9.298).

The above definitions allow us to immediately write the relativistic Lagrangian equations of hydrodynamics in a Cartesian coordinate system. The three components of the position equation (9.240) are

$$\frac{\partial x}{\partial \tau} = u_x, \tag{9.301}$$

$$\frac{\partial y}{\partial \tau} = u_y, \tag{9.302}$$

$$\frac{\partial z}{\partial \tau} = u_z. \tag{9.303}$$

The three components of the momentum equation (9.286) are

$$\Delta\varrho\,\frac{\partial}{\partial\tau}\left[\frac{\Lambda^2(\varrho c^2 + E_m + P_m)\,u_x + F_x}{\Delta\varrho c^2}\right] + D_x P_m$$

$$+ D_x\left(p_{xx} - \frac{1}{c^2}\,u_x F_x\right) + D_y\left(p_{yx} - \frac{1}{c^2}\,u_y F_x\right) + D_z\left(p_{zx} - \frac{1}{c^2}\,u_z F_x\right) = 0, \tag{9.304}$$

$$\Delta\varrho\,\frac{\partial}{\partial\tau}\left[\frac{\Lambda^2(\varrho c^2 + E_m + P_m)\,u_y + F_y}{\Delta\varrho c^2}\right] + D_y P_m$$

$$+ D_x\left(p_{xy} - \frac{1}{c^2}\,u_x F_y\right) + D_y\left(p_{yy} - \frac{1}{c^2}\,u_y F_y\right) + D_z\left(p_{zy} - \frac{1}{c^2}\,u_z F_y\right) = 0, \tag{9.305}$$

$$\Delta\varrho\,\frac{\partial}{\partial\tau}\left[\frac{\Lambda^2(\varrho c^2 + E_m + P_m)\,u_z + F_z}{\Delta\varrho c^2}\right] + D_z P_m$$

$$+ D_x\left(p_{xz} - \frac{1}{c^2}\,u_x F_z\right) + D_y\left(p_{yz} - \frac{1}{c^2}\,u_y F_z\right) + D_z\left(p_{zz} - \frac{1}{c^2}\,u_z F_z\right) = 0. \tag{9.306}$$

The energy equation (9.287) is written

$$\Delta\varrho\,\frac{\partial}{\partial\tau}\left[\frac{\Lambda(\Lambda-1)\,\varrho c^2 + \Lambda^2(E_m + P_m) - P_m + E_r}{\Delta\varrho}\right] + D_x[F_x + (P_m - E_r)\,u_x]$$

$$+ D_y[F_y + (P_m - E_r)\,u_y] + D_z[F_z + (P_m - E_r)\,u_z] = W. \tag{9.307}$$

By letting c approach infinity in the material terms in Eqs. (9.301) through (9.307) we obtain the nonrelativistic Lagrangian equations of hydrodynamics in a Cartesian coordinate system. The position equations (9.301) through (9.303) remain the same. The momentum equations (9.304) through (9.306) become

$$\varrho\frac{\partial}{\partial\tau}\left(u_x+\frac{F_x}{\varrho c^2}\right)+D_xP_m+D_x\left(p_{xx}-\frac{1}{c^2}\,u_xF_x\right)$$

$$+D_y\left(p_{yx}-\frac{1}{c^2}\,u_yF_x\right)+D_z\left(p_{zx}-\frac{1}{c^2}\,u_zF_x\right)=0, \tag{9.308}$$

$$\varrho\frac{\partial}{\partial\tau}\left(u_y+\frac{F_y}{\varrho c^2}\right)+D_yP_m+D_x\left(p_{xy}-\frac{1}{c^2}\,u_xF_y\right)$$

$$+D_y\left(p_{yy}-\frac{1}{c^2}\,u_yF_y\right)+D_z\left(p_{zy}-\frac{1}{c^2}\,u_zF_y\right)=0, \tag{9.309}$$

$$\varrho\frac{\partial}{\partial\tau}\left(u_z+\frac{F_z}{\varrho c^2}\right)+D_zP_m+D_x\left(p_{xz}-\frac{1}{c^2}\,u_xF_z\right)$$

$$+D_y\left(p_{yz}-\frac{1}{c^2}\,u_yF_z\right)+D_z\left(p_{zz}-\frac{1}{c^2}\,u_zF_z\right)=0. \tag{9.310}$$

The energy equation (9.307) becomes

$$\varrho\frac{\partial}{\partial\tau}\left[\frac{1}{2}u^2+\frac{1}{\varrho}\,(E_m+E_r)\right]+D_x[F_x+(P_m-E_r)\,u_x]$$

$$+D_y[F_y+(P_m-E_r)\,u_y]+D_z[F_z+(P_m-E_r)\,u_z]=W. \tag{9.311}$$

Equations (9.308) through (9.311) are the Cartesian form of the nonrelativistic Lagrangian momentum and energy equations (9.292) and (9.293).

We now turn to a cylindrical system with coordinates r, θ, and z, defined in terms of the Cartesian coordinates x, y, and z by the standard relationships Eqs. (9.103) through (9.105). The r, θ, and z components of the Eulerian gradient of any scalar A in cylindrical coordinates are given by Eq. (9.112). To obtain the Lagrangian gradient, we need only compute $\partial(\ )/\partial r$, $\partial(\ )/\partial\theta$, and $\partial(\ )/\partial z$ in Lagrangian form. Using methods identical to those used in the Cartesian analysis to obtain $\partial(\ )/\partial x$, $\partial(\ )/\partial y$, and $\partial(\ )/\partial z$, Eqs. (9.279) through (9.281), we find

$$\frac{\partial(\ )}{\partial r}=\frac{1}{K}\begin{vmatrix}\dfrac{\partial(\ )}{\partial r_0} & \dfrac{\partial(\ )}{\partial\theta_0} & \dfrac{\partial(\ )}{\partial z_0}\\[1.2em]\dfrac{\partial\theta}{\partial r_0} & \dfrac{\partial\theta}{\partial\theta_0} & \dfrac{\partial\theta}{\partial z_0}\\[1.2em]\dfrac{\partial z}{\partial r_0} & \dfrac{\partial z}{\partial\theta_0} & \dfrac{\partial z}{\partial z_0}\end{vmatrix}\equiv D_r(\ ), \tag{9.312}$$

$$\frac{\partial(\ )}{\partial\theta}=\frac{1}{K}\begin{vmatrix}\dfrac{\partial r}{\partial r_0} & \dfrac{\partial r}{\partial\theta_0} & \dfrac{\partial r}{\partial z_0}\\[1.2em]\dfrac{\partial(\ )}{\partial r_0} & \dfrac{\partial(\ )}{\partial\theta_0} & \dfrac{\partial(\ )}{\partial z_0}\\[1.2em]\dfrac{\partial z}{\partial r_0} & \dfrac{\partial z}{\partial\theta_0} & \dfrac{\partial z}{\partial z_0}\end{vmatrix}\equiv D_\theta(\ ), \tag{9.313}$$

$$\frac{\partial(\ \ )}{\partial z} = \frac{1}{K} \begin{vmatrix} \dfrac{\partial r}{\partial r_0} & \dfrac{\partial r}{\partial \theta_0} & \dfrac{\partial r}{\partial z_0} \\[2ex] \dfrac{\partial \theta}{\partial r_0} & \dfrac{\partial \theta}{\partial \theta_0} & \dfrac{\partial \theta}{\partial z_0} \\[2ex] \dfrac{\partial(\ \ )}{\partial r_0} & \dfrac{\partial(\ \ )}{\partial \theta_0} & \dfrac{\partial(\ \ )}{\partial z_0} \end{vmatrix} \equiv D_z(\ \ ). \tag{9.314}$$

Here K is the determinant

$$K = \begin{vmatrix} \dfrac{\partial r}{\partial r_0} & \dfrac{\partial r}{\partial \theta_0} & \dfrac{\partial r}{\partial z_0} \\[2ex] \dfrac{\partial \theta}{\partial r_0} & \dfrac{\partial \theta}{\partial \theta_0} & \dfrac{\partial \theta}{\partial z_0} \\[2ex] \dfrac{\partial z}{\partial r_0} & \dfrac{\partial z}{\partial \theta_0} & \dfrac{\partial z}{\partial z_0} \end{vmatrix}. \tag{9.315}$$

The variables r_0, θ_0, and z_0 in Eqs. (9.312) through (9.315) are, of course, the labels in cylindrical coordinates which define a fluid particle, and give the position of this particle at the initial time $\tau = 0$. The last equality in Eqs. (9.312) through (9.314) is a defining equality for the Lagrangian operators $D_r(\ \ )$, $D_\theta(\ \ )$, and $D_z(\ \ )$. It then follows from Eqs. (9.112) and (9.283) that the r, θ, and z components of the Lagrangian gradient of any scalar A are given by

$$(\mathbf{D}A)_r = D_r A; \quad (\mathbf{D}A)_\theta = \frac{1}{r} D_\theta A; \quad (\mathbf{D}A)_z = D_z A. \tag{9.316}$$

It should be emphasized that Eq. (9.316) gives the components of the Lagrangian gradient with respect to a local orthogonal coordinate system at the point $\mathbf{r}$, not $\mathbf{r}_0$. This is so in spite of the fact that r_0, θ_0, and z_0 are the independent spatial variables in the Lagrangian equations. Throughout the analysis we shall consider the cylindrical components of all vectors to be with respect to a local orthogonal system at the point $\mathbf{r}$.

The Eulerian divergence of any vector $\mathbf{B}$ with cylindrical components B_r, B_θ, and B_z is given in cylindrical coordinates by Eq. (9.113). This equation, together with Eq. (9.284), gives the Lagrangian divergence of a vector $\mathbf{B}$ in cylindrical coordinates as

$$\mathbf{D} \cdot \mathbf{B} = \frac{1}{r} D_r(rB_r) + \frac{1}{r} D_\theta B_\theta + D_z B_z, \tag{9.317}$$

where the operators $D_r(\ \ )$, $D_\theta(\ \ )$, and $D_z(\ \ )$ are again given by Eqs. (9.312) through (9.315). Similarly, Eqs. (9.123) through (9.125) and Eq. (9.285) give for the r, θ, and z components of the Lagrangian divergence of any dyad $\mathbf{BC}$,

$$[\mathbf{D} \cdot (\mathbf{BC})]_r = \frac{1}{r} D_r(rB_r C_r) + \frac{1}{r} D_\theta(B_\theta C_r) + D_z(B_z C_r) - \frac{1}{r} B_\theta C_\theta, \tag{9.318}$$

$$[\mathbf{D} \cdot (\mathbf{BC})]_\theta = \frac{1}{r} D_r(rB_r C_\theta) + \frac{1}{r} D_\theta(B_\theta C_\theta) + D_z(B_z C_\theta) + \frac{1}{r} B_\theta C_r, \tag{9.319}$$

$$[\mathbf{D} \cdot (\mathbf{BC})]_z = \frac{1}{r} D_r(rB_r C_z) + \frac{1}{r} D_\theta(B_\theta C_z) + D_z(B_z C_z). \tag{9.320}$$

Finally, we consider the Lagrangian time derivative of a vector $\mathbf{B}$ in cylindrical coordinates. We have

$$\frac{\partial \mathbf{B}}{\partial \tau} = \frac{\partial}{\partial \tau}[B_r\mathbf{e}_r+B_\theta\mathbf{e}_\theta+B_z\mathbf{e}_z] = \mathbf{e}_r\frac{\partial B_r}{\partial \tau}+\mathbf{e}_\theta\frac{\partial B_\theta}{\partial \tau}+\mathbf{e}_z\frac{\partial B_z}{\partial \tau}+B_r\frac{\partial \mathbf{e}_r}{\partial \tau}+B_\theta\frac{\partial \mathbf{e}_\theta}{\partial \tau}. \tag{9.321}$$

Here $\mathbf{e}_r$, $\mathbf{e}_\theta$, and $\mathbf{e}_z$ are unit vectors, at the point $\mathbf{r}$, along the r, θ, and z directions, respectively. In writing Eq. (9.321) we have used $\partial \mathbf{e}_z/\partial \tau = 0$ since $\mathbf{e}_z$ is fixed in space and hence independent of τ. Now, from Eqs. (9.106) through (9.108), which give the Cartesian components of any vector in terms of the cylindrical components, we find

$$\mathbf{e}_r = \mathbf{i} \cos \theta+\mathbf{j} \sin \theta, \tag{9.322}$$

$$\mathbf{e}_\theta = -\mathbf{i} \sin \theta+\mathbf{j} \cos \theta, \tag{9.323}$$

where $\mathbf{i}$ and $\mathbf{j}$ are unit vectors along the Cartesian x and y axes (these vectors are fixed in space and hence independent of τ). Equation (9.321) can then be written

$$\frac{\partial \mathbf{B}}{\partial \tau} = \mathbf{e}_r\frac{\partial B_r}{\partial \tau}+\mathbf{e}_\theta\frac{\partial B_\theta}{\partial \tau}+\mathbf{e}_z\frac{\partial B_z}{\partial \tau}$$
$$+[B_r(-\mathbf{i} \sin \theta+\mathbf{j} \cos \theta)+B_\theta(-\mathbf{i} \cos \theta-\mathbf{j} \sin \theta)]\frac{\partial \theta}{\partial \tau}. \tag{9.324}$$

From Eqs. (9.109) through (9.111), which give the cylindrical components of any vector in terms of its Cartesian components, we find

$$\mathbf{i} = \mathbf{e}_r \cos \theta-\mathbf{e}_\theta \sin \theta, \tag{9.325}$$

$$\mathbf{j} = \mathbf{e}_r \sin \theta+\mathbf{e}_\theta \cos \theta. \tag{9.326}$$

Use of Eqs. (9.325) and (9.326) in Eq. (9.324) gives the final result for the Lagrangian time derivative of an arbitrary vector $\mathbf{B}$ in a cylindrical coordinate system. This result is

$$\frac{\partial \mathbf{B}}{\partial \tau} = \left(\frac{\partial B_r}{\partial \tau}-B_\theta\frac{\partial \theta}{\partial \tau}\right)\mathbf{e}_r+\left(\frac{\partial B_\theta}{\partial \tau}+B_r\frac{\partial \theta}{\partial \tau}\right)\mathbf{e}_\theta+\left(\frac{\partial B_z}{\partial \tau}\right)\mathbf{e}_z. \tag{9.327}$$

The position vector in cylindrical coordinates is given by

$$\mathbf{r} = r\mathbf{e}_r+z\mathbf{e}_z. \tag{9.328}$$

If we apply Eq. (9.327) to the vector $\mathbf{r}$, and let u_r, u_θ, and u_z denote the three cylindrical components of the fluid velocity $\mathbf{u}$, we find the position equation (9.240) becomes, upon equating components of this vector equation,

$$\frac{\partial r}{\partial \tau} = u_r, \tag{9.329}$$

$$r\frac{\partial \theta}{\partial \tau} = u_\theta, \tag{9.330}$$

$$\frac{\partial z}{\partial \tau} = u_z. \tag{9.331}$$

Using Eq. (9.330) to eliminate $\partial\theta/\partial\tau$, we can rewrite Eq. (9.327) as

$$\frac{\partial \mathbf{B}}{\partial \tau} = \left(\frac{\partial B_r}{\partial \tau} - \frac{1}{r} B_\theta u_\theta\right) \mathbf{e}_r + \left(\frac{\partial B_\theta}{\partial \tau} + \frac{1}{r} B_r u_\theta\right) \mathbf{e}_\theta + \left(\frac{\partial B_z}{\partial \tau}\right) \mathbf{e}_z, \tag{9.332}$$

a form we shall find useful later when we consider the momentum equation.

To obtain the Jacobian J in a cylindrical coordinate system, we apply the divergence expression, Eq. (9.317), to the fluid velocity $\mathbf{u}$, interpreting the result as $(\partial J/\partial \tau)/J$ [see Eq. (9.268)]. We thus have

$$\frac{1}{J}\frac{\partial J}{\partial \tau} = \frac{1}{r} D_r(r u_r) + \frac{1}{r} D_\theta u_\theta + D_z u_z, \tag{9.333}$$

or, using Eqs. (9.329) through (9.331) and recognizing that $D_\theta r = 0$,

$$\frac{1}{J}\frac{\partial J}{\partial \tau} = \frac{1}{r} D_r\left(r\,\frac{\partial r}{\partial \tau}\right) + D_\theta\left(\frac{\partial \theta}{\partial \tau}\right) + D_z\left(\frac{\partial z}{\partial \tau}\right). \tag{9.334}$$

Using the defining equations for the operators $D_r(\ )$, $D_\theta(\ )$, and $D_z(\ )$, Eqs. (9.312) through (9.314), we find that Eq. (9.334) simplifies to

$$\frac{1}{J}\frac{\partial J}{\partial \tau} = \frac{1}{rK}\frac{\partial(rK)}{\partial \tau}, \tag{9.335}$$

where K is the determinant given by Eq. (9.315). Integration of Eq. (9.335) gives

$$J = CrK, \tag{9.336}$$

where the integration "constant" C may depend upon r_0, θ_0, and z_0, but not τ. We can evaluate this constant by considering $\tau = 0$. Clearly, since at $\tau = 0$ we have $\mathbf{r} = \mathbf{r}_0$, we find $J(\tau = 0) = K(\tau = 0) = 1$ [see Eqs. (9.255) and (9.315)]. This gives $C = 1/r_0$, and hence

$$J = \frac{r}{r_0} K. \tag{9.337}$$

Using Eq. (9.315) for K, we find that in a cylindrical coordinate system the Jacobian J is given by

$$J = \frac{r}{r_0}\begin{vmatrix} \dfrac{\partial r}{\partial r_0} & \dfrac{\partial r}{\partial \theta_0} & \dfrac{\partial r}{\partial z_0} \\[2ex] \dfrac{\partial \theta}{\partial r_0} & \dfrac{\partial \theta}{\partial \theta_0} & \dfrac{\partial \theta}{\partial z_0} \\[2ex] \dfrac{\partial z}{\partial r_0} & \dfrac{\partial z}{\partial \theta_0} & \dfrac{\partial z}{\partial z_0} \end{vmatrix}. \tag{9.338}$$

To write the Lagrangian momentum and energy equations in a cylindrical coordinate system, we let F_r, F_θ, and F_z denote the three cylindrical components of the radiative flux $\mathbf{F}_r$. We denote by p_{ij} $(i, j = r,\, \theta,\, z)$ the nine cylindrical components of the radiation pressure tensor $\mathbf{P}_r$ [see Eqs. (9.126) and (9.127)].

With these definitions and the expressions we have derived for the differential Lagrangian operators in cylindrical coordinates, the three components of the relativistic momentum equation (9.286) can immediately be written as

$$\Lambda\varrho\,\frac{\partial}{\partial\tau}\left[\frac{\Lambda^2(\varrho c^2+E_m+P_m)u_r+F_r}{\Lambda\varrho c^2}\right]-\frac{1}{r}\,u_\theta\left[\frac{\Lambda^2(\varrho c^2+E_m+P_m)u_\theta+F_\theta}{c^2}\right]$$

$$+D_rP_m+\frac{1}{r}\,D_r\left[r\left(p_{rr}-\frac{1}{c^2}\,u_rF_r\right)\right]+\frac{1}{r}\,D_\theta\left(p_{\theta r}-\frac{1}{c^2}\,u_\theta F_r\right)$$

$$+D_z\left(p_{zr}-\frac{1}{c^2}\,u_zF_r\right)-\frac{1}{r}\left(p_{\theta\theta}-\frac{1}{c^2}\,u_\theta F_\theta\right)=0,\tag{9.339}$$

$$\Lambda\varrho\,\frac{\partial}{\partial\tau}\left[\frac{\Lambda^2(\varrho c^2+E_m+P_m)u_\theta+F_\theta}{\Lambda\varrho c^2}\right]+\frac{1}{r}\,u_\theta\left[\frac{\Lambda^2(\varrho c^2+E_m+P_m)\,u_r+F_r}{c^2}\right]$$

$$+\frac{1}{r}\,D_\theta P_m+\frac{1}{r}\,D_r\left[r\left(p_{r\theta}-\frac{1}{c^2}\,u_rF_\theta\right)\right]+\frac{1}{r}\,D_\theta\left(p_{\theta\theta}-\frac{1}{c^2}\,u_\theta F_\theta\right)$$

$$+D_z\left(p_{z\theta}-\frac{1}{c^2}\,u_zF_\theta\right)+\frac{1}{r}\left(p_{\theta r}-\frac{1}{c^2}\,u_\theta F_r\right)=0,\tag{9.340}$$

$$\Lambda\varrho\,\frac{\partial}{\partial\tau}\left[\frac{\Lambda^2(\varrho c^2+E_m+P_m)\,u_z+F_z}{\Lambda\varrho c^2}\right]+D_zP_m+\frac{1}{r}\,D_r\left[r\left(p_{rz}-\frac{1}{c^2}\,u_rF_z\right)\right]$$

$$+\frac{1}{r}\,D_\theta\left(p_{\theta z}-\frac{1}{c^2}\,u_\theta F_z\right)+D_z\left(p_{zz}-\frac{1}{c^2}\,u_zF_z\right)=0.\tag{9.341}$$

The relativistic energy equation (9.287) in cylindrical coordinates is written

$$\Lambda\varrho\,\frac{\partial}{\partial\tau}\left[\frac{\Lambda(\Lambda-1)\varrho c^2+\Lambda^2(E_m+P_m)-P_m+E_r}{\Lambda\varrho}\right]+\frac{1}{r}\,D_r[rF_r+r(P_m-E_r)u_r]$$

$$+\frac{1}{r}\,D_\theta[F_\theta+(P_m-E_r)u_\theta]+D_z[F_z+(P_m-E_r)u_z]=W.\tag{9.342}$$

The nonrelativistic limit of Eqs. (9.339) through (9.342) is found by letting c approach infinity in the fluid terms. The momentum equations (9.339) through (9.341) become

$$\varrho\,\frac{\partial}{\partial\tau}\left(u_r+\frac{F_r}{\varrho c^2}\right)-\varrho\,\frac{u_\theta}{r}\left(u_\theta+\frac{F_\theta}{\varrho c^2}\right)+D_rP_m+\frac{1}{r}\,D_r\left[r\left(p_{rr}-\frac{1}{c^2}\,u_rF_r\right)\right]$$

$$+\frac{1}{r}\,D_\theta\left(p_{\theta r}-\frac{1}{c^2}\,u_\theta F_r\right)+D_z\left(p_{zr}-\frac{1}{c^2}\,u_zF_r\right)-\frac{1}{r}\left(p_{\theta\theta}-\frac{1}{c^2}\,u_\theta F_\theta\right)=0,\tag{9.343}$$

$$\varrho\,\frac{\partial}{\partial\tau}\left(u_\theta+\frac{F_\theta}{\varrho c^2}\right)+\varrho\,\frac{u_\theta}{r}\left(u_r+\frac{F_r}{\varrho c^2}\right)+\frac{1}{r}\,D_\theta P_m+\frac{1}{r}\,D_r\left[r\left(p_{r\theta}-\frac{1}{c^2}\,u_rF_\theta\right)\right]$$

$$+\frac{1}{r}\,D_\theta\left(p_{\theta\theta}-\frac{1}{c^2}\,u_\theta F_\theta\right)+D_z\left(p_{z\theta}-\frac{1}{c^2}\,u_zF_\theta\right)+\frac{1}{r}\left(p_{\theta r}-\frac{1}{c^2}\,u_\theta F_r\right)=0,\tag{9.344}$$

$$\varrho\,\frac{\partial}{\partial\tau}\left(u_z+\frac{F_z}{\varrho c^2}\right)+D_zP_m+\frac{1}{r}\,D_r\left[r\left(p_{rz}-\frac{1}{c^2}\,u_rF_z\right)\right]$$

$$+\frac{1}{r}\,D_\theta\left(p_{\theta z}-\frac{1}{c^2}\,u_\theta F_z\right)+D_z\left(p_{zz}-\frac{1}{c^2}\,u_zF_z\right)=0.\tag{9.345}$$

The Equations of Radiation Hydrodynamics

The energy equation (9.342) becomes

$$\varrho \frac{\partial}{\partial \tau}\left[\frac{1}{2}u^2 + \frac{1}{\varrho}(E_m + E_r)\right] + \frac{1}{r}D_r[rF_r + r(P_m - E_r)u_r]$$

$$+ \frac{1}{r}D_\theta[F_\theta + (P_m - E_r)u_\theta] + D_z[F_z + (P_m - E_r)u_z] = W. \tag{9.346}$$

Equations (9.343) through (9.346) are the cylindrical coordinate forms of the nonrelativistic Lagrangian momentum and energy equations (9.292) and (9.293).

We now turn to a spherical system with coordinates r, θ, and φ, defined in terms of the Cartesian coordinates x, y, and z by the standard relationships Eqs. (9.138) through (9.140). The r, θ, and φ components of the Eulerian gradient of any scalar A in spherical coordinates are given by Eq. (9.147). To obtain the Lagrangian gradient, we need only compute $\partial(\ )/\partial r$, $\partial(\ )/\partial\theta$, and $\partial(\ )/\partial\varphi$ in Lagrangian form. Using methods identical to those used in the Cartesian analysis to obtain $\partial(\ )/\partial x$, $\partial(\ )/\partial y$, and $\partial(\ )/\partial z$, Eqs. (9.279) through (9.281), we find

$$\frac{\partial(\ )}{\partial r} = \frac{1}{K}\begin{vmatrix} \dfrac{\partial(\ )}{\partial r_0} & \dfrac{\partial(\ )}{\partial\theta_0} & \dfrac{\partial(\ )}{\partial\varphi_0} \\[2ex] \dfrac{\partial\theta}{\partial r_0} & \dfrac{\partial\theta}{\partial\theta_0} & \dfrac{\partial\theta}{\partial\varphi_0} \\[2ex] \dfrac{\partial\varphi}{\partial r_0} & \dfrac{\partial\varphi}{\partial\theta_0} & \dfrac{\partial\varphi}{\partial\varphi_0} \end{vmatrix} \equiv D_r(\ ), \tag{9.347}$$

$$\frac{\partial(\ )}{\partial\theta} = \frac{1}{K}\begin{vmatrix} \dfrac{\partial r}{\partial r_0} & \dfrac{\partial r}{\partial\theta_0} & \dfrac{\partial r}{\partial\varphi_0} \\[2ex] \dfrac{\partial(\ )}{\partial r_0} & \dfrac{\partial(\ )}{\partial\theta_0} & \dfrac{\partial(\ )}{\partial\varphi_0} \\[2ex] \dfrac{\partial\varphi}{\partial r_0} & \dfrac{\partial\varphi}{\partial\theta_0} & \dfrac{\partial\varphi}{\partial\varphi_0} \end{vmatrix} \equiv D_\theta(\ ), \tag{9.348}$$

$$\frac{\partial(\ )}{\partial\varphi} = \frac{1}{K}\begin{vmatrix} \dfrac{\partial r}{\partial r_0} & \dfrac{\partial r}{\partial\theta_0} & \dfrac{\partial r}{\partial\varphi_0} \\[2ex] \dfrac{\partial\theta}{\partial r_0} & \dfrac{\partial\theta}{\partial\theta_0} & \dfrac{\partial\theta}{\partial\varphi_0} \\[2ex] \dfrac{\partial(\ )}{\partial r_0} & \dfrac{\partial(\ )}{\partial\theta_0} & \dfrac{\partial(\ )}{\partial\varphi_0} \end{vmatrix} \equiv D_\varphi(\ ). \tag{9.349}$$

Here K is the determinant

$$K = \begin{vmatrix} \dfrac{\partial r}{\partial r_0} & \dfrac{\partial r}{\partial\theta_0} & \dfrac{\partial r}{\partial\varphi_0} \\[2ex] \dfrac{\partial\theta}{\partial r_0} & \dfrac{\partial\theta}{\partial\theta_0} & \dfrac{\partial\theta}{\partial\varphi_0} \\[2ex] \dfrac{\partial\varphi}{\partial r_0} & \dfrac{\partial\varphi}{\partial\theta_0} & \dfrac{\partial\varphi}{\partial\varphi_0} \end{vmatrix}. \tag{9.350}$$

The variables r_0, θ_0, and φ_0 in Eqs. (9.347) through (9.350) are, of course, the labels in spherical coordinates which define a fluid particle, and give the position of this particle at the initial time $\tau = 0$. The last equality in Eqs. (9.347) through (9.349) is a defining equality for the Lagrangian operators $D_r(\ )$, $D_\theta(\ )$, and $D_\varphi(\ )$. It then follows from Eqs. (9.147) and (9.283) that the r, θ, and φ components of the Lagrangian gradient of any scalar A are given by

$$(\mathbf{D}A)_r = D_r A; \quad (\mathbf{D}A)_\theta = \frac{1}{r} D_\theta A; \quad (\mathbf{D}A)_\varphi = \frac{1}{r \sin \theta} D_\varphi A. \tag{9.351}$$

It should be emphasized that Eq. (9.351) gives the components of the Lagrangian gradient with respect to a local orthogonal coordinate system at the point $\mathbf{r}$, not $\mathbf{r}_0$. This is so in spite of the fact that r_0, θ_0, and φ_0 are the independent spatial variables in the Lagrangian equations. Throughout the analysis we shall consider the spherical components of all vectors to be with respect to a local orthogonal system at the point $\mathbf{r}$.

The Eulerian divergence of any vector $\mathbf{B}$ with spherical components B_r, B_θ, and B_φ is given in spherical coordinates by Eq. (9.148). This equation, together with Eq. (9.284), gives the Lagrangian divergence of a vector $\mathbf{B}$ in spherical coordinates as

$$\mathbf{D} \cdot \mathbf{B} = \frac{1}{r^2} D_r(r^2 B_r) + \frac{1}{r \sin \theta} D_\theta(\sin \theta B_\theta) + \frac{1}{r \sin \theta} D_\varphi B_\varphi, \tag{9.352}$$

where the operators $D_r(\ )$, $D_\theta(\ )$, and $D_\varphi(\ )$ are again given by Eqs. (9.347) through (9.349). Similarly, Eqs. (9.149) through (9.151) and Eq. (9.285) give for the r, θ, and φ components of the Lagrangian divergence of any dyad $\mathbf{BC}$,

$$[\mathbf{D} \cdot (\mathbf{BC})]_r = \frac{1}{r^2} D_r(r^2 B_r C_r) + \frac{1}{r \sin \theta} D_\theta(\sin \theta B_\theta C_r)$$

$$+ \frac{1}{r \sin \theta} D_\varphi(B_\varphi C_r) - \frac{1}{r}(B_\theta C_\theta + B_\varphi C_\varphi), \tag{9.353}$$

$$[\mathbf{D} \cdot (\mathbf{BC})]_\theta = \frac{1}{r^2} D_r(r^2 B_r C_\theta) + \frac{1}{r \sin \theta} D_\theta(\sin \theta B_\theta C_\theta)$$

$$+ \frac{1}{r \sin \theta} D_\varphi(B_\varphi C_\theta) + \frac{1}{r} B_\theta C_r - \frac{1}{r} \cot \theta B_\varphi C_\varphi, \tag{9.354}$$

$$[\mathbf{D} \cdot (\mathbf{BC})]_\varphi = \frac{1}{r^2} D_r(r^2 B_r C_\varphi) + \frac{1}{r \sin \theta} D_\theta(\sin \theta B_\theta C_\varphi)$$

$$+ \frac{1}{r \sin \theta} D_\varphi(B_\varphi C_\varphi) + \frac{1}{r} B_\varphi C_r + \frac{1}{r} \cot \theta B_\varphi C_\theta. \tag{9.355}$$

Finally, we consider the Lagrangian time derivative of a vector $\mathbf{B}$ in spherical coordinates. We have

$$\frac{\partial \mathbf{B}}{\partial \tau} = \frac{\partial}{\partial \tau} [B_r \mathbf{e}_r + B_\theta \mathbf{e}_\theta + B_\varphi \mathbf{e}_\varphi]$$

$$= \mathbf{e}_r \frac{\partial B_r}{\partial \tau} + \mathbf{e}_\theta \frac{\partial B_\theta}{\partial \tau} + \mathbf{e}_\varphi \frac{\partial B_\varphi}{\partial \tau} + B_r \frac{\partial \mathbf{e}_r}{\partial \tau} + B_\theta \frac{\partial \mathbf{e}_\theta}{\partial \tau} + B_\varphi \frac{\partial \mathbf{e}_\varphi}{\partial \tau}. \tag{9.356}$$

The Equations of Radiation Hydrodynamics

Here e_r, e_θ, and e_φ are unit vectors, at the point r, along the r, θ, and φ directions, respectively. Now, from Eqs. (9.141) through (9.143), which give the Cartesian components of any vector in terms of the spherical components, we find

$$e_r = i \sin\theta\cos\varphi + j\sin\theta\sin\varphi + k\cos\theta, \tag{9.357}$$

$$e_\theta = i\cos\theta\cos\varphi + j\cos\theta\sin\varphi - k\sin\theta, \tag{9.358}$$

$$e_\varphi = -i\sin\varphi + j\cos\varphi, \tag{9.359}$$

where i, j, and k are unit vectors along the Cartesian x, y, and z axes. Since these axes are fixed in space, the vectors i, j, and k are independent of τ. Equation (9.356) can then be written

$$\frac{\partial B}{\partial \tau} = e_r \frac{\partial B_r}{\partial \tau} + e_\theta \frac{\partial B_\theta}{\partial \tau} + e_\varphi \frac{\partial B_\varphi}{\partial \tau}$$

$$+ B_r(i\cos\theta\cos\varphi + j\cos\theta\sin\varphi - k\sin\theta)\frac{\partial\theta}{\partial\tau}$$

$$+ B_r(-i\sin\theta\sin\varphi + j\sin\theta\cos\varphi)\frac{\partial\varphi}{\partial\tau}$$

$$+ B_\theta(-i\sin\theta\cos\varphi - j\sin\theta\sin\varphi - k\cos\theta)\frac{\partial\theta}{\partial\tau}$$

$$+ B_\theta(-i\cos\theta\sin\varphi + j\cos\theta\cos\varphi)\frac{\partial\varphi}{\partial\tau}$$

$$+ B_\varphi(-i\cos\varphi - j\sin\varphi)\frac{\partial\varphi}{\partial\tau}. \tag{9.360}$$

From Eqs. (9.144) through (9.146), which give the spherical components of any vector in terms of its Cartesian components, we find

$$i = e_r\sin\theta\cos\varphi + e_\theta\cos\theta\cos\varphi - e_\varphi\sin\varphi, \tag{9.361}$$

$$j = e_r\sin\theta\sin\varphi + e_\theta\cos\theta\sin\varphi + e_\varphi\cos\varphi, \tag{9.362}$$

$$k = e_r\cos\theta - e_\theta\sin\theta. \tag{9.363}$$

Use of Eqs. (9.361) through (9.363) in Eq. (9.360) gives the final result for the Lagrangian time derivative of an arbitrary vector B in a spherical coordinate system. This result is

$$\frac{\partial B}{\partial\tau} = \left(\frac{\partial B_r}{\partial\tau} - B_\theta\frac{\partial\theta}{\partial\tau} - B_\varphi\sin\theta\frac{\partial\varphi}{\partial\tau}\right)e_r$$

$$+ \left(\frac{\partial B_\theta}{\partial\tau} + B_r\frac{\partial\theta}{\partial\tau} - B_\varphi\cos\theta\frac{\partial\varphi}{\partial\tau}\right)e_\theta$$

$$+ \left[\frac{\partial B_\varphi}{\partial\tau} + (B_r\sin\theta + B_\theta\cos\theta)\frac{\partial\varphi}{\partial\tau}\right]e_\varphi. \tag{9.364}$$

The position vector in spherical coordinates is given by

$$r = re_r. \tag{9.365}$$

If we apply Eq. (9.364) to the vector $\mathbf{r}$, and let u_r, u_θ, and u_φ denote the three spherical components of the fluid velocity $\mathbf{u}$, we find the position equation (9.240) becomes, upon equating components of this vector equation,

$$\frac{\partial r}{\partial \tau} = u_r, \tag{9.366}$$

$$r\frac{\partial \theta}{\partial \tau} = u_\theta, \tag{9.367}$$

$$r\sin\theta\,\frac{\partial \varphi}{\partial \tau} = u_\varphi. \tag{9.368}$$

Using Eqs. (9.367) and (9.368) to eliminate $\partial\theta/\partial\tau$ and $\partial\varphi/\partial\tau$, we can rewrite Eq. (9.364) as

$$\begin{aligned}
\frac{\partial \mathbf{B}}{\partial \tau} = {}& \left(\frac{\partial B_r}{\partial \tau} - \frac{1}{r}B_\theta u_\theta - \frac{1}{r}B_\varphi u_\varphi\right)\mathbf{e}_r \\
&+ \left(\frac{\partial B_\theta}{\partial \tau} + \frac{1}{r}B_r u_\theta - \frac{1}{r}\cot\theta\, B_\varphi u_\varphi\right)\mathbf{e}_\theta \\
&+ \left(\frac{\partial B_\varphi}{\partial \tau} + \frac{1}{r}B_r u_\varphi + \frac{1}{r}\cot\theta\, B_\theta u_\varphi\right)\mathbf{e}_\varphi,
\end{aligned} \tag{9.369}$$

a form we shall find useful later when we consider the momentum equation.

To obtain the Jacobian J in a spherical coordinate system, we apply the divergence expression, Eq. (9.352), to the fluid velocity $\mathbf{u}$, interpreting the result as $(\partial J/\partial\tau)/J$ [see Eq. (9.268)]. We thus have

$$\frac{1}{J}\frac{\partial J}{\partial \tau} = \frac{1}{r^2}D_r(r^2 u_r) + \frac{1}{r\sin\theta}D_\theta(\sin\theta u_\theta) + \frac{1}{r\sin\theta}D_\varphi u_\varphi, \tag{9.370}$$

or, using Eqs. (9.366) through (9.368) and recognizing that $D_\theta r = D_\varphi(r\sin\theta) = 0$,

$$\frac{1}{J}\frac{\partial J}{\partial \tau} = \frac{1}{r^2}D_r\left(r^2\frac{\partial r}{\partial \tau}\right) + \frac{1}{\sin\theta}D_\theta\left(\sin\theta\frac{\partial \theta}{\partial \tau}\right) + D_\varphi\left(\frac{\partial \varphi}{\partial \tau}\right). \tag{9.371}$$

Using the defining equations for the operators $D_r(\ )$, $D_\theta(\ )$, and $D_\varphi(\ )$, Eqs. (9.347) through (9.349), we find that Eq. (9.371) simplifies to

$$\frac{1}{J}\frac{\partial J}{\partial \tau} = \frac{1}{Kr^2\sin\theta}\frac{\partial(Kr^2\sin\theta)}{\partial \tau}, \tag{9.372}$$

where K is the determinant given by Eq. (9.350). Integration of Eq. (9.372) gives

$$J = CKr^2\sin\theta, \tag{9.373}$$

where the integration "constant" C may depend upon r_0, θ_0, and φ_0, but not τ. We can evaluate this constant by considering $\tau = 0$. Clearly, since at $\tau = 0$ we have $\mathbf{r} = \mathbf{r}_0$, we find $J(\tau = 0) = K(\tau = 0) = 1$ [see Eqs. (9.255) and (9.350)]. This gives $C = 1/(r_0^2\sin\theta_0)$, and hence

$$J = \frac{r^2\sin\theta}{r_0^2\sin\theta_0}K. \tag{9.374}$$

The Equations of Radiation Hydrodynamics

Using Eq. (9.350) for K, we find that in a spherical coordinate system the Jacobian J is given by

$$J = \frac{r^2 \sin \theta}{r_0^2 \sin \theta_0} \begin{vmatrix} \dfrac{\partial r}{\partial r_0} & \dfrac{\partial r}{\partial \theta_0} & \dfrac{\partial r}{\partial \varphi_0} \\[2mm] \dfrac{\partial \theta}{\partial r_0} & \dfrac{\partial \theta}{\partial \theta_0} & \dfrac{\partial \theta}{\partial \varphi_0} \\[2mm] \dfrac{\partial \varphi}{\partial r_0} & \dfrac{\partial \varphi}{\partial \theta_0} & \dfrac{\partial \varphi}{\partial \varphi_0} \end{vmatrix}. \tag{9.375}$$

To write the Lagrangian momentum and energy equations in a spherical coordinate system, we let F_r, F_θ, and F_φ denote the three spherical components of the radiative flux $\mathbf{F}_r$. We denote by p_{ij} $(i, j = r, \theta, \varphi)$ the nine spherical components of the radiation pressure tensor $\mathbf{P}_r$ [see Eqs. (9.152) and (9.153)].

With these definitions and the expressions we have derived for the differential Lagrangian operators in spherical coordinates, the three components of the relativistic momentum equation (9.286) can immediately be written as

$$\begin{aligned}
&\Lambda\varrho\,\frac{\partial}{\partial \tau}\left[\frac{\Lambda^2(\varrho c^2 + E_m + P_m)u_r + F_r}{\Lambda\varrho c^2}\right] - \frac{u_\theta}{r}\left[\frac{\Lambda^2(\varrho c^2 + E_m + P_m)u_\theta + F_\theta}{c^2}\right] \\[2mm]
&\quad - \frac{u_\varphi}{r}\left[\frac{\Lambda^2(\varrho c^2 + E_m + P_m)u_\varphi + F_\varphi}{c^2}\right] + D_r P_m + \frac{1}{r^2}\,D_r\left[r^2\left(p_{rr} - \frac{1}{c^2}u_r F_r\right)\right] \\[2mm]
&\quad + \frac{1}{r\sin\theta}\,D_\theta\left[\sin\theta\left(p_{\theta r} - \frac{1}{c^2}u_\theta F_r\right)\right] + \frac{1}{r\sin\theta}\,D_\varphi\left(p_{\varphi r} - \frac{1}{c^2}u_\varphi F_r\right) \\[2mm]
&\quad - \frac{1}{r}\left[p_{\theta\theta} + p_{\varphi\varphi} - \frac{1}{c^2}(u_\theta F_\theta + u_\varphi F_\varphi)\right] = 0,
\end{aligned} \tag{9.376}$$

$$\begin{aligned}
&\Lambda\varrho\,\frac{\partial}{\partial \tau}\left[\frac{\Lambda^2(\varrho c^2 + E_m + P_m)u_\theta + F_\theta}{\Lambda\varrho c^2}\right] + \frac{u_\theta}{r}\left[\frac{\Lambda^2(\varrho c^2 + E_m + P_m)u_r + F_r}{c^2}\right] \\[2mm]
&\quad - \frac{u_\varphi}{r}\cot\theta\left[\frac{\Lambda^2(\varrho c^2 + E_m + P_m)u_\varphi + F_\varphi}{c^2}\right] + \frac{1}{r}\,D_\theta P_m + \frac{1}{r^2}\,D_r\left[r^2\left(p_{r\theta} - \frac{1}{c^2}u_r F_\theta\right)\right] \\[2mm]
&\quad + \frac{1}{r\sin\theta}\,D_\theta\left[\sin\theta\left(p_{\theta\theta} - \frac{1}{c^2}u_\theta F_\theta\right)\right] + \frac{1}{r\sin\theta}\,D_\varphi\left(p_{\varphi\theta} - \frac{1}{c^2}u_\varphi F_\theta\right) \\[2mm]
&\quad + \frac{1}{r}\left(p_{\theta r} - \frac{1}{c^2}u_\theta F_r\right) - \frac{1}{r}\cot\theta\left(p_{\varphi\varphi} - \frac{1}{c^2}u_\varphi F_\varphi\right) = 0,
\end{aligned} \tag{9.377}$$

$$\begin{aligned}
&\Lambda\varrho\,\frac{\partial}{\partial \tau}\left[\frac{\Lambda^2(\varrho c^2 + E_m + P_m)u_\varphi + F_\varphi}{\Lambda\varrho c^2}\right] + \frac{u_\varphi}{r}\left[\frac{\Lambda^2(\varrho c^2 + E_m + P_m)u_r + F_r}{c^2}\right] \\[2mm]
&\quad + \frac{u_\varphi}{r}\cot\theta\left[\frac{\Lambda^2(\varrho c^2 + E_m + P_m)u_\theta + F_\theta}{\Lambda\varrho c^2}\right] + \frac{1}{r\sin\theta}\,D_\varphi P_m \\[2mm]
&\quad + \frac{1}{r^2}\,D_r\left[r^2\left(p_{r\varphi} - \frac{1}{c^2}u_r F_\varphi\right)\right] + \frac{1}{r\sin\theta}\,D_\theta\left[\sin\theta\left(p_{\theta\varphi} - \frac{1}{c^2}u_\theta F_\varphi\right)\right] \\[2mm]
&\quad + \frac{1}{r\sin\theta}\,D_\varphi\left(p_{\varphi\varphi} - \frac{1}{c^2}u_\varphi F_\varphi\right) + \frac{1}{r}\left(p_{\varphi r} - \frac{1}{c^2}u_\varphi F_r\right) \\[2mm]
&\quad + \frac{1}{r}\cot\theta\left(p_{\varphi\theta} - \frac{1}{c^2}u_\varphi F_\theta\right) = 0.
\end{aligned} \tag{9.378}$$

The relativistic energy equation (9.287) in spherical coordinates is written

$$\Lambda\varrho\,\frac{\partial}{\partial\tau}\left[\frac{\Lambda(\Lambda-1)\,\varrho c^2+\Lambda^2(E_m+P_m)-P_m+E_r}{\Lambda\varrho}\right]+\frac{1}{r^2}\,D_r[r^2F_r+r^2(P_m-E_r)\,u_r]$$

$$+\frac{1}{r\sin\theta}\,D_\theta[\sin\theta F_\theta+\sin\theta(P_m-E_r)\,u_\theta]+\frac{1}{r\sin\theta}\,D_\varphi[F_\varphi+(P_m-E_r)\,u_\varphi]=W.\qquad(9.379)$$

The nonrelativistic limit of Eqs. (9.376) through (9.379) is found by letting c approach infinity in the fluid terms. The momentum equations (9.376) through (9.378) become

$$\varrho\,\frac{\partial}{\partial\tau}\left(u_r+\frac{F_r}{\varrho c^2}\right)-\varrho\,\frac{u_\theta}{r}\left(u_\theta+\frac{F_\theta}{\varrho c^2}\right)-\varrho\,\frac{u_\varphi}{r}\left(u_\varphi+\frac{F_\varphi}{\varrho c^2}\right)+D_rP_m$$

$$+\frac{1}{r^2}\,D_r\left[r^2\left(p_{rr}-\frac{1}{c^2}\,u_rF_r\right)\right]+\frac{1}{r\sin\theta}\,D_\theta\left[\sin\theta\left(p_{\theta r}-\frac{1}{c^2}\,u_\theta F_r\right)\right]$$

$$+\frac{1}{r\sin\theta}\,D_\varphi\left(p_{\varphi r}-\frac{1}{c^2}\,u_\varphi F_r\right)-\frac{1}{r}\left[p_{\theta\theta}+p_{\varphi\varphi}-\frac{1}{c^2}\,(u_\theta F_\theta+u_\varphi F_\varphi)\right]=0,\qquad(9.380)$$

$$\varrho\,\frac{\partial}{\partial\tau}\left(u_\theta+\frac{F_\theta}{\varrho c^2}\right)+\varrho\,\frac{u_\theta}{r}\left(u_r+\frac{F_r}{\varrho c^2}\right)-\varrho\,\frac{u_\varphi}{r}\cot\theta\left(u_\varphi+\frac{F_\varphi}{\varrho c^2}\right)$$

$$+\frac{1}{r}\,D_\theta P_m+\frac{1}{r^2}\,D_r\left[r^2\left(p_{r\theta}-\frac{1}{c^2}\,u_rF_\theta\right)\right]+\frac{1}{r\sin\theta}\,D_\theta\left[\sin\theta\left(p_{\theta\theta}-\frac{1}{c^2}\,u_\theta F_\theta\right)\right]$$

$$+\frac{1}{r\sin\theta}\,D_\varphi\left(p_{\varphi\theta}-\frac{1}{c^2}\,u_\varphi F_\theta\right)+\frac{1}{r}\left(p_{\theta r}-\frac{1}{c^2}\,u_\theta F_r\right)$$

$$-\frac{1}{r}\cot\theta\left(p_{\varphi\varphi}-\frac{1}{c^2}\,u_\varphi F_\varphi\right)=0,\qquad(9.381)$$

$$\varrho\,\frac{\partial}{\partial\tau}\left(u_\varphi+\frac{F_\varphi}{\varrho c^2}\right)+\varrho\,\frac{u_\varphi}{r}\left(u_r+\frac{F_r}{\varrho c^2}\right)+\varrho\,\frac{u_\varphi}{r}\cot\theta\left(u_\theta+\frac{F_\theta}{\varrho c^2}\right)$$

$$+\frac{1}{r\sin\theta}\,D_\varphi P_m+\frac{1}{r^2}\,D_r\left[r^2\left(p_{r\varphi}-\frac{1}{c^2}\,u_rF_\varphi\right)\right]+\frac{1}{r\sin\theta}\,D_\theta\left[\sin\theta\left(p_{\theta\varphi}-\frac{1}{c^2}\,u_\theta F_\varphi\right)\right]$$

$$+\frac{1}{r\sin\theta}\,D_\varphi\left(p_{\varphi\varphi}-\frac{1}{c^2}\,u_\varphi F_\varphi\right)+\frac{1}{r}\left(p_{\varphi r}-\frac{1}{c^2}\,u_\varphi F_r\right)$$

$$+\frac{1}{r}\cot\theta\left(p_{\varphi\theta}-\frac{1}{c^2}\,u_\varphi F_\theta\right)=0.\qquad(9.382)$$

The energy equation (9.379) becomes

$$\varrho\,\frac{\partial}{\partial\tau}\left[\frac{1}{2}\,u^2+\frac{1}{\varrho}\,(E_m+E_r)\right]+\frac{1}{r^2}\,D_r[r^2F_r+r^2(P_m-E_r)u_r]$$

$$+\frac{1}{r\sin\theta}\,D_\theta[\sin\theta F_\theta+\sin\theta(P_m-E_r)u_\theta]+\frac{1}{r\sin\theta}\,D_\varphi[F_\varphi+(P_m-E_r)u_\varphi]=W.$$

$$(9.383)$$

Equations (9.380) through (9.383) are the spherical coordinate form of the nonrelativistic Lagrangian momentum and energy equations (9.292) and (9.293).

APPENDIX

The Lorentz Transformation of the Equation of Transfer

We consider the equation of radiative transfer as seen by an observer in an inertial frame of reference. We call this the zero frame and subscript all quantities with a zero. The equation of transfer is, rewriting Eq. (2.167) with zero subscripts,

$$\frac{1}{c}\frac{\partial I_0(\nu_0, \Omega_0)}{\partial t_0} + \Omega_0 \cdot \nabla_0 I_0(\nu_0, \Omega_0)$$

$$= \left[1 + \frac{c^2}{2h\nu_0^3} I_0(\nu_0, \Omega_0)\right] S_0(\nu_0, \Omega_0) - \sigma_{a0}(\nu_0, \Omega_0) I_0(\nu_0, \Omega_0)$$

$$+ \int_0^\infty d\nu_0' \int_{4\pi} d\Omega_0' \sigma_{s0}(\nu_0' \to \nu_0, \Omega_0' \to \Omega_0) \frac{\nu_0}{\nu_0'} I_0(\nu_0', \Omega_0') \left[1 + \frac{c^2}{2h\nu_0^3} I_0(\nu_0, \Omega_0)\right]$$

$$- \int_0^\infty d\nu_0' \int_{4\pi} d\Omega_0' \sigma_{s0}(\nu_0 \to \nu_0', \Omega_0 \to \Omega_0') I_0(\nu_0, \Omega_0) \left[1 + \frac{c^2}{2h\nu_0'^3} I_0(\nu_0', \Omega_0')\right]. \tag{A.1}$$

Equation (A.1) is a slight generalization of Eq. (2.167) in that the source S and absorption coefficient σ_a are allowed an Ω dependence, and the scattering kernel σ_s can depend upon Ω and Ω' separately. The same generalization of Eq. (2.170) is

$$\frac{1}{c}\frac{\partial I_0(\nu_0, \Omega_0)}{\partial t_0} + \Omega_0 \cdot \nabla_0 I_0(\nu_0, \Omega_0) = \sigma_{a0}'(\nu_0, \Omega_0)[B_0(\nu_0, \Omega_0) - I_0(\nu_0, \Omega_0)]$$

$$+ \int_0^\infty d\nu_0' \int_{4\pi} d\Omega_0' \sigma_{s0}(\nu_0' \to \nu_0, \Omega_0' \to \Omega_0) \frac{\nu_0}{\nu_0'} I_0(\nu_0', \Omega_0') \left[1 + \frac{c^2}{2h\nu_0^3} I_0(\nu_0, \Omega_0)\right]$$

$$- \int_0^\infty d\nu_0' \int_{4\pi} d\Omega_0' \sigma_{s0}(\nu_0 \to \nu_0', \Omega_0 \to \Omega_0') I_0(\nu_0, \Omega_0) \left[1 + \frac{c^2}{2h\nu_0'^3} I_0(\nu_0', \Omega_0')\right]. \tag{A.2}$$

In both Eqs. (A.1) and (A.2) we have not explicitly written the spatial and temporal arguments of all quantities, but these dependences are understood. The remainder of this

The Equations of Radiation Hydrodynamics

Appendix is devoted to the development of the Lorentz transformations of the various terms appearing in Eqs. (A.1) and (A.2).

For this purpose, we consider a second inertial frame of reference moving with velocity $\mathbf{v}$ with respect to the zero frame. For convenience we choose a coordinate system so that the vector $\mathbf{v}$ points along the positive x_0 axis as seen by an observer in the zero frame. In this second frame we leave all quantities unadorned. Hence to an observer in this frame, Eq. (A.1) or (A.2), with all zero subscripts dropped, constitutes the description of radiative transfer. We undertake the task of relating all quantities occurring in the equation of transfer in the zero frame to those in the unadorned frame. At the particular space point and time in question, we take the two frames to coincide. Hence

$$\mathbf{r}_0 = \mathbf{r}, \tag{A.3}$$

$$t_0 = t; \tag{A.4}$$

i.e., only the identity transformation is involved for the space and time coordinates. All other transformations follow from the basic Lorentz transformation of four vectors (see any standard physics text). Denote by $\mathbf{F}_0$ a four vector with components $F_0^{(1)}, F_0^{(2)}, F_0^{(3)}, F_0^{(4)}$ in the zero frame, and let $\mathbf{F}$ with components $F^{(1)}, F^{(2)}, F^{(3)}, F^{(4)}$ be the corresponding four vector in the unadorned frame. If the unadorned frame moves with speed v in the direction of the positive x_0 axis, $\mathbf{F}_0$ and $\mathbf{F}$ are related by

$$\mathbf{F} = \mathbf{T}\mathbf{F}_0, \tag{A.5}$$

where $\mathbf{T}$ is the Lorentz transformation matrix given by

$$\mathbf{T} = \begin{bmatrix} \lambda & 0 & 0 & i\lambda v/c \\ 0 & 1 & 0 & 0 \\ 0 & 0 & 1 & 0 \\ -i\lambda v/c & 0 & 0 & \lambda \end{bmatrix}, \tag{A.6}$$

with $i = \sqrt{-1}$ and

$$\lambda = (1 - v^2/c^2)^{-1/2}. \tag{A.7}$$

Written out in component form, Eq. (A.5) becomes

$$F^{(1)} = \lambda(F_0^{(1)} + iv F_0^{(4)}/c), \tag{A.8}$$

$$F^{(2)} = F_0^{(2)}, \tag{A.9}$$

$$F^{(3)} = F_0^{(3)}, \tag{A.10}$$

$$F^{(4)} = \lambda(F_0^{(4)} - iv F_0^{(1)}/c). \tag{A.11}$$

This transformation can be inverted to give

$$F_0^{(1)} = \lambda(F^{(1)} - iv F^{(4)}/c), \tag{A.12}$$

$$F_0^{(2)} = F^{(2)}, \tag{A.13}$$

$$F_0^{(3)} = F^{(3)}, \tag{A.14}$$

$$F_0^{(4)} = \lambda(F^{(4)} + iv F^{(1)}/c). \tag{A.15}$$

The fundamental four vector has components $\Delta x, \Delta y, \Delta z,$ and $ic\,\Delta t$, where $\Delta x, \Delta y, \Delta z,$

and Δt measure the distances in space and time from the point at which the two frames coincide, i.e., from $\mathbf{r}_0 = \mathbf{r}$ and $t_0 = t$. For notational convenience we shall drop the Δ prefix and take x, y, z, and ict as the space–time four vector, but it must be remembered that these quantities represent distances and time measured from the space–time point at which the two frames coincide. For this four vector, Eqs. (A.8) through (A.11) give

$$x = \lambda(x_0 - vt_0), \tag{A.16}$$

$$y = y_0, \tag{A.17}$$

$$z = z_0, \tag{A.18}$$

$$t = \lambda(t_0 - vx_0/c^2). \tag{A.19}$$

From Eqs. (A.16) through (A.19) we can easily determine how the time derivative and gradient operator transform. We have, by the chain rule of differentiation,

$$\frac{\partial}{\partial t_0} = \left(\frac{\partial t}{\partial t_0}\right)\frac{\partial}{\partial t} + \left(\frac{\partial x}{\partial t_0}\right)\frac{\partial}{\partial x} + \left(\frac{\partial y}{\partial t_0}\right)\frac{\partial}{\partial y} + \left(\frac{\partial z}{\partial t_0}\right)\frac{\partial}{\partial z}. \tag{A.20}$$

Equations (A.16) through (A.19) then yield

$$\frac{\partial}{\partial t_0} = \lambda\left(\frac{\partial}{\partial t} - v\frac{\partial}{\partial x}\right). \tag{A.21}$$

Similarly, we find

$$\frac{\partial}{\partial x_0} = \lambda\left(\frac{\partial}{\partial x} - \frac{v}{c^2}\frac{\partial}{\partial t}\right), \tag{A.22}$$

$$\frac{\partial}{\partial y_0} = \frac{\partial}{\partial y}, \tag{A.23}$$

$$\frac{\partial}{\partial z_0} = \frac{\partial}{\partial z}, \tag{A.24}$$

as the transformation of the gradient.

We now consider the transformation of ν and $\boldsymbol{\Omega}$. We know p_x, p_y, p_z, and iE/c form the components of a four vector, where $\mathbf{p}$ is the photon momentum and E is the photon energy. We have

$$\mathbf{p} = \frac{h\nu}{c}\boldsymbol{\Omega}, \tag{A.25}$$

$$E = h\nu, \tag{A.26}$$

and hence $\nu\Omega_x$, $\nu\Omega_y$, $\nu\Omega_z$, and $i\nu$ form a four vector. Equations (A.12) through (A.15) give in this case

$$\nu_0\Omega_{x0} = \lambda\nu(\Omega_x + v/c), \tag{A.27}$$

$$\nu_0\Omega_{y0} = \nu\Omega_y, \tag{A.28}$$

$$\nu_0\Omega_{z0} = \nu\Omega_z, \tag{A.29}$$

$$\nu_0 = \lambda\nu(1 + \Omega_x v/c). \tag{A.30}$$

If we define

$$D = 1 + \Omega_x v/c, \tag{A.31}$$

The Equations of Radiation Hydrodynamics

Eqs. (A.27) through (A.30) can be rewritten as

$$v_0 = \lambda D v, \tag{A.32}$$

$$\Omega_{x0} = \frac{1}{D}(\Omega_x + v/c), \tag{A.33}$$

$$\Omega_{y0} = \frac{1}{\lambda D}\Omega_y, \tag{A.34}$$

$$\Omega_{z0} = \frac{1}{\lambda D}\Omega_z. \tag{A.35}$$

From Eqs. (A.21) through (A.24) and (A.33) through (A.35) we can easily obtain the transformation of the streaming operator. After a bit of straightforward algebra, we find the simple result

$$\frac{1}{c}\frac{\partial}{\partial t_0} + \Omega_0 \cdot \nabla_0 = \frac{1}{\lambda D}\left[\frac{1}{c}\frac{\partial}{\partial t} + \Omega \cdot \nabla\right]. \tag{A.36}$$

We now consider the transformation of the specific intensity $I(v, \Omega)$. We first show that the distribution function per unit volume and per unit momentum for particles of rest mass m_0 is a Lorentz invariant. Since the number of particles is a Lorentz invariant, we have

$$\psi_0(\mathbf{p}_0)\, d\mathbf{p}_0\, d\mathbf{r}_0 = \psi(\mathbf{p})\, d\mathbf{p}\, d\mathbf{r}, \tag{A.37}$$

where $\psi(\mathbf{p})$ is the distribution function defined above. The volume elements $d\mathbf{r}_0$ and $d\mathbf{r}$ in Eq. (A.37) are the values of a volume element moving with the particles as observed from the two frames. We first transform the momentum differential in Eq. (A.37). Since p_x, p_y, p_z, and iE/c form a four vector, Eqs. (A.12) through (A.15) give

$$p_{x0} = \lambda(p_x + vE/c^2), \tag{A.38}$$

$$p_{y0} = p_y, \tag{A.39}$$

$$p_{z0} = p_z, \tag{A.40}$$

$$E_0 = \lambda(E + vp_x). \tag{A.41}$$

The three components of $\mathbf{p}$ and E are not independent, but are related according to

$$E^2 = p^2 c^2 + m_0^2 c^4, \tag{A.42}$$

with a similar relationship valid in the zero frame. The differentials in momentum are related according to

$$d\mathbf{p}_0 = |J(\mathbf{p}_0; \mathbf{p})|\, d\mathbf{p}, \tag{A.43}$$

where $J(\mathbf{p}_0; \mathbf{p})$ is the Jacobian of the transformation. Because of Eqs. (A.39) and (A.40), the three by three determinant representing the Jacobian reduces to a single term and we find

$$J(\mathbf{p}_0; \mathbf{p}) = \frac{\partial p_{x0}}{\partial p_x} = \lambda\left(1 + \frac{v}{c^2}\frac{\partial E}{\partial p_x}\right) = \lambda(1 + vp_x/E). \tag{A.44}$$

Use of Eq. (A.41) then gives

$$J(\mathbf{p}_0; \mathbf{p}) = E_0/E, \tag{A.45}$$

and hence

$$d\mathbf{p}_0 = (E_0/E)\, d\mathbf{p}. \tag{A.46}$$

Use of this result in Eq. (A.37) gives

$$E_0\psi_0(\mathbf{p}_0)\, d\mathbf{r}_0 = E\psi(\mathbf{p})\, d\mathbf{r}. \tag{A.47}$$

We now transform the volume elements in Eq. (A.47). Let dV denote the volume element as seen by an observer moving with the particles. An observer in the zero frame will see a Lorentz contraction in one dimension (that dimension parallel to $\mathbf{p}_0$) and no contraction in the two perpendicular dimensions. Hence

$$d\mathbf{r}_0 = (1-v_0^2/c^2)^{1/2}\, dV, \tag{A.48}$$

where v_0 is the speed of the particles as seen by an observer in the zero frame. Similarly,

$$d\mathbf{r} = (1-v_1^2/c^2)^{1/2}\, dV, \tag{A.49}$$

where v_1 is the speed of the particles as seen by an observer in the unadorned frame. Hence the volume elements transform as

$$d\mathbf{r}_0 = \left[\frac{1-v_0^2/c^2}{1-v_1^2/c^2}\right]^{1/2} d\mathbf{r}. \tag{A.50}$$

Now, the energy E_0 of a particle in the zero frame is

$$E_0 = \frac{m_0 c^2}{(1-v_0^2/c^2)^{1/2}}, \tag{A.51}$$

and in the unadorned frame

$$E = \frac{m_0 c^2}{(1-v_1^2/c^2)^{1/2}}. \tag{A.52}$$

Thus we find

$$d\mathbf{r}_0 = (E/E_0)\, d\mathbf{r}. \tag{A.53}$$

Use of this result in Eq. (A.47) gives the desired result

$$\psi_0(\mathbf{p}_0) = \psi(\mathbf{p}); \tag{A.54}$$

that is, the distribution function per unit volume and per unit momentum for particles of any rest mass m_0 is a Lorentz invariant. If we set $m_0 = 0$, Eq. (A.54) is then valid for photons.

We can easily derive the transformation law for the specific intensity $I(\nu, \Omega)$ from Eq. (A.54). To change from $\psi(\mathbf{p})$, a photon number distribution function, to $I(\nu, \Omega)$ we need multiply by $ch\nu$ and change from unit momentum to unit frequency and solid angle. Thus we have

$$I(\nu, \Omega)\, d\nu\, d\Omega = ch\nu\psi(\mathbf{p})\, d\mathbf{p}. \tag{A.55}$$

We write

$$d\mathbf{p} = p^2\, dp\, d\Omega, \tag{A.56}$$

and since for photons

$$p = h\nu/c, \tag{A.57}$$

we have

$$dp = (h/c)^3 \nu^2 \, d\nu \, d\Omega. \tag{A.58}$$

Then Eq. (A.55) gives

$$I(\nu, \Omega) = (h^4/c^2)\nu^3\psi(\mathbf{p}). \tag{A.59}$$

Since $\psi(\mathbf{p})$ is a Lorentz invariant, Eq. (A.59) shows that $I(\nu, \Omega)/\nu^3$ is also an invariant, and thus

$$I_0(\nu_0, \Omega_0) = (\nu_0/\nu)^3 I(\nu, \Omega). \tag{A.60}$$

Use of Eq. (A.32) then gives the transformation law for the specific intensity as

$$I_0(\nu_0, \Omega_0) = (\lambda D)^3 I(\nu, \Omega). \tag{A.61}$$

Combining Eqs. (A.36) and (A.61), we see

$$\frac{1}{c}\frac{\partial I_0(\nu_0, \Omega_0)}{\partial t} + \Omega_0 \cdot \nabla_0 I_0(\nu_0, \Omega_0) = (\lambda D)^2 \left[\frac{1}{c}\frac{\partial I(\nu, \Omega)}{\partial t} + \Omega \cdot \nabla I(\nu, \Omega)\right]. \tag{A.62}$$

Hence the right hand sides of Eqs. (A.1) and (A.2) must also transform as $(\lambda D)^2$. Let us show this directly.

We consider any one of the individual source (or loss) terms on the right hand sides of Eqs. (A.1) and (A.2). Each term represents a gain or loss of energy due to emission, absorption, or scattering of photons, measured per unit interval of volume, time, frequency, and solid angle. Division of any one of these terms by $h\nu$ to convert from energy to photon number and multiplication of the result by all the differentials involved gives the number of particles gained or lost, which is a Lorentz invariant. Thus

$$\frac{A_0(\nu_0, \Omega_0)}{\nu_0} \, d\nu_0 \, d\Omega_0 \, d\mathbf{r}_0 \, dt_0 = \frac{A(\nu, \Omega)}{\nu} \, d\nu \, d\Omega \, d\mathbf{r} \, dt, \tag{A.63}$$

where $A(\nu, \Omega)$ denotes any one of the individual terms mentioned above. Here $d\mathbf{r}_0$ and $d\mathbf{r}$ are volume elements fixed in space and are not to be confused with those involved in Eqs. (A.48) and (A.49). The Jacobian between the variables $\mathbf{r}_0$, t_0 and $\mathbf{r}$, t is unity, and hence $d\mathbf{r} \, dt$ is a Lorentz invariant, i.e.,

$$d\mathbf{r}_0 \, dt_0 = d\mathbf{r} \, dt. \tag{A.64}$$

Further, from Eq. (A.46) we have

$$\frac{d\mathbf{p}_0}{E_0} = \frac{d\mathbf{p}}{E}. \tag{A.65}$$

Use of Eqs. (A.26) and (A.58) in Eq. (A.65) yields

$$\nu_0 \, d\nu_0 \, d\Omega_0 = \nu \, d\nu \, d\Omega. \tag{A.66}$$

Using Eqs. (A.64) and (A.66) in Eq. (A.63), we find

$$A_0(\nu_0, \Omega_0) = (\nu_0/\nu)^2 A(\nu, \Omega), \tag{A.67}$$

and finally, making use of Eq. (A.32), we obtain

$$A_0(\nu_0, \Omega_0) = (\lambda D)^2 A(\nu, \Omega). \tag{A.68}$$

272

Hence any given source or loss term on the right hand sides of Eqs. (A.1) and (A.2) transforms as $(\lambda D)^2$. In particular, setting $A(\nu, \Omega) = S(\nu, \Omega)$, we find

$$S_0(\nu_0, \Omega_0) = (\lambda D)^2 S(\nu, \Omega). \tag{A.69}$$

Also, setting $A(\nu, \Omega) = \sigma_a(\nu, \Omega)I(\nu, \Omega)$ and using Eq. (A.61) we obtain

$$\sigma_{a0}(\nu_0, \Omega_0) = \frac{1}{\lambda D}\sigma_a(\nu, \Omega). \tag{A.70}$$

From the transformation formulae for ν, S, and σ_a, Eqs. (A.32), (A.69), and (A.70), and the definitions of B and σ'_a, Eqs. (2.168) and (2.169), we immediately find

$$B_0(\nu_0, \Omega_0) = (\lambda D)^3 B(\nu, \Omega), \tag{A.71}$$

$$\sigma'_{a0}(\nu_0, \Omega_0) = \frac{1}{\lambda D}\sigma'_a(\nu, \Omega). \tag{A.72}$$

The final quantity we need consider is the scattering kernel. Setting $A(\nu, \Omega)$ equal to the outscattering term in Eq. (A.1) or (A.2), Eq. (A.68) gives

$$\sigma_{s0}(\nu_0 \to \nu'_0, \Omega_0 \to \Omega'_0)\, I_0(\nu_0, \Omega_0) \left[1 + \frac{c^2}{2h\nu'^3_0} I_0(\nu'_0, \Omega'_0)\right] d\nu'_0\, d\Omega'_0$$

$$= (\lambda D)^2 \sigma_s(\nu \to \nu', \Omega \to \Omega')\, I(\nu, \Omega) \left[1 + \frac{c^2}{2h\nu'^3} I(\nu', \Omega')\right] d\nu'\, d\Omega'. \tag{A.73}$$

Making use of the transformation laws for ν, I, and $\nu\, d\nu\, d\Omega$, Eqs. (A.32), (A.61) and (A.66), we find Eq. (A.73) reduces to

$$\sigma_{s0}(\nu_0 \to \nu'_0, \Omega_0 \to \Omega'_0) = \frac{D'}{D}\sigma_s(\nu \to \nu', \Omega \to \Omega'), \tag{A.74}$$

where D is still defined by Eq. (A.31) and

$$D' = 1 + \Omega'_x v/c. \tag{A.75}$$

The same result is obtained by considering the inscattering term in Eq. (A.1) or (A.2).

At this point we have specified the transformation formulae for all quantities in the equation of transfer. It must be remembered that the unadorned frame was assumed to move with speed v in the direction of the positive x_0 axis. We can allow for a general direction of the velocity between frames by simply introducing vector notation. For all scalar quantities, i.e., all quantities except Ω, the transformations are generalized to a general velocity by merely writing λ, D, and D' in vector notation. Since λ, defined by Eq. (A.7), is a function of v^2 only, its definition is unchanged. The quantities D and D', defined by Eqs. (A.31) and (A.75), generalize to

$$D = 1 + \mathbf{\Omega} \cdot \mathbf{v}/c, \tag{A.76}$$

$$D' = 1 + \mathbf{\Omega'} \cdot \mathbf{v}/c, \tag{A.77}$$

where $\mathbf{v}$ is the velocity of the unadorned frame as seen by an observer in the zero frame. The

The Equations of Radiation Hydrodynamics

generalization of the transformation law for Ω, Eqs. (A.33) through (A.35), to an arbitrary velocity, is given by

$$\Omega_0 = \frac{1}{\lambda D}\left[\Omega + \frac{\lambda}{\lambda+1}\,(\lambda D+1)\,\frac{\mathbf{v}}{c}\right].\tag{A.78}$$

It is easily verified that Eq. (A.78) reduces to Eqs. (A.33) through (A.35) if $\mathbf{v}$ is in the direction of increasing x_0. Another transformation which we shall find useful is that for $\Omega\cdot\Omega'$. The transformation for this scalar product turns out to be remarkably simple. Use of Eq. (A.78) gives, after a fair amount of straightforward algebra,

$$1-\Omega_0\cdot\Omega_0' = \frac{1}{\lambda^2 DD'}\,(1-\Omega\cdot\Omega').\tag{A.79}$$

We now summarize the results we have obtained in this Appendix.

Summary

Let an inertial frame of reference, the unadorned frame, move with velocity $\mathbf{v}$ as seen by an observer in another frame of reference, the zero frame. We define

$$\lambda = (1-v^2/c^2)^{-1/2},\tag{A.80}$$

$$D = 1+\Omega\cdot\mathbf{v}/c,\tag{A.81}$$

$$D' = 1+\Omega'\cdot\mathbf{v}/c.\tag{A.82}$$

The Lorentz transformation of the equation of transfer then gives the following transformation laws between the two frames:

$$\nu_0 = \lambda D\nu,\tag{A.83}$$

$$\Omega_0 = \frac{1}{\lambda D}\left(\Omega + \frac{\lambda}{\lambda+1}\,(\lambda D+1)\frac{\mathbf{v}}{c}\right),\tag{A.84}$$

$$\frac{1}{c}\frac{\partial}{\partial t_0}+\Omega_0\cdot\nabla_0 = \frac{1}{\lambda D}\left[\frac{1}{c}\frac{\partial}{\partial t}+\Omega\cdot\nabla\right],\tag{A.85}$$

$$1-\Omega_0\cdot\Omega_0' = \frac{1}{\lambda^2 DD'}\,(1-\Omega\cdot\Omega'),\tag{A.86}$$

$$I_0(\nu_0,\,\Omega_0) = (\lambda D)^3 I(\nu,\,\Omega),\tag{A.87}$$

$$S_0(\nu_0,\,\Omega_0) = (\lambda D)^2 S(\nu,\,\Omega),\tag{A.88}$$

$$\sigma_{a0}(\nu_0,\,\Omega_0) = \frac{1}{\lambda D}\,\sigma_a(\nu,\,\Omega),\tag{A.89}$$

$$B_0(\nu_0,\,\Omega_0) = (\lambda D)^3 B(\nu,\,\Omega),\tag{A.90}$$

$$\sigma_{a0}'(\nu_0,\,\Omega_0) = \frac{1}{\lambda D}\,\sigma_a'(\nu,\,\Omega),\tag{A.91}$$

$$\sigma_{s0}(\nu_0 \to \nu_0',\,\Omega_0 \to \Omega_0') = \frac{D'}{D}\,\sigma_s(\nu \to \nu',\,\Omega \to \Omega'),\tag{A.92}$$

$$d\nu_0\,d\Omega_0 = \frac{1}{\lambda D}\,d\nu\,d\Omega.\tag{A.93}$$

References

ABRAMOWITZ, M., and STEGUN, I. (editors) (1964) *Handbook of Mathematical Functions*, Dover, New York, pp. 332–53.

AIR FORCE WEAPONS LABORATORY CONFERENCE ON OPACITIES (1964) *J. Quant. Spectros. Rad. Trans.* **4**, 583–761.

AIR FORCE WEAPONS LABORATORY CONFERENCE ON OPACITIES (1965) *J. Quant. Spectros. Rad. Trans.* **5**, 3–271.

AMBARTSUMYAN, V. (1956) *Theoretical Astrophysics*, Pergamon Press, New York, chapter 11.

BATES, D. R. (1962) *Atomic and Molecular Processes*, Academic Press, New York.

BAUER, G. N., and BROOKE, W. E. (1932) *Plane and Spherical Trigonometry*, D. C. Heath, New York, pp. 189–95.

BELL, G. I., HANSEN, G. E., and SANDMEIER, H. A. (1967) *Nucl. Sci. Eng.* **28**, 376.

CAMPBELL, P. M. (1969) *An Introduction to High Temperature Radiation Gas Dynamics*, Air Force Weapons Laboratory Report No. AFWL-TR-69-10, Albuquerque, New Mexico, pp. 39–82.

CARLSON, B., and LATHROP, K. (1965) *Trans. Am. Nucl. Soc.* **8**, 487.

CARLSON, B., and LATHROP, K. (1968) In *Computing Methods in Reactor Physics* (H. Greenspan, C. Kelber, and D. Okrent, editors), Gordon & Breach, New York.

CASE, K., DE HOFFMANN, F., and PLACZEK, G. (1953a) *Introduction to the Theory of Neutron Diffusion*, Los Alamos Scientific Laboratory, Los Alamos, New Mexico, pp. 153–62.

CASE, K., DE HOFFMANN, F., and PLACZEK, G. (1953b) *Ibid.*, p. 55.

CHANDRASEKHAR, S. (1957) *Stellar Structure*, Dover, New York, pp. 357–411.

CHANDRASEKHAR, S. (1960a) *Radiative Transfer*, Dover, New York, pp. 54–88.

CHANDRASEKHAR, S. (1960b) *Ibid.*, pp. 24–44.

CHANDRASEKHAR, S. (1960c) *Ibid.*, pp. 171–7.

CHANDRASEKHAR, S. (1960d) *Ibid.*, pp. 35–37.

CONDON, E. U., and SHORTLY, G. H. (1957) *The Theory of Atomic Spectra*, Cambridge University Press, chapter IV.

COX, A. N. (1965a) In *Stellar Structure* (L. Aller and D. McLaughlin, editors), University of Chicago Press, p. 214.

COX, A. N. (1965b) *Ibid.*, chapter III.

COX, J. P., and GIULI, R. T. (1968) *Principles of Stellar Structure*, vol. I, Gordon & Breach, New York, pp. 103–5.

DAVISON, B. (1957a) *Neutron Transport Theory*, Clarendon Press, Oxford, pp. 126–7.

DAVISON, B. (1957b) *Ibid.*, pp. 127–8.

DAVISON, B. (1957c) *Ibid.*, pp. 129–30.

DAVISON, B. (1957d) *Ibid.*, pp. 158–9.

DAVISON, B. (1957e) *Ibid.*, pp. 247–8.

DAVISON, B. (1957f) *Ibid.*, pp. 166–9.

DIRAC, P. A. M. (1925) *Mon. Not. R. Astr. Soc.* **85**, 825.

EISBERG, L. (1961) *Fundamentals of Modern Physics*, Wiley, New York.

ELSASSER, W. M. (1938) *Phys. Rev.* **54**, 126.

ERDELYI, A., MAGNUS, W., OBERHETTINGER, F., and TRICOMI, F. G. (1953a) *Higher Transcendental Functions*, vol. 1, McGraw-Hill, New York, pp. 120–81.

ERDELYI, A., MAGNUS, W., OBERHETTINGER, F., and TRICOMI, F. G. (1953b) *Ibid.*, vol. 2, pp. 43–44, 82.

EVANS, R. D. (1955) *The Atomic Nucleus*, McGraw-Hill, New York, p. 682.

FEDERIGHI, F. (1964) *Nukleonik* **6**, 277.

FEYNMAN, R. P., LEIGHTON, R. B., and SANDS, M. (1966a) *The Feynman Lectures on Physics*, Addison-Wesley, Reading, Massachusetts, vol. III, pp. 4–7.

References

FEYNMAN, R. P., LEIGHTON, R. B., and SANDS, M. (1966b) *The Feynman Lectures on Physics*, Addison-Wesley, Reading, Massachusetts, vol. I, p. 31.

FOWLER, R. H. (1936) *Statistical Mechanics*, Cambridge University Press.

FOWLER, R. H., and GUGGENHEIM, E. A. (1939) *Statistical Thermodynamics*, Cambridge University Press.

FRANK-KAMENETSKII, D. (1962) *Physical Processes in Stellar Interiors*, National Science Foundation, Washington, DC, p. 89.

FRASER, A. R. (1966) *The Fundamental Equations of Radiation Hydrodynamics*, Report No. AWRE O-82/65, Atomic Weapons Research Establishment, Berkshire, England.

GAUNT, J. A. (1930) *Phil. Trans. R. Soc. London* A **229**, 163.

GRIEM, H. R. (1960) *Astrophys. J.* **132**, 883.

HARRIS, E. G. (1965) *Phys. Rev.* **138**, B479.

HEITLER, W. (1954) *The Quantum Theory of Radiation*, Oxford University Press.

HERZBERG, G. (1950) *Spectra of Diatomic Molecules*, van Nostrand, Princeton, New Jersey.

HUNT, B. and SIBULKIN, M. (1967) *J. Quant. Spectros. Rad. Trans.* **7**, 951.

HURWITZ, H., and ZWEIFEL, P. F. (1955) *J. Appl. Phys.* **26**, 923.

KRAMERS, H. A. (1923) *Phil. Mag.* **46**, 836.

LANDAU, L. D., and LIFSHITZ, E. M. (1965) *Quantum Mechanics, Non-Relativistic Theory*, 2nd edn., Addison-Wesley, Reading, Massachusetts, pp. 164, 241.

LATHROP, K. D. (1968) *Nucl. Sci. Eng.* **32**, 357.

LEIGHTON, R. B. (1959) *Principles of Modern Physics*, McGraw-Hill, New York.

MAYER, H. (1947a) *Methods of Opacity Calculations*, Report No. AECD-1870, Los Alamos Scientific Laboratory, Los Alamos, New Mexico.

MAYER, H. (1947b) *Ibid.*, chapter IV.

MAYER, H. (1947c) *Ibid.*, chapter V.

MAYER, J. E., and MAYER, M. G. (1940) *Statistical Mechanics*, Wiley, New York.

McINERNEY, J. J. (1964) *Nucl. Sci. Eng.* **19**, 458.

MENZEL, D. H., and PEKERIS, C. H. (1935) *Royal Astronom. Soc. Mon. Not.* **96**, 77.

MIKA, J. R. (1961) *Nucl. Sci. Eng.* **11**, 415.

MILNE, E. A. (1924) *Proc. Phys. Soc.* **36**, 100.

MORSE, P. M., and FESHBACH, H. (1961) *Methods of Theoretical Physics*, McGraw-Hill, New York, p. 1494.

PAULING, L., and WILSON, E. (1935) *Introduction to Quantum Mechanics*, McGraw-Hill, New York, p. 144.

PENNER, S. (1959) *Quantitative Molecular Spectroscopy and Gas Emissivities*, Addison-Wesley, Reading, Massachusetts.

POMRANING, G. C. (1964) *Annals of Physics* **27**, 193.

POMRANING, G. C. (1965a) *Nukleonik* **6**, 348.

POMRANING, G. C. (1965b) *Nucl. Sci. Eng.* **22**, 328.

POMRANING, G. C. (1965c) *A Method of Solution for Particle Transport Problems*, Report No. GA-6497, General Atomic, La Jolla, California, pp. 108–22.

POMRANING, G. C. (1965d) *Ibid.*, pp. 44–63.

POMRANING, G. C. (1968a) *J. Quant. Spectros. Rad. Trans.* **8**, 1087.

POMRANING, G. C. (1968b) *Astrophys. J.* **152**, 809.

POMRANING, G. C. (1969a) *J. Quant. Spectros. Rad. Trans.* **9**, 1011.

POMRANING, G. C. (1969b) *Ibid.*, **9**, 407.

POMRANING, G. C. (1969c) *Int. J. Heat Mass. Trans.* **12**, 81.

POMRANING, G. C. (1971) *J. Quant. Spectros. Rad. Trans.* **11**, 597.

POMRANING, G. C., and CLARK, M. (1963) *Nucl. Sci. Eng.* **17**, 227.

POMRANING, G. C., and FROEHLICH, R. (1969) *Astron. Astrophys.* **1**, 286.

REED, W., and LATHROP, K. D. (1970) *Nucl. Sci. Eng.* **41**, 237.

RICHTMYER, F., KENNARD, E., and LAURITSEN, T. (1955a) *Introduction to Modern Physics*, McGraw-Hill, New York.

RICHTMYER, F., KENNARD, E., and LAURITSEN, T. (1955b) *Ibid.*, p. 168.

RUSHBROOKE, G. S. (1949) *Introduction to Statistical Mechanics*, Oxford University Press.

SELENGUT, D. (1962) *Trans. Am. Nucl. Soc.* **5**, 40.

SLATER, J. C. (1930) *Phys. Rev.* **36**, 57.

SLATER, J. C. (1960) *Quantum Theory of Atomic Structure*, vols. 1 and 2, McGraw-Hill, New York.

SLATER, J. C., and FRANK, N. H. (1947) *Electromagnetism*, McGraw-Hill, New York, pp. 161–5.

STIX, T. H. (1962) *The Theory of Plasma Waves*, McGraw-Hill, New York, pp. 55–60.

STONE, S. and NELSON, R. G. (1966) *Compton Scattering from Relativistic Electrons*, Report No. UCRL-14918-T, Lawrence Radiation Laboratory, Livermore, California.

ter Haar, D. (1954) *Elements of Statistical Mechanics*, Rinehart, New York.

Tolman, R. C. (1938) *Principles of Statistical Mechanics*, Oxford University Press.

Volterra, V. (1959) *Theory of Functionals and of Integral and Integro-Differential Equations*, Dover, New York.

Weinberg, S. (1962) *Phys. Rev.* **126**, 1899.

Wilkins, E., Hellens, R., and Zweifel, P. F. (1955) *Proc. Int. Conf. Peaceful Uses Atomic Energy* 5, P/597.

Winslow, A. (1968) *Nucl. Sci. Eng.* **32**, 101.

Woolley, R. and Stibbs, D. (1953) *The Outer Layers of a Star*, Oxford University Press, chapter VI.

Zel'dovich, Ya., and Raizer, Yu. (1966) *Physics of Shock Waves and High Temperature Hydrodynamic Phenomena*, vol. 1, Academic Press, New York, p. 194.

Bibliography

In this bibliography we list some general reading references which should be useful to persons interested in radiative transfer, fluid flow, and the interaction of the radiation field with the fluid. For the most part, these references are readily available books. The reader is referred to the bibliography of P. M. Campbell *(An Annotated Bibliography on High Temperature Radiation Gas Dynamics*, Air Force Weapons Laboratory Technical Report No. AFWL-TR-68-119, Kirtland Air Force Base, New Mexico, 1968) for a guide to selected journal articles and unpublished reports in the general area of radiation hydrodynamics.

ALDER, B., FERNBACH, S., and ROTENBERG, M. (editors), *Methods in Computational Physics*, Academic Press New York, continuing series, started in 1963.

ALLER, L. H., *Astrophysics: The Atmospheres of the Sun and Stars*, Ronald Press, New York, 1963.

AMBARTSUMYAN, V. A., *Theoretical Astrophysics*, Pergamon Press, London, 1958.

BATES, D. R. (editor), *Atomic and Molecular Processes*, Academic Press, New York, 1962.

BELLMAN, R. E., KALABA, R. E., and PRESTRUD, M. C., *Invariant Imbedding and Radiative Transfer in Slabs of Finite Thickness*, American Elsevier, New York, 1962.

BELLMAN, R. E., KAGIWADA, H. H., KALABA, R. E., and PRESTRUD, M. C., *Invariant Imbedding and Time Dependent Transport Processes*, American Elsevier, New York, 1963.

BIRD, R. B., STEWART, W. E., and LIGHTFOOT, E. N., *Transport Phenomena*, Wiley, New York, 1960.

BOND, J. W., WATSON, K. M., and WELCH, J. A., Jr., *Atomic Theory of Gas Dynamics*, Addison-Wesley, Reading, Massachusetts, 1965.

BUSBRIDGE, I. W., *The Mathematics of Radiative Transfer*, Cambridge University Press, 1960.

CARSLAW, H. S., and JAEGER, J. C., *Conduction of Heat in Solids*, Oxford University Press, London, 1959.

CASE, K., DE HOFFMANN, F., and PLACZEK, G., *Introduction to the Theory of Neutron Diffusion*, Los Alamos Scientific Laboratory, Los Alamos, New Mexico, 1953.

CASE, K. M., and ZWEIFEL, P. F., *Linear Transport Theory*, Addison-Wesley, Reading, Massachusetts, 1967.

CASHWELL, E. D., and EVERETT, C. J., *A Practical Manual on the Monte Carlo Method for Random Walk Problems*, Pergamon Press, New York, 1959.

CHANDRASEKHAR, S., *Stellar Structure*, Dover, New York, 1957.

CHANDRASEKHAR, S., *Radiative Transfer*, Dover, New York, 1960.

COLLATZ, L., *The Numerical Treatment of Differential Equations*, Springer-Verlag, Berlin, 1960.

COMPTON, A. H., Jr., and ALLISON, S. K., *X-rays in Theory and Experiment*, van Nostrand, New York, 1935.

COURANT, R., and FRIEDRICHS, K. O., *Supersonic Flow and Shock Waves*, Interscience, New York, 1948.

COX, A. N., Stellar absorption coefficients and opacities, in *Stellar Structure* (L. Aller and D. McLaughlin, editors), University of Chicago Press, Chicago, 1965.

COX, J. P., and GIULI, R. T., *Principles of Stellar Structure*, vols. 1 and 2, Gordon & Breach, New York, 1968.

DAVISON, B., *Neutron Transport Theory*, Clarendon Press, Oxford, 1957.

FANO, U., SPENCER, L., and BERGER, M., Penetration and diffusion of X-rays, in *Encyclopedia of Physics* (S. Flugge, editor), Springer-Verlag, Berlin, 1959.

FORSYTHE, G. E., and WASOW, W. R., *Finite Difference Methods for Partial Differential Equations*, Wiley, New York, 1960.

FRASER, A. R., *The Fundamental Equations of Radiation Hydrodynamics*, Atomic Weapons Research Establishment Report No. AWRE O-82/65, Berkshire, England, 1966.

Bibliography

GOODJOHN, A. J., and POMRANING, G. C. (editors), *Reactor Physics in the Resonance and Thermal Regions*, vols. 1 and 2, MIT Press, Cambridge, Massachusetts, 1966.

GOODY, R. M., *Atmospheric Radiation*, Clarendon Press, Oxford, 1964.

GREENSPAN, H., KELBER, C., and OKRENT, D. (editors), *Computing Methods in Reactor Physics*, Gordon & Breach, New York, 1968.

HEITLER, W., *The Quantum Theory of Radiation*, Clarendon Press, Oxford, 1954.

HOPF, E., *Mathematical Problems of Radiative Equilibrium*, Stechert-Hafner, New York, 1964.

HOTTEL, H. C., and SAROFIM, A. F., *Radiative Transfer*, McGraw-Hill, New York, 1967.

INONU, E., and ZWEIFEL, P. F. (editors), *Developments in Transport Theory*, Academic Press, London, 1967.

JEFFRIES, J. T., *Spectral Line Formation*, Blaisdell, Waltham, Massachusetts, 1968.

KONDRAT'YEV, K. YA., *Radiative Heat Exchange in the Atmosphere*, Pergamon Press, New York, 1965 (translated by O. Tedder).

KOURGANOFF, V., *Basic Methods in Transfer Problems*, Dover, New York, 1963.

KREITH, F., *Radiation Heat Transfer*, International Textbook Company, Scranton, Pennsylvania, 1962.

LAMB, H., *Hydrodynamics*, Dover, New York, 1879.

LOVE, T. J., *Radiative Heat Transfer*, Charles E. Merrill, Columbus, Ohio, 1968.

MENZEL, D. H. (editor), *Selected Papers on Physical Processes in Ionized Plasmas*, Dover, New York, 1962.

MENZEL, D. H. (editor), *Selected Papers of the Transfer of Radiation*, Dover, New York, 1966.

MORSE, P. M., and FESHBACH, H., *Methods of Theoretical Physics*, vols. 1 and 2, McGraw-Hill, New York, 1953.

PAI, S. I., *Radiation Gas Dynamics*, Springer-Verlag, New York, 1966.

PENNER, S. S., *Quantitative Molecular Spectroscopy and Gas Emissivities*, Addison-Wesley, Reading, Massachusetts, 1959.

PENNER, S. S., and OLFE, D. B., *Radiation and Reentry*, Academic Press, New York, 1968.

PLANCK, M., *The Theory of Heat Radiation*, Dover, New York, 1959.

PREISENDORFER, R. W., *Radiative Transfer on Discrete Spaces*, Pergamon Press, New York, 1965.

Proceedings of the First Air Force Weapons Laboratory Conference on Opacities, *Journal of Quantitative Spectroscopy and Radiative Transfer*, vol. 4, 1964.

Proceedings of the Second Air Force Weapons Laboratory Conference on Opacities, *Journal of Quantitative Spectroscopy and Radiative Transfer*, vol. 5, 1965.

Proceedings of the Atlas Computer Laboratory Conference on Transport Theory, *Journal of Quantitative Spectroscopy and Radiative Transfer*, vol. 11, 1971.

RALSTON, A., and WILF, H. S. (editors), *Mathematical Methods for Digital Computers*, Wiley, New York, 1960.

RICHTMYER, R. D., *Difference Methods for Initial Value Problems*, Interscience, New York, 1957.

SAMPSON, D. H., *Radiative Contributions to Energy and Momentum Transport in a Gas*, Interscience, New York, 1965.

SANGREN, W. C., *Digital Computers and Nuclear Reactor Calculations*, Wiley, New York, 1960.

SCHMIDT, G., *Physics of High Temperature Plasmas, An Introduction*, Academic Press, New York, 1966.

SLATER, J. C., *Quantum Theory of Atomic Structure*, vols. 1 and 2, McGraw-Hill, New York, 1960.

SOBEL'MAN, I. N., *Introduction to the Theory of Atomic Spectra*, Moscow, 1963.

SOBOLEV, V. V., *A Treatise on Radiative Transfer*, D. van Nostrand, Princeton, New Jersey, 1963 (translated by S. I. Gaposchkin).

SPANIER, J., and GELBARD, E. M., *Monte Carlo Principles and Neutron Transport Problems*, Addison-Wesley, Reading, Massachusetts, 1966.

SYNGE, J. L., *The Relativistic Gas*, North-Holland, Amsterdam, 1957.

TAIT, J. H., *An Introduction to Neutron Transport Theory*, American Elsevier, New York, 1964.

THOMAS, R. N., *Some Aspects of Non-Equilibrium Thermodynamics in the Presence of a Radiation Field*, University of Colorado Press, Boulder, Colorado, 1965.

UNSÖLD, A., *Physik der Sternatmosphären*, Springer-Verlag, Berlin, 1955.

VAN DE HULST, H. C., *Light Scattering by Small Particles*, Wiley, New York, 1957.

WING, G. M., *An Introduction to Transport Theory*, Wiley, New York, 1962.

ZEL'DOVICH, YA. B., and RAIZER, YU. P., *Physics of Shock Waves and High Temperature Hydrodynamic Phenomena*, vols. 1 and 2, Academic Press, New York, 1966, 1967.

ZHELEZNYAKOV, V. V., *Radioemission of the Sun and Planets*, Pergamon, Oxford, 1970.

Index

Index

A CATALOG OF SELECTED
DOVER BOOKS
IN SCIENCE AND MATHEMATICS

Physics

OPTICAL RESONANCE AND TWO-LEVEL ATOMS, L. Allen and J. H. Eberly. Clear, comprehensive introduction to basic principles behind all quantum optical resonance phenomena. 53 illustrations. Preface. Index. 256pp. 5⅜ x 8½.   65533-4

QUANTUM THEORY, David Bohm. This advanced undergraduate-level text presents the quantum theory in terms of qualitative and imaginative concepts, followed by specific applications worked out in mathematical detail. Preface. Index. 655pp. 5⅜ x 8½.   65969-0

ATOMIC PHYSICS (8th edition), Max Born. Nobel laureate's lucid treatment of kinetic theory of gases, elementary particles, nuclear atom, wave-corpuscles, atomic structure and spectral lines, much more. Over 40 appendices, bibliography. 495pp. 5⅜ x 8½.   65984-4

A SOPHISTICATE'S PRIMER OF RELATIVITY, P. W. Bridgman. Geared toward readers already acquainted with special relativity, this book transcends the view of theory as a working tool to answer natural questions: What is a frame of reference? What is a "law of nature"? What is the role of the "observer"? Extensive treatment, written in terms accessible to those without a scientific background. 1983 ed. xlviii+172pp. 5⅜ x 8½.   42549-5

AN INTRODUCTION TO HAMILTONIAN OPTICS, H. A. Buchdahl. Detailed account of the Hamiltonian treatment of aberration theory in geometrical optics. Many classes of optical systems defined in terms of the symmetries they possess. Problems with detailed solutions. 1970 edition. xv + 360pp. 5⅜ x 8½.   67597-1

PRIMER OF QUANTUM MECHANICS, Marvin Chester. Introductory text examines the classical quantum bead on a track: its state and representations; operator eigenvalues; harmonic oscillator and bound bead in a symmetric force field; and bead in a spherical shell. Other topics include spin, matrices, and the structure of quantum mechanics; the simplest atom; indistinguishable particles; and stationary-state perturbation theory. 1992 ed. xiv+314pp. 6⅛ x 9¼.   42878-8

LECTURES ON QUANTUM MECHANICS, Paul A. M. Dirac. Four concise, brilliant lectures on mathematical methods in quantum mechanics from Nobel Prize-winning quantum pioneer build on idea of visualizing quantum theory through the use of classical mechanics. 96pp. 5⅜ x 8½.   41713-1

THIRTY YEARS THAT SHOOK PHYSICS: The Story of Quantum Theory, George Gamow. Lucid, accessible introduction to influential theory of energy and matter. Careful explanations of Dirac's anti-particles, Bohr's model of the atom, much more. 12 plates. Numerous drawings. 240pp. 5⅜ x 8½.   24895-X

ELECTRONIC STRUCTURE AND THE PROPERTIES OF SOLIDS: The Physics of the Chemical Bond, Walter A. Harrison. Innovative text offers basic understanding of the electronic structure of covalent and ionic solids, simple metals, transition metals and their compounds. Problems. 1980 edition. 582pp. 6⅛ x 9¼.
66021-4

HYDRODYNAMIC AND HYDROMAGNETIC STABILITY, S. Chandrasekhar. Lucid examination of the Rayleigh-Benard problem; clear coverage of the theory of instabilities causing convection. 704pp. 5⅜ x 8¼. 64071-X

INVESTIGATIONS ON THE THEORY OF THE BROWNIAN MOVEMENT, Albert Einstein. Five papers (1905–8) investigating dynamics of Brownian motion and evolving elementary theory. Notes by R. Fürth. 122pp. 5⅜ x 8½. 60304-0

THE PHYSICS OF WAVES, William C. Elmore and Mark A. Heald. Unique overview of classical wave theory. Acoustics, optics, electromagnetic radiation, more. Ideal as classroom text or for self-study. Problems. 477pp. 5⅜ x 8½. 64926-1

PHYSICAL PRINCIPLES OF THE QUANTUM THEORY, Werner Heisenberg. Nobel Laureate discusses quantum theory, uncertainty, wave mechanics, work of Dirac, Schroedinger, Compton, Wilson, Einstein, etc. 184pp. 5⅜ x 8½. 60113-7

ATOMIC SPECTRA AND ATOMIC STRUCTURE, Gerhard Herzberg. One of best introductions; especially for specialist in other fields. Treatment is physical rather than mathematical. 80 illustrations. 257pp. 5⅜ x 8½. 60115-3

AN INTRODUCTION TO STATISTICAL THERMODYNAMICS, Terrell L. Hill. Excellent basic text offers wide-ranging coverage of quantum statistical mechanics, systems of interacting molecules, quantum statistics, more. 523pp. 5⅜ x 8½. 65242-4

THEORETICAL PHYSICS, Georg Joos, with Ira M. Freeman. Classic overview covers essential math, mechanics, electromagnetic theory, thermodynamics, quantum mechanics, nuclear physics, other topics. First paperback edition. xxiii + 885pp. 5⅜ x 8½. 65227-0

PROBLEMS AND SOLUTIONS IN QUANTUM CHEMISTRY AND PHYSICS, Charles S. Johnson, Jr. and Lee G. Pedersen. Unusually varied problems, detailed solutions in coverage of quantum mechanics, wave mechanics, angular momentum, molecular spectroscopy, more. 280 problems plus 139 supplementary exercises. 430pp. 6½ x 9¼. 65236-X

THEORETICAL SOLID STATE PHYSICS, Vol. 1: Perfect Lattices in Equilibrium; Vol. II: Non-Equilibrium and Disorder, William Jones and Norman H. March. Monumental reference work covers fundamental theory of equilibrium properties of perfect crystalline solids, non-equilibrium properties, defects and disordered systems. Appendices. Problems. Preface. Diagrams. Index. Bibliography. Total of 1,301pp. 5⅜ x 8½. Two volumes. Vol. I: 65015-4   Vol. II: 65016-2

WHAT IS RELATIVITY? L. D. Landau and G. B. Rumer. Written by a Nobel Prize physicist and his distinguished colleague, this compelling book explains the special theory of relativity to readers with no scientific background, using such familiar objects as trains, rulers, and clocks. 1960 ed. vi+72pp. 5⅜ x 8½. 42806-0

A TREATISE ON ELECTRICITY AND MAGNETISM, James Clerk Maxwell. Important foundation work of modern physics. Brings to final form Maxwell's theory of electromagnetism and rigorously derives his general equations of field theory. 1,084pp. 5⅜ x 8½. Two-vol. set. Vol. I: 60636-8   Vol. II: 60637-6

QUANTUM MECHANICS: Principles and Formalism, Roy McWeeny. Graduate student-oriented volume develops subject as fundamental discipline, opening with review of origins of Schrödinger's equations and vector spaces. Focusing on main principles of quantum mechanics and their immediate consequences, it concludes with final generalizations covering alternative "languages" or representations. 1972 ed. 15 figures. xi+155pp. 5⅜ x 8½. 42829-X

INTRODUCTION TO QUANTUM MECHANICS With Applications to Chemistry, Linus Pauling & E. Bright Wilson, Jr. Classic undergraduate text by Nobel Prize winner applies quantum mechanics to chemical and physical problems. Numerous tables and figures enhance the text. Chapter bibliographies. Appendices. Index. 468pp. 5⅜ x 8½. 64871-0

METHODS OF THERMODYNAMICS, Howard Reiss. Outstanding text focuses on physical technique of thermodynamics, typical problem areas of understanding, and significance and use of thermodynamic potential. 1965 edition. 238pp. 5⅜ x 8½. 69445-3

TENSOR ANALYSIS FOR PHYSICISTS, J. A. Schouten. Concise exposition of the mathematical basis of tensor analysis, integrated with well-chosen physical examples of the theory. Exercises. Index. Bibliography. 289pp. 5⅜ x 8½. 65582-2

THE ELECTROMAGNETIC FIELD, Albert Shadowitz. Comprehensive undergraduate text covers basics of electric and magnetic fields, builds up to electromagnetic theory. Also related topics, including relativity. Over 900 problems. 768pp. 5⅜ x 8¼. 65660-8

GREAT EXPERIMENTS IN PHYSICS: Firsthand Accounts from Galileo to Einstein, edited by Morris H. Shamos. 25 crucial discoveries: Newton's laws of motion, Chadwick's study of the neutron, Hertz on electromagnetic waves, more. Original accounts clearly annotated. 370pp. 5⅜ x 8½. 25346-5

RELATIVITY, THERMODYNAMICS AND COSMOLOGY, Richard C. Tolman. Landmark study extends thermodynamics to special, general relativity; also applications of relativistic mechanics, thermodynamics to cosmological models. 501pp. 5⅜ x 8½. 65383-8

STATISTICAL PHYSICS, Gregory H. Wannier. Classic text combines thermodynamics, statistical mechanics and kinetic theory in one unified presentation of thermal physics. Problems with solutions. Bibliography. 532pp. 5⅜ x 8½. 65401-X